Cambridge Archaeological and Ethnological Series

THE PLACE-NAMES

OF

NORTHUMBERLAND

AND

DURHAM

THE PLACE-NAMES

OF

NORTHUMBERLAND

AND

DURHAM

by

ALLEN MAWER, M.A.

Joseph Cowen Professor of English in Armstrong College,
University of Durham.
Late Fellow of Gonville and Caius College

CAMBRIDGE

AT THE UNIVERSITY PRESS

1920

CAMBRIDGE UNIVERSITY PRESS
Cambridge, New York, Melbourne, Madrid, Cape Town,
Singapore, São Paulo, Delhi, Tokyo, Mexico City

Cambridge University Press
The Edinburgh Building, Cambridge CB2 8RU, UK

Published in the United States of America by Cambridge University Press, New York

www.cambridge.org
Information on this title: www.cambridge.org/9781107608344

First published 1920
First paperback edition 2011

A catalogue record for this publication is available from the British Library

ISBN 978-1-107-60834-4 Paperback

PREFACE

THE study of English place-names is steadily advancing in its methods and extent and in the present volume an attempt is made to deal with two more counties. Work on them has taken much the greater part of eight years, sadly interrupted by the war and other circumstances. These interruptions have had their bad influence in making it more difficult than usual to secure that uniformity of handling and presentation which is desirable in a theme of this kind. On the other hand, they have enabled the author to take advantage of the ever-growing literature—English and Scandinavian—that deals with these matters. As one reads it, the unhappy conviction is more and more brought home to one that no single county can be dealt with satisfactorily apart from a survey of the field of English place-nomenclature as a whole. In the disorganised condition of English place-name study it is impossible to look for those happy results that have already been attained in Norway, and are fast becoming possible in Sweden and Denmark, as the result of organised research, aided by the State. Those who recognise the importance of these studies must labour to secure similar co-operation in England, and, in the meantime, endeavour to keep interest alive by such single-handed efforts as they can individually make.

The present volume follows the general lines of study laid down by Skeat, Wyld, and Moorman, but there are a few points in which it endeavours to work on newer and

b

more definite lines than those hitherto followed. They are roughly as follows :—

1. An attempt is made to deal with all names found in documents dating from before 1500, which can be identified on the modern map, and the study is, with some half dozen insignificant exceptions, rigidly confined to such. Books which deal with undocumented names on the same lines as documented ones stultify themselves and the newer methods of place-name study generally.

2. Topographical conditions have been carefully studied by the aid of maps, guide-books, personal observation, and local enquiry. Explanations, satisfactory from the philological point of view, which do not harmonise with these conditions have been rejected.

3. Special attention has been given to sixteenth, seventeenth, and eighteenth century spellings, which are of interest as suggesting peculiar local pronunciations. A rich harvest of phonetic spellings has been gathered from the Parish Registers, and one's only regret is that only too often pedantic misunderstanding has brought it about that names are now pronounced as spelled, rather than spelled as pronounced. Again and again local enquiry has failed to discover any trace of some perfectly legitimate pronunciation which must at one time have prevailed.

4. Many parallel names are quoted from other counties, but these have been rigidly limited to those in which old forms justify the parallel. Identity of modern form is often only misleading. For this and for certain details in Part II a detailed study has been made of all names which can now be identified in (1) Birch, Kemble, and Earle, (2) D.B., (3) Charter Rolls, (4) Index to Charters in British Museum, (5) Feudal Aids. In some counties other documents also have been studied.

Early documents are not plentiful, especially for Co. Durham. There are no Anglo-Saxon charters for these counties. *Boldon Buke*, with its thirteenth or fourteenth century spellings is, for the philologist, a poor substitute for Domesday Book, and for a large number of Durham place-names no forms have been found earlier than Bishop Hatfield's Survey (1382). This has made much of the interpretation uncertain, at least in the case of these last names, and the comments should perhaps have been seasoned with " probably " and " possibly " a good deal more frequently than they have been.

Work on a book of this kind means an ever increasing sense of indebtedness to the labours of other writers and scholars. This indebtedness is in some measure indicated by the lists of books on pp. xxviii.-xxxvi., but one must mention in particular the *New English Dictionary*, Wright's *English Dialect Dictionary* and *Grammar*, Hodgson's *History of Northumberland*, Surtees' *History of Durham*, Raine's *History of North Durham*, the publications of the Newcastle Society of Antiquaries (*Archaeologia Aeliana*), the Surtees Society, and the Northumberland and Durham Church Register Society. Without the aid of these the book could hardly have been written, and would certainly have lost any merit it may now possess.

Very cordial thanks are due to the Newcastle Society of Antiquaries for the use of the transcript of the *Lay Subsidy Roll* in their possession; to the Northumberland County History Committee for the use of His Grace the Duke of Northumberland's transcripts of the *Coram Rege*, *De Banco*, and *Placita Forestæ* Rolls temporarily in their possession, and for the use of the transcript of the *Feet of Fines* (1514-1603) ; to Mr M. H. Wood, Secretary of the Northumberland and Durham Parish Register Society, for the use of the invaluable series of unprinted transcripts of registers in his possession and for the most part copied by him personally; to Dr J. A. Smythe, of Armstrong College, whose unique knowledge of Northumbrian topography was most helpful in checking conclusions reached on philological grounds alone; to Mr T. W. Moles, of Rutherford College, for untiring efforts in gleaning local pronunciations; to scholars and friends too numerous to mention who have been most helpful in elucidating problems referred to them; to the many clergymen and others who so readily answered enquiries about local pronunciation.

Finally, the author feels himself peculiarly fortunate in having had help in the proof stage from the Rev. Canon Fowler, of whom all Northern scholars and antiquaries are so justly proud, and Mr Hamilton Thompson, whose antiquarian and topographical knowledge have alike been most helpful.

ALLEN MAWER.

NEWCASTLE-UPON-TYNE,
October 1920.

CONTENTS

INTRODUCTION

§ 1. *The Names of the Counties and their Divisions.*

Northumberland is one of those counties which, like Surrey, Essex, and Sussex, have taken their rise from the ruins of an ancient kingdom. The Anglian settlements of England north of the Humber—*Norðhymbraland*—were originally grouped in two kingdoms, Bernicia and Deira, of which Bernicia extended roughly from the Tees to the Forth and was bounded by the Pennines on the west. Of the name Bernicia not a trace remains, and the application of the term Northumberland was gradually restricted. The kingdom was reduced to an earldom, with its centre at Bamburgh, Lothian passed to the Scots, Durham developed into a palatinate under the rule of the Bishops of Durham, and from the 11th century on the term is used with increasing definiteness of an area corresponding, apart from certain notable exceptions (*v. infra*), to the modern county.

The Durham area, when first distinguished from the rest of the earldom of Northumberland, was known as *Haliwer(es)folc* or *Haliwersocn*=the people or soke (i.e. jurisdiction) of the holy man or saint, a term which is the equivalent of the common Latin expression *terra* or *patrimonium Sancti Cuthberti*. This term is found in the forged charter of Bishop William in 1093, and, technically at any rate, did not at that time include certain parts of the present county (e.g. Wirralshire *infra*), and did include considerable areas of territory (e.g. Norhamshire), which now lie within the county of Northumberland. It went out of use in the 15th century, but was revived by historians of an etymologising term of mind in the form *Haliwerk-folk*,[1]

---

[1] For this and other points see the clear and full account in Lapsley *The County Palatine of Durham*, pp. 22-4. An alternative name is *Cuthbert folk* (*v. Metrical Life of St Cuthbert, c.* 1430).

a form found sporadically as early as the 14th century, and then explained as "people of the holy work," i.e. people whose tenure depended on their fulfilling the duty of defending the body of St Cuthbert. There is no doubt that this is simply an antiquarian blunder.

Gradually the territory now known as County Durham came to be more and more completely in the hands of the Bishops of Durham, and their lands in Northumberland were known as North Durham, in contrast to these more southerly possessions. It is only since the abolition of the Palatinate jurisdiction in 1836 that the latter term has gone out of use.

Within the two counties there are several smaller districts to which the term *shire* has at various times been applied. These shires are of varied origin :—

(1) *Hexhamshire* is probably identical with the district originally granted to St Wilfrid for the endowment of the bishopric of Hexham. Authority within it, both ecclesiastical and civil, was long a matter of dispute between the Bishops of Lindisfarne and their successors the Bishops of Durham on the one hand and the Archbishop of York on the other, claiming as the successor of St Wilfrid. Ultimately the district became a regality under the jurisdiction of the Archbishops of York. The regality came to an end in the 16th century, but the district is still commonly known in S. Northumberland as "the Shire."

(2) *Bedlingtonshire, Islandshire, Norhamshire* were outlying portions of the palatinate of Durham, but the names are now only used archaically.

(3) *Bamburghshire*, called sometimes in medieval times the wapentake of Bamburgh, was applied to a district around Bamburgh of no very definite limits (N. i. 1). The term is no longer in use.

(4) *Tynemouthshire* was the name of a district around Tynemouth in which the monks of that priory had certain rights and privileges. The name has left its trace in Shire Moor, between Tynemouth and Newcastle.

(5) *Bywellshire* and *Feltonshire* are used sporadically (N. vi. 180, vii. 230) with no very definite connotation.

(6) In the charter of Bp. William (1093) the monks of Durham are granted certain lands in territory between the Tyne and the Wear, including the two Heworths, Headworth, Jarrow, Hebburn, Monkton, Preston, Westoe, Harton. This district is called *Werhale*, and the name must be a compound of the name *Wear* and *hale* (Part II), the name meaning "the corner of land by the Wear." Later the district is called *Weralshire, Werehal(f)shir*, or *Wirralshire*, but the name has now passed out of use.

(8) *Staindropshire* is applied to a district round Staindrop, conferred by Bishop Flambard on the monks of Durham. *Quarringtonshire* and *Billinghamshire* are also found occasionally.

§ 2. *The Celtic Element.*

The Celtic element in the place-names of Northumberland and Durham is certainly no stronger than in most English counties, and a good deal weaker than in those on the Welsh Border. There is no increase in the frequency of such names on the north-west and west borders of these counties such as might suggest an unsubdued Celtic element in the hill-country. The Anglian conquest was complete.

Here, as elsewhere, the river-names, except for some of the smaller burns, are uniformly Celtic. Allen, Alne, Ayle, Blyth, Derwent, Devil's (Water), Don, Eden (Burn), Glen Lyne, Ouse (Burn), Team, Tyne are found elsewhere in the same or similar forms. Others stand alone : Bowmont (Water), Breamish, Cong (Burn), Coquet, Deerness, Erring (Burn), Gaunless, Irthing, Nanny, Poltross (Burn), Pont (twice), Tees, Till, Tweed, Wansbeck, Warren (Burn), Wear. Other natural features are seldom mentioned in ancient documents and the only noteworthy Celtic hill-name recorded is Cheviot.

Of town-, village-, and farm-names that must be Celtic there are a good number. We may note Alwent (ultimately a river-name), Amble, Cambois, Carraw, Cocken, Glendue,

Jarrow, Kielder (probably a river-name), Lampart, Lindisfarne, Maughan, Mindrum, Painshaw, Plenmeller, Ross,
Tecket, Teppermoor, Trewhitt, Troughend, Wardrew,
Yeavering. Only in a very few cases is the meaning of
these names at all clear, e.g. Glendue, Ross. In some names
a Celtic element has been compounded with an English one :
Carrick, Cockerton, Ottercaps, Gloster Hill, Wooperton,
Wrekin Dike are examples of this. An interesting name of this
type is Kirkley (v. *infra*), where English *law* (later *ley*) has been
added to Celtic *cric*=hill, in explanation of an unfamiliar
term.[1] Carham and Crag Shiel contain Celtic elements which
had been naturalised in English speech. Corsenside would
seem to contain the Gaelic personal name *Crossan*. If that
is the case, the name could best be explained as due to some
settler of Hiberno-Scandinavian origin or connexion.

In dealing with many names in these counties which,
so far as the evidence goes, can be readily explained
as of English origin, one has the uneasy feeling that
these apparently genuine English names may really be
etymological perversions of Celtic names. O.E. *Eoforwic*
(York) and *Searoburh* (Salisbury) might be quite convinc-
ingly explained as " boar-dwelling " and " fort of trickery,"
did we not happen to have record of the earlier Romano-
Celtic forms—*Eburacum* and *Sorbiodunum*—which prove
quite clearly that the Old English names are due to folk-
etymology. There is good reason to believe that in the
counties under consideration, Auckland, Gateshead, Hexham show this process at an early date, and it is quite
possible there may be others which cannot be detected
on the evidence we have. Very few Northumbrian names
have been preserved in their O.E., let alone in their
British form. Such etymologising is clearly present at a
later date in Carrycoats, Hebburn, Heddon, Sowershope
Hill, Painshaw.[2]

---

[1] *Cf.* the addition of -*beck* after -*burn* (§ 4 *infra*.)

[2] In the case of two English names—Bamburgh and Tynemouth—
literary tradition in the one case and antiquarian research in the other
have preserved the ancient Celtic names which have no connexion with
the modern name (v. *infra*.)

Roman occupation has left hardly a trace in place-nomenclature, except indirectly in the fairly numerous *chesters*, so named from O.E. *ceaster* (< L. *castra*), a term applied by the English to any Roman fort or the ruins of such.

§ 3. *The English Element.*

The vast majority of the names both in Northumberland and Durham are of English or, more strictly speaking, of Anglian [1] origin. Many of them are doubtless of comparatively late date, and in some this is shown beyond doubt by their phonology, but the problem of the relative chronology of O.E. place-nomenclature is as yet an almost entirely untouched one.

Probably one of the oldest strata of names is that formed by the names in *-ingham*. It is shown below (pp. xxiv.-xxvii.) that these in all probability go back in every case to O.E. names in *-ingaham*, i.e. " homestead of the sons of —." If this is the case, all these names are probably old, representing as they do a primitive type of settlement. This view is further strengthened when we find that all such names, except Ealingham, belong to places lying in well-watered, fertile valleys, such as we may imagine would early be seized upon by our Anglian forbears.

The *-ington* names in Northumberland have already been the subject of comment. Dr Woolacott, writing in the *Geographical Journal* [2] in an article on the early settlement of Northumbria, says : " Another important effect of the Glacial period was that nearly the whole of the country was clothed in a mantle of drift. . . . The surface deposits lie thickest along the ancient washes. . . . On the higher ground some of the escarpments rise like islands from

---

[1] Throughout the book, however, O.E. words and names have been given in West Saxon rather than in Anglian form, unless some important phonological distinction was involved. Anglian forms would serve no useful purpose in the majority of cases, would not be so readily understood by readers of limited linguistic attainments in O.E., and could only be justified on pedantic grounds.

[2] July 1907, pp. 48-51.

beneath the cloak of superficial deposits, and in Northumberland especially, on account of the ease with which water could be obtained at these places, this has had considerable influence in determining the position of the minor places of settlement. The villages of Northumberland (especially those with the syllable *ing* in their names) are, as Topley pointed out many years ago, old settlements, and either stand on sand and gravel hillocks, lying on the boulder clay or on exposures of sandstone which rise above the uniform level of the surface formations. A large number of the pit villages, which are in many cases merely enlargements of the ancient settlements belong to the latter class, e.g. Killingworth, Widdrington, Earsdon."

With a view to testing this theory with regard to the -*ing* names, a fresh survey of the topography of all the genuine examples has been made, with the help of Dr J. A. Smythe and Dr Woolacott himself. Of the long line of -*ington* names, the theory would seem definitely to hold good of Acklington, Bedlington, Cramlington, Easington, Edington, Riplington, Shilvington, Widdrington. It is true also of Killingworth. All alike stand on high ground, where geological circumstance would favour the finding of springs. It is of doubtful application in Chevington, partly because it is difficult to be certain where the original settlement was. It can hardly be applied to Choppington, Tritlington, or Willington, for these stand on or near streams from which water could in any case be easily obtained.

In the west of the county geological conditions are different. There is not so much of the land surface covered with glacial drift. The water supply is dependent on different factors. Wallington, Kirkwhelpington, and Little Whittington stand on streams sufficient to give a water supply. Grottington, Bavington, Thockerington, and Great Whittington stand high on comparatively waterless country, and it is difficult to see how conditions of water supply can have had much to do with determining their position. To the north-west of Newcastle lie Dinnington, Dissington, and Woolsington. The first-named is a good illustration of the theory, but the other two lie low in well-watered country. Similarly there is the group Eslington, Titlington, Yetlington. The first two lie in well-watered valleys, the last is in much the same kind of country as the Grottington group. Finally Doddington, in the Till valley, and Hetherington and Shitlington, in North Tyndale, are well supplied from the rivers themselves.[1]

Taking all the evidence into consideration, it would seem that the theory can only be established at all for Eastern Northumberland. There it does seem that the proportion of -ington names on ground of the type indicated is too large to be due to coincidence alone.

In County Durham, Farrington, Herrington, Merrington, and Easington fulfil the required conditions. Washington is a doubtful case. Heighington and Lutterington are definitely against the theory, and Stillington stands in a well-watered plain.

English suffixes which are specially common in Northumberland are haugh, law, hope, peth or path, shiel. Of these all except haugh are fairly common in Durham also : shiel is confined to these two counties.

§ 4. *The Scandinavian Element.*

Northumberland and Durham are not counties in which the evidence for Scandinavian settlement is strong. In Nthb. there are no examples of *by, beck* (Wansbeck is deceptive), *toft, thwaite, garth, scale*, suffixes specially characteristic of that settlement. In Co. Durham there is only one *toft* and one *garth*, and no example of *beck* has been found in any early document. Such names as Euden Beck, Thornhope Beck all originally show *burn*, and the same is true of such pleonastic forms as Linburn Beck. The use of *beck* in modern times must be due to the influence of neighbouring Yorkshire custom. There are, however, some eight examples of *by*; of these four are in and near Teesdale—Aislaby, Ulnaby, Killerby, Raby; two—Ornsby

[1] Rennington does not come into consideration as it would seem to be of ninth-cent. origin (*v. s.n. infra*).

and Rumby—are in the north-west of the county; and two—Follingsby and Raceby—in the north-east and east.

Names which can only have been given by men of Scandinavian birth (i.e. names built up from elements which, so far as we know, were never naturalised in England) are few and far between. In Nthb. we have Akeld and Copeland in the Till and Glen valleys, North Sunderland and Lucker near Bamburgh, Howick near Embleton, Tosson and Snitter in Coquetdale, Dingbell Hill near Whitfield, Knaresdale on the S. Tyne, in Durham we have Hurbuck (with Ornsby) near Lanchester and Hutton (Henry).

Such names, when they occur in isolation, are probably due to direct transference by some Scandinavian settler of a name familiar to him in his own country, not necessarily with any very precise reference to the topographical conditions prevailing in the English place.

When they occur in a group, they may point to a more or less extensive settlement in which an Anglo-Scandinavian dialect was at one time spoken. One such is clearly that of Sadberge, Skerningham, and the river Skerne, including possibly (Newton) Ketton, all in the *wapentake* of Sadberge, a Scandinavian unit of territory found nowhere else north of the Tees. Another such group is found in Gainford and district. Here Dyance, Staindrop, Raby, Killerby, Ulnaby are purely Scandinavian, Eggleston and Ingleton probably so, Coniscliffe has been modified from its earlier Anglian form under the influence of Scandinavian speech, Gainford is probably named from a Norse settler, and Cleatlam is a hybrid such as might well arise in a district of this type. Possibly a third group is found in a well-marked series of names round Rothbury. Tosson and Snitter are purely Scandinavian, Cartington is named after a settler bearing the rare Hiberno-Scandinavian name Kiartan, Bickerton is an Anglo-Scandinavian hybrid, Brinkburn may be such, while Plainfield and Rothbury would seem to be named from their Scandinavian settlers.

The presence of a Scandinavian personal-name as the first element in a place-name is not decisive as to

Scandinavian settlement unless the name is very rare in Scandinavia or one which, so far as our knowledge goes, was never naturalised in England. Thus it is probable that Carp Shiel, Cartington, Claxton, Eltringham, Farrow Shiel, Henshaw, Hisehope, Nafferton, Offerton, Pandon, Glanton, Glantlees, Scrainwood, Stirkscleugh, Slingley, Thrundle, Thrunton, Tranwell, Trewitley, Whessoe take their names from men of Scandinavian birth as well as Scandinavian name, but one cannot be so sure in names like Blakeston, Brancepeth, Brotherwick, Ilderton, Lumley, Ouston, Stoney Burn, Stannington, Swainston, Thrislington, Ushaw. If the name, though naturalised, is found in a district in which Scandinavian influence may be suspected on other grounds, then the probability is in favour of an actual Scandinavian settlement. Examples of this are " Routh " in Rothbury, and " Ingald " in Ingleton. Further, these names are sometimes found in such definitely marked groups that we may naturally infer that we have to do with settlements made by men of the same origin. Such groups are Dotland and Eshells in Hexhamshire, Ouston by Dingbell Hill in Whitfield, Bolt's Law, Hisehope, Carp Shiel, together with Waskerley and Nookton in N.W. Durham ; Thrislington, Tursdale, Thrundle, and Raceby, near Ferryhill ; Amerston, Swainston, Blakeston, and Claxton, together with Carlton and Thorpe (Thewles) near Wynyard. It is not suggested that these groups were of the same intimate character as the Sadberge, Gainford, and Rothbury ones indicated above. There is no evidence here that we have to do with districts in which an Anglo-Scandinavian dialect was ever actually the current speech.

A considerable number of Scandinavian loan-words are found in the dialects of Northumberland and Durham. Such words do not necessarily prove Scandinavian settlement when found either there or in place-names. They may well have found their way here through contact with dialects of a more definitely Scandinavian complexion. We have clear evidence of this in the case of *beck* (*v. supra*) and *dale* (cf. Harsondale, Tursdale, with earlier *-den*), and it is probably true of *-biggin* in the numerous Newbiggins. The chief of these dialect words are *car* or *ker* in Byker, Walker ; *carl* in Carlbury ; *clints* in Clints Wood ; *crook*, in Crawcrook, Crook, Crookham, Crookhouse, Crookburn ; *haining*, *hagg*, in Hagg Wood, ; *felling* (found in North M.E.), *flat* in Shortflatt and Oxney-flat ; *stain*, in Fourstones, Stonecroft and *Stainshaw*, for Stagshaw ; *wham* in Wham and Whitwham.

Hybrid formations, consisting of a Scandinavian first element (other than a personal name) which was never naturalised and an English second element, are improbable except in a district where a mixed Anglo-Scandinavian speech prevailed. This makes the assumption of a Scandinavian element in Nookton, Satley, and Waskerley somewhat doubtful. It is more likely in Plainfield.

School Aycliffe is of special interest in that we can with fair certainty identify the Skúli, from whom it takes its name, with the Scula who, together with Onlafbald, divided the patrimony of St Cuthbert (*c.* 920).[1] Bulbeck in Northumberland reminds us that the Norman-French lords were ultimately Scandinavian in origin and speech.

The general distribution of these names compels us to believe that such Scandinavian settlements as there may have been, were made by men arriving either from the sea and moving up the great river-valleys, or, to some extent at least, by men moving up from more southerly and more distinctively Scandinavian districts. There is no evidence which could support the idea of an influx from the west, either from Cumberland or from S.W. Scotland, Dumfries, and Galloway. Further, there may have been very extensive ravagings of the two counties by Viking invaders, but there was, on the other hand, no definite and permanent parcelling out of the land of these two counties among alien settlers who had ousted the old Anglian population from their farms.

§ 5. *The French Element.*

The number of place-names purely French, or rather Anglo-French, in their form is not large. There are some

---

[1] Hist. Regum, § 83.

nine examples of the common type of French name in *Beau-* or *Bel-*, viz. Beamish, Bear Park, Bewdley, Bewley, Beaumont (2), Bellasis (3). *Carriteth* and *Plessey* are names formed from elements never naturalised in English, though there is evidence for the common use of the latter in Anglo-French. *Landieu* and *Blanchland* are purely French formations. *The Hermitage, Close House, Pallion* are examples of English loan-words from French. There are several examples of *le* (O.Fr. *les*) meaning " near, by the side of," viz. Chester-le-Street, Dalton-le-Dale, Haughton-le-Skerne, Houghton-le-Side, Witton-le-Wear, and the suffix was sufficiently common to give rise to illegitimate forms such as Hetton-le-Hole, Houghton-le-Spring (*v. infra*).

At one time a good many manors were distinguished by the addition of the names of their Norman or French holders, but many of these have now lost this distinctive addition. It is retained in Seaton Delaval, Coatham Mundeville, Dalton Piercy, Newton Hansard, but lost in Brunton (Bataill) in Embleton, Callerton (Delaval and Valence), Dissington (Delaval), Horton (Guiscard) in Blyth, Horton (Turberville) in Doddington. In Darras the name of the holder alone is retained, and the same is true of Feugar House, Puncherton, and Gubeon. In Battle Shield and Whisker Shiel the French name is found as the first element, so also in Heron's Close and Barnard Castle. Guyzance is probably of the same type as Darras, though it may be an example of direct transference of a French to an English place-name.

A few names show the influence of Anglo-French spelling and pronunciation in their ultimate forms, leading at times to strange transformations. This is specially true of Dissington, Dyance, Sedgefield, and Sessinghope.

NOTE

Names in -ing.

One of the most difficult problems in English place-nomenclature is the interpretation of the element -*ing*- in early forms.

In O.E. we have three types of -*ing*- names, viz. :—

(I) Gen. pl. in -inga-, e.g. *Ricingahaam* (B.C.S. 81), *billingabyrig* (ib. 144).

(II) Gen. sg. in -inges-, e.g. *heringesleah* (ib. 543).

(III) Simple -ing-, e.g. *lavingtun* (ib. 144), *tucincgnæs* (227), *dunincglond* (254), *sceofingdene* (370).

Types I and II clearly represent gen. pl. and gen. sg. respectively of patronymics in -*ing*. By far the commonest suffix found after *inga* is *ham*, but other suffixes, such as *burh*, *leah*, *weald*, *burna*, are found sporadically, the only suffix which is unexpectedly rare is the common *tun*. In all the *Cartularium*,[1] only two examples have been noted—*wassingatun* (834), *hwessingatun* (1131). Of the names in -*inges*, four are boundaries (cf. Nos. 5, 428, 784, 801), one is compounded with *hlaw*, referring probably to a barrow (No. 777), only the remaining four are, strictly speaking, place-names, viz., *fyrdingeslea* (27, 391, 690), *liabingescota* (518), *heringesleah* (543), *scyllinges broc* (505).[2]

The rareness of names of Type II is what we might expect, for personal names of patronymic form are very rare in O.E. Examination of the pages of Searle yields very few examples, even if we count examples of the type *Aelfred Aeþelwulfing*, where the patronymic is not independently

[1] Here and elsewhere no attempt is made to deal with charters in purely M.E. language. Their evidence is not contradictory to the conclusions here arrived at, but is valueless.

[2] In our counties Muggleswick (*v. infra*) is the only name which is unequivocally of this type.

used as a personal name. Redin (pp. 165-174) gives some forty in all.

Type I is evidently of older date and represents a stage in the development of Teutonic society when the clan-feeling was still strong, and a whole body of settlers might be named, not necessarily after their immediate father, but after some more remote common ancestor. Thus the Danish royal house of the *Scyldingas* were not so called because they were sons of Scyld, but because the whole house was thought to be descended from this eponymous ancestor. Similarly we should probably be correct in interpreting such a name as Edlingham *infra* as " homestead of the *descendants* of Eadwulf," rather than of the more strictly limited *sons* of Eadwulf.

A similar type of place-name is that which in O.E. documents is represented by the nom. pl. *-ingas*, named from such a family group without the addition of any suffix at all, e.g. *Geddingas* (265), *Mallingas* (421), and the familiar Reading from *Readingas*. Birling *infra* is clearly an example of this kind.

Type III is by far the commonest, and is the one about which controversy has chiefly arisen. Bradley endeavoured to explain the difference between names in *-ing* and *-inga* as due to the number of syllables in the personal name, *-ing* being favoured in the case of polysyllabic names; but Alexander, after further examination of the evidence (*Essays and Studies by Members of the English Association*, vol. i. p. 7, vol. ii. pp. 158-182), showed this not to be the case. Others have suggested that all names in *-ing* really represent earlier *-inga* names with loss of the inflexional suffix. Ekblom (*Place-Names of Wiltshire*, pp. 6, 7) takes this view strongly, but it is impossible to agree. Examples of these names are found in some of the earliest of our charters, preserved in original or contemporary copies (e.g. B.C.S., 81, 144, 227, 254, 289, 293, 332, 335), where inflexional loss is out of the question.[1]

There are, however, graver and more serious objections to such an interpretation. It would compel us to believe in

---

[1] Only two examples of names spelled alternatively with *-inga* and *-ing* have been noted, neither in early documents.

a far wider prevalence of a clan system, or at least a system of inheritance between the sons of a family, than English social history gives any warrant for. Rather we must believe, with the late Professor Moorman (*Place-Names of the West Riding*, Introd. p. xli), that the -*ing* in these names is not a patronymic suffix at all, but has possessive force, e.g. *dunincglond*=Dunna's Land.

Examination of the O.E. evidence confirms this theory beyond the possibility of doubt. For

(1) What can we make of names like *werburgingwic* (373), *cyneburgincgtun* (1436) on the patronymic-theory? Patronymics are not formed from women's names.

(2) Still more impossible as a patronymic is the first element in *bisceopincgdene* (378). This can only be interpreted as " Bishop's *dene*."

(3) There is definite evidence for equating such names with possessive forms. Birch No. 97 is an original 7th century charter dealing with a grant of land at *wieghelmestun*. This charter is endorsed in a late 10th or early 11th century hand,[1] and the name of the land is given as " nunc *wigelmignctun* " (*sic*).[2] Rennington *infra* is by Simeon of Durham (i. 80) explained as named after Reingualdus—" a quo illa quam condiderat villa Reiningtun est appellata."

With this explanation we see at once how it comes about that -*ing*- names are far more numerous than names in -*inga*- or -*inges*-, for it is no longer a question of the comparatively rare patronymic names. Such place-names are simply the farm, clearing, or whatever it may be, of or belonging to a man bearing a certain name. Probably we should be right in interpreting all genuine -*ington* names in Nthb. and Durh. as representing O.E. -*ingtun* rather than the very rare -*ingatun*, but in the absence of any O.E. forms for Nthb. and Durh., the alternative possibility has been allowed for in the explanations of such name given below.

In M.E. there are two other possible sources for -*ing*-names :—

---

[1] This information is due to the kindness of Mr J. P. Gibson of the British Museum.

[2] To be identified with Wilmington in Sellinge.

(1) O.E. *-an*, gen. sg. of a personal name of weak form. This becomes M.E. *-en, -in,* and later *-ing,* as in the familiar example of *Abbandun* (=Abba's Hill) > M.E. *Abbendon* > *Abingdon.* This explains a great many place-names in *-ing,* but cannot apply to any names in Nthb. and Durham, for here the *n* of the suffix *an* was lost long before the M.E. period, e.g. *Tunnacæstir* (Bede, iv. 22), named after *Tunna,* priest and abbot. In Merc. and West Saxon we should have *Tunnanceaster.*[1] Forms with persistent *en* in M.E. forms must then be given some other explanation, at least in these counties, e.g. Beadnell *infra* must be from *Bedwine* rather than from *Bedan.*

(2) There is a M.E. *ing,* a Scand. loan-word meaning "grass-land," which is in common use in certain dialects. It is found in Nthb. and Durham, though it is not common in the former county. In place-names it is certainly found in Skirningham, and possibly in Broomyholm, Elrington, and Stannington *infra.*

In Mod. Eng., under the influence of the analogy of the very common *-ing-* type of name, certain names have received *-ing* forms quite unjustifiably. Examples in Nthb. and Durham are Errington, Follingsby, Hallington, Hartington, Hollingside, Lemmington, Yeavering.

---

[1] The form *Brincaburch,* given under Brinkburn *infra,* is probably an archaic survival of an *n*-less gen. sg.

BIBLIOGRAPHY

i. *Primary Authorities cited for place-name forms in the two counties, and, where marked with an asterisk, only for counties other than these two.*

	Abbreviated reference.
Placitorum Abbreviatio, 1811.	Abbr.
Durham Account Rolls. (Surtees Soc., vols. xcix., c., ciii.)	Acct.
Tracts Printed at the Darlington Press of G. Allan, Esq., relating to the County of Durham, 1777.	Allan
Catalogue of Ancient Deeds in the Public Record Office. 6 vols. In progress. 1890–.	Anc.D.
Archaeologia Aeliana. 1st Series, 4 vols. ; 2nd Series, 25 vols. ; 3rd Series (in progress). Newcastle-on-Tyne.	Arch.
Two of the Saxon Chronicles. Ed. Plummer and Earle. 2 vols. Oxford, 1892–9.	A.S.C.
Three Early Assize Rolls for the County of Northumberland. (Surtees Soc., vol. lxxxviii.)	Ass.
Injunctions of Bishop Barnes. (Ib. vol. xxii.)	Barnes
Bedae Opera Historica. Ed. Plummer. 2 vols. Oxford, 1896.	Bede
Birch, W. de G. Cartularium Saxonicum, 1885–9.	B.C.S.
Boldon Buke. (B. Bodleian MS., C. Chapter MS.) (Surtees Soc., vol. xxv.)	B.B.
Black Book of Hexham. (v. Hexham Priory *infra.*)	B.B.H.
Index to Charters and Rolls in the British Museum. 2 vols. 1900–12.	B.M.
Border Papers. Ed. Bain. 2 vols. 1894–6.	Bord.
Nicholson, Wm. (Bp.) *Leges Marchiarum,* or *Border Laws.* 1747.	Bord. Laws.
Survey of the East and Middle Marches, by Sir Robert Bowes and Sir Ralph Ellerker. (H. 3. 2. 171–248)	Bord. Surv.
Chartulary of Brinkburn Priory. (Surtees Soc., vol. xc.)	Brkb.

Abbreviated
reference.

Richard d'Aungerville of Bury. *Fragments of* Bury
his Register and other Documents. (Surtees
Soc., vol. cxix.)

Camden, Wm. *Britain* (tr. P. Holland). 1637. Camden

Calendar of Charter Rolls. In progress. 1903–. Ch.

Calendar of Close Rolls. In progress. 1892–. Cl.

Coldingham Documents. (v. Raine, *History of* Colding.
North Durham.)

Acts of the High Commission Court within the Comm.
Diocese of Durham. (Surtees Soc., vol.
xxxiv.)

Records of the Committee for Compounding in Comps.
Durham and Northumberland. (Surtees
Soc., vol. cxi.)

Coram Rege Rolls (Transcripts). Coram

*Crawford Charters. Ed. Napier and Stevenson. Crawf.
1895.

Two 13th Century Ass. Rolls for Co. Durham. D.Ass.
(Surtees Soc., vol. cxxvii.)

De Banco Rolls (Transcripts). De Banco

*Domesday Book. 4 vols. 1783–1816.[1] D.B.

Historiae Dunelmensis Scriptores Tres. (Surtees D.S.T.
Soc., vol. ix.)

Dugdale, Sir Wm. *Monasticon Anglicanum.* Dugdale
6 vols. 1817–30.

Depositions and other Ecclesiastical Proceedings Eccl.
from the Courts of Durham. (Surtees Soc.,
vol. xxi.)

Registers of Baptisms, etc., in Elsdon. 1903. Elsdon

Register of Christenings, etc., at the Ancient Esh
Chapel of Esh. 1896.

Inquisitions and Assessments relating to Feudal F.A.
Aids. In progress. 1899.

Feet of Fines, 1514–1603. Transcribed by Miss F.F.
Emma Welford.

Charters, etc., of the Priory of Finchale. (Surtees Finch.
Soc., vol. v.)

Calendar of Fine Rolls. In progress. 1911–. Fine

Feodarium Prioratus Dunelmensis. (Surtees F.P.D.
Soc., vol. lviii.)

List of Freeholders of Northumberland in 1628 Freeh.
and 1638–9. (Arch. Ael. i. 2., pp. 316–25.)

[1] Transcripts in the *Victoria County Histories* have also been freely
used.

	Abbreviated reference.
Index to the Gainford Registers. Parts I–III. 1884–90.	Gainf.
Geoffrey of Coldingham. (*v.* D.S.T.)	Geoffr.
The Registers of Walter Giffard, Archbishop of York. (Surtees Soc., vol. cix.)	Giff.
Libellus de vita et miraculis S. Godrici. (Surtees Soc., vol. xx.)	Godr.
Register of Walter Gray, Archbishop of York. (Surtees Soc., vol. lvi.)	Gray.
Halmota Prioratus Dunelmensis. (Surtees Soc., vol. lxxxii.)	Halm.
Bishop Hatfield's Survey. (Surtees Soc., vol. xxxii.)	Hatf.
Priory of Hexham. (Surtees Soc., vols. xliv. and xlvi.)	Hexh. Pr.
Survey of Hexham Manor, 1547 and 1608. (*v.* N. iii. 66–104.)	Hexh. Surv.
Historia Regum. (*v. Symeonis Monachi Opera infra.*)	Hist. Reg.
Historia de Sancto Cuthberto (ib.)	H.S.C.
Calendarium R.C. et Inquisitionum ad quod damnum. 1803.	Inq. a.q.d.
Calendar of Inquisitions Post Mortem. In progress. 1904–.	Ipm.
Inquisitions Post Mortem in 45th Report of Deputy Keeper of Records.	Ipm.
Feudal and Military Antiquities of Northumberland. Ed. Hartshorne. 1858. App. pp. 9–68. *Iter of Wark.*	Iter
Inventories and Account Rolls of Jarrow and Monk-Wearmouth. (Surtees Soc., vol. xxix.)	J. and W.
John of Hexham. (*v.* Hexh. Pr., vol. I.)	Joh. Hexh,
Jordan Fantosme. *Chronicle of the War between the English and the Scots.* (Surtees Soc., vol. xi.)	Jord.
Kemble, J. M. Codex diplomaticus ævi Saxonici. 6 vols. 1839–48.	K.C.D.
Registers of Lanchester. 1909.	Lanch.
Leland, J. *Itinerary.* Ed. L. T. Smith. 5 vols. 1906–10.	Léland
Joannis Lelandi Collectanea. Ed. Hearne. 6 vols. 1744.	Leland
List of Knights in the Liberty of Durham at the Time of the Battle of Lewes. (*v.* Hatf. Surv.)	Lewes
The Lincolnshire Survey.	Lincs. Surv.

Abbreviated
reference.

Liber Vitae Ecclesiae Dunelmensis. (Surtees
Soc., vol. xiii.) [1]   L.V.D.

Ancient Maps in the Public Library, Newcastle-   Maps
on-Tyne ; The University Library, Durham ;
The Library of the Newcastle Society of
Antiquaries ; and the Library of the
Literary and Philosophical Society, New-
castle-on-Tyne.

Parish Registers of Muggleswick. 1906.   Muggles.

Musters for Northumberland in MDXXXVIII.   Must.
(Arch. i. 4. 157–206.)

North Country Diaries. (Surtees Soc., vol. cxxiv.)   N.C.D.

North Country Wills. (Surtees Soc., vol. cxxi.)   N.C.W.

Chartularium Abbathiae de Novo Monasterio.   Newm.
(Surtees Soc., vol. lxvi.)

The Old English Version of Bede's Ecclesiastical   O.E. Bede
History. E.E.T.S. Orig. Series. (Vols.
xcv., xcvi.)

Rotulorum Originalium Abbreviatio. 2 vols.   Orig.
1810.

Papal Registers and Papal Letters. In progress.   Pap.
1893–.

Calendar of Patent Rolls. In progress. 1891–.   Pat.

Extracts from Patent Rolls. (Arch. i. 3. 51–75.)   Pat.

Percy Chartulary. (Surtees Soc., vol. cxvii.)   Perc.

Great Roll of the Pipe. (Pipe Rolls Soc.). In   Pipe
progress. 1884–.

Magnus Rotulus Pipae. (*v.* H. 3. 3. 1–306.)   Pipe

Pipe Rolls of 1st, 2nd, and 3rd Edw. I. (Arch.   Pipe
I. 4. 207–60.)

The Pipe Rolls for the Counties of Cumberland,   Pipe
Westmoreland, and Durham. 1847.

Placita Forestae (Transcripts).   Plac. For.

Placita de Quo Warranto. 1818.   Q.W.

Raine, J. *History of North Durham.* 1852.   Raine

Red Book of the Exchequer. 3 vols. 1896.   R.B.E.

Calendarium Rotulorum Chartarum, etc. 1803.   R.C.

A Rental of the Ancient Principality of Redesdale.   Redesd.
(Arch. I. 2. 326–338.)

Reginaldi Monachi Dunelmensis Libellus, etc.   Reg. Dun.
(Surtees Soc., vol. i.)

Rentals and Rates for Northumberland in 1663.   Rental
(H. 3. 1. 243–347.)

[1] Sweet's ed. of part of this in *Oldest English Texts* has also been used.

	Abbreviated reference.
Rotuli Hundredorum. London, 1818.	R.H.
Richard of Hexham. (*v.* Hexh. Pr.)	Ric. Hexh.
Robert of Graystanes. (*v.* D.S.T.)	Robt. de Grayst.
Registrum Palatinum Dunelmense. 4 vols. 1813–8.	R.P.D.
Calendar of Documents relating to Scotland. 4 vols. 1881–8.	Sc.
Historia Dunelmensis Ecclesiae. (*v. Symeonis Monachi Opera Omnia.* 2 vols. 1882–5.)	S.D.
Lay Subsidy Roll. (Tr. by G. G. Baker Cresswell.)	S.R.
Speed, John. *Maps of Northumberland and the Bishopric of Durham.* 1608.	Speed.
Charters in the possession of Sir J. E. Swinburne. (H 3. 1. 1–25.)	Swinb.
Taxatio Ecclesiastica. 1802.	Tax.
Testa de Nevill. 1807.	T.N.
Appendix of Original Documents in Gibson, M.S., *History of the Monastery of Tynemouth.* 2 vols. 1846–7.	Ty.
The Parish Registers of Tynemouth. Vol. i.	Tyne.
Calendar of Various Chancery Rolls. 1912.	Var.
Registrum Cart. Conv. de Holne. (*v. Hartshorne, Feudal and Military Antiquities u.s.* App. pp. 69–109.)	Vescy
The Visitation of Northumberland in 1618. 1878.	V.N.
Visitations of the North. Part I. (Surtees Soc., vol. cxxii.)	V.N.
Registers of Warkworth Parish. 1899.	Warkw.
MSS. of Louisa, Marchioness of Waterford. *Hist. MSS. Comm.* 11th Report. App. Part VII.	Waterf.
Whellan & Co. *History, Topography, and Directory of Northumberland.* 1855.	Whellan.
The Register of William Wickwane, Lord Archbishop of York. (Surtees Soc., vol. cxiv.)	Wickw.
Wills and Inventories. (Surtees Soc., vols. ii. and xxxviii.)	Wills
31st, 32nd, 33rd, 34th, 35th, 36th, 45th Annual Report of the Deputy Keeper of Records.	31, 32, 33, 34, 35, 36, 45.

Abbreviated
reference.

Northumberland and Durham Parish Register
Society. (1) Whickham (*Whickh.*), (2)
Eglingham (*Egling.*), (3) Ebchester (*Ebch.*),
(4) Stanhope, (5) Bothal with Hebburn
(*Bothal*), (6) Hebburn, (7) Ryton, (8)
Ingram, (9) Edlingham (*Edling.*), (10) St
Margaret's Durham (*St Marg.*), (11) Whit-
burn (*Whitb.*), (13) Middleton St George,
(14) Bp. Middleham, (15) Alnham, (19)
Coniscliffe (*Coniscl.*), (22) Whalton, (25)
Corbridge (*Corbr.*), (27) St Mary le Bow,
Durham (*St Mary le B.*), (30) Castle Eden
(*Castle E.*), (31) Sherburn Hospital (*Sherb.*),
(32) Chatton, (34) Meldon, (35) Ilderton.

Transcripts of Registers in the possession of
M. H. Wood, Esq.: Alwinton (*Alw.*), Bel-
ford, Long Benton, Embleton, Ford, Hart-
burn, Haydon, Horton, Houghton-le-Spring
(*H. le Spr.*), Knaresdale, Lambley, Lan-
chester (*Lanch.*), Long Framlington, Lowick,
Mitford, Netherwitton (*Netherw.*), Newburn
(*Newb.*), Norham, Ovingham, Ponteland,
Redmarshall (*Redm.*), Ryton, Sedgefield
(*Sedgf.*), Shotley, Staindrop, Tweedmouth,
Bishop Wearmouth (*Bp. Wearm.*), Whitfield
(*Whitf.*), Witton Gilbert, Woodhorn.

ii. *Books on Place- and Personal Names.*

ALEXANDER, H. Place-names of Oxfordshire. 1912.

BADDELEY, W. St C. Place-names of Gloucestershire.
1913.

BANNISTER, A. T. Place-names of Herefordshire.
1916.

BARDSLEY, C. W. Dictionary of English and Welsh
Surnames. 1901.

BJÖRKMAN, E. Nordische Personennamen in England.   N.P.
1910.

BJÖRKMAN, E. Zur Englischen Namenkunde. 1912.   Z.E.N.

DUIGNAN W. H. Notes on Staffordshire Place-names.
1902.

DUIGNAN, W. H. Warwickshire Place-names. 1912.

DUIGNAN, W. H. Worcestershire Place-names. 1905.

EKBLOM, E. Place-names of Wiltshire. Uppsala,
1907.

Abbreviated
reference.

EKWALL, E.   Scandinavians and Celts in the North-
West of England. Lund, 1918.

FABRICIUS, A.   Danske Minder i Normandiet.   1897.

FALKMAN, A.   Ortnamen i Skåne.   Lund, 1877.

FORSSNER, T.   Continental-Germanic Personal Names
in England. Uppsala, 1914.

FÖRSTEMANN, E.   Altdeutsches namenbuch.   Vol. 2.
Ortsnamen.   Ed. Jellinghaus.   3rd ed.   1913.   14
Parts. In progress.

GILLIES, H. C.   Place-names of Argyll.   1906.

GOODALL. A.   Place-names of S.W. Yorkshire.   1914.

HEINTZE, R.   Die Deutschen Familien-namen.   1903.

HELLQUIST, E.   Svenska Ortnamnen pa -inge, -unge,
ock-unga.   Göteborg, 1904.

HOGAN, E.   Onomasticon Goidelicum.   1910.

HOLDER, A. T.   Alt-celtisches sprachschatz.   In progress.
Leipzig, 1896–.

JACKSON, C. E.   Place-names of Durham.   1916.

JAKOBSEN, J.   Shetlandsøerne Stednavne.   (Aarb. for
Nord. Oldkynd. og Hist.   1901.   pp. 55-258).

JAKOBSEN, J.   Stednavne og Personnavne i Normandiet.   Danske Studier.   1911.   pp. 59–84.

JELLINGHAUS, H.   Die   westfälischen   Ortsnamen.
1902.

JOHNSTON, J. B.   Place-names of England and Wales.
1915.

JOHNSTON, J. B.   Place-names of Scotland.   1903.

JÓNSSON, F.   Tilnavne i den islandske oldlitteratur.
(Aarb. f. Nord. Olldkynd. og Hist.   1907).

JÓNSSON, F.   Bæjanöfni á Islandi.   1911.

JOYCE, P. W.   Origin and History of Irish Names of
Places.   Two Series.   1870, 1875.

KAHLE, B.   Die Altwestnordischen Beinamen.   (Arkiv.
f. Nordisk Filologi.   New Series.   Vol. xl.   pp.
142–202, 227–268.)

KÅLUND, P. E. K.   Bidrag til en historisk-topografisk
Beskrivelse af Island.   2 vols.   1877–82.

LIND, E. H.   Norsk-isländska dopnamn ock fingerade
namn från medeltiden.   1905-15.

Abbreviated
reference.

LINDROTH, Hj. Bohusläns Härads ock Sockenamn.
1918.

LINDKVIST, H. Middle-English Place-names of Scandinavian Origin. 1912.

M'CLURE, EDMUND. British Place-names in their
Historical Setting. London, 1910.

MATHESON, D. Place-names of Elginshire. 1905.

MAXWELL, Sir H. E. Studies in the Topography of
Galloway. 1887.

MIDDENDORF, H. Altenglisches Flurnamenbuch. 1902.

MILNE, J. Celtic Place-names in Aberdeenshire. 1912.

MOORE, A. W. Manx names. 1903.

MOORMAN, F. W. Place-names of the West Riding of
Yorkshire. Thoresby Society, 1910.

MORGAN, T. Place-names of Wales. 1912.

MUTSCHMANN, H. Place-names of Nottinghamshire.
1913.

NAMN OG BYGD. Tidskrift för Nordisk Ortnamnsforsk-
ning. 1913. In progress.   No. B.

NAUMANN, H. Altnordische Namenstudien. 1912.

NIELSEN, O. Olddanske Personnavne. 1883.

REDIN, M. Uncompounded Personal Names in Old
English. 1919.

ROBERTS, R. G. Place-names of Sussex. 1914.

RYGH, O. Norske Gaardnavne. Forord og Indledning.   Indl.
1898.

RYGH, O. Gamle Personnavne i Norske Stednavne.   G.P.

RYGH, O., and OTHERS, ed. Norske Gaardnavne. In
progress. 1898–.   N.G.

SCHÖNFELD, M. Wörterbuch der Altgermanischen
Personen- und Völkernamen. 1912.

SEARLE, W. G. Onomasticon Anglo-Saxonicum. 1897.

SEPHTON, J. A Handbook of Lancashire Place-names.
1913.

SKEAT, W. W. Place-names of Bedfordshire, 1906.

SKEATS, W. W. Place-names of Berkshire. 1911.

,,   ,,   Place-names of Cambridgeshire. 1911.

,,   ,,   Place-names of Hertfordshire. 1904.

,,   ,,   Place-names of Huntingdonshire.
(Camb. Antiq. Soc.'s Comm., vol. x. pp. 317–60.)

Abbreviated
reference.

SKEAT, W. W.   Place-names of Suffolk.   1913.

STEENSTRUP, J.   Indledende Studier over de ældste Danske Stednavnes Bygning.   1909.

STENTON, F. M.   Place-names of Berkshire.   1911.

STEVENSON, W. H.   Unexplained O.E. words.   (Phil. Soc. Trans., 1895-8, pp. 528-42.)

STURMFELS, W.   Die ortsnamen Hessens.   Second edition.   1910.

Sveriges Ortnamn.   Älvsborgs Län.   In progress. 1906–.

TAYLOR, J.   Words and Places.   Revised ed.

Victoria History of the Counties of England.   In pro-   V.C.H. gress.   1900–.

WALKER, B.   Place-names of Derbyshire.   (Derbys. Arch. and Nat. Hist. Soc.'s Journal.   Vol. xxxvi. 1914–15.)

WATSON, W. J.   Place-names of Ross and Cromarty. 1904.

WEEKLEY, E.   Surnames.   1916.

WINKLER, J.   Friesche naamlijst.   1898.

WYLD, H. C., and HIRST, T. O.   Place-names of Lancashire.   1911.

ZACHRISSON, R. E.   Anglo-Norman Influence on English Place-names.   1909.

ZACHRISSON, R. E.   Some Instances of Latin Influence on English Place-Nomenclature.   1910.

iii. *Books on the History, Topography, and Dialect of Northumberland and Durham.*

BATES, C. J.   Border Holds of Northumberland.   1891.

BOYLE, J. R.   The County of Durham.   1892.

DIXON, D. D.   Upper Coquetdale.   1903.

EGGLESTON, W. M.   Weardale Names.   1886.

HESLOP, R. O.   Northumberland Words.   1892–4.

HODGSON, J., and HINDE, J. H.   History of Northumberland.   6 vols.   Pt. I, Pt. II, vols. i.–iii.; Pt. III, vols, i.–iii.   1820–58.   H.

History of Northumberland.   In progress.   1893–.   N.

RAINE, J.   History of North Durham.   1852.

SURTEES, R.   History of the County Palatine of Durham.   4 vols.   1816–40.

Abbreviated
reference.

TATE, G. History of Alnwick. 2 vols. 1846–9.

TOMLINSON, W. W. Guide to Northumberland. *n.d.*

WALLIS, J. Nat. Hist. and Antiquities of Northumberland. 2 vols. 1759.

iv. *Dictionaries, Grammars, etc.*

BJÖRKMAN, E. Scandinavian Loan-Words in Middle
English. 1900–2.

BÜLBRING, K. D. Altenglisches Elementarbuch. 1902.

English Dialect Dictionary. Ed. Joseph Wright. 6
vols. 1898–1905.   E.D.D.

FALK, H., and TORP, A. Etymologisk Ordbog. 1903–6.

FRITZNER, J. Ordbog over det gamle norske sprog. 3
vols. 1886–96.

HORN, W. Historische Neuenglische Grammatik. 1908.

JESPERSEN, O. A. Modern English Grammar. Part I.
1909.

MORSBACH, L. Mittelenglische Grammatik. Erste
Hälfte. 1896.

A New English Dictionary. In progress. 1888–.   N.E.D.

WRIGHT, J. English Dialect Grammar. 1909.   E.D.G.

ABBREVIATIONS.

(Other than those detailed above.)

Angl.	Anglian.	Norw.	Norwegian.
A.F.	Anglo-French.	O.E.	Old English.
A.N.	Anglo-Norman.	O.Fr.	Old French.
Dan.	Danish.	O.H.G.	Old High German.
E.M.E.	Early Middle English.	O.Sw.	Old Swedish.
Gael.	Gaelic.	O.W.Sc.	Old West Scandinavian.
Ir.	Irish.	O.N.	Old Norse.
L.O.E.	Late Old English.	O.N.F.	Old Norman French.
M.E.	Middle English.	St. Eng.	Standard English.

SYMBOLS USED IN PHONETIC SCRIPT.

a	North Country *a*.	ʌ	*u* of *but* (St. Eng.)
ai	=a+i, *i* of *mine*.	ə	*e*(*r*) of *better*.
au	=a+u, *ou* of *house*.	ɔ	*o* of *hot*.
ε	*e* of *there*.	ɑ	*a* of *father*.
ei	=e+i, *a* of *fate*.	i·	*ee* of *feed*.
ou	*o* of *note*.	u·	*u* of *rule*.
u	*u* of *pull* (St. Eng.).	ɔ·	*aw* of *raw*.
ʃ	*sh* of *shut*.	θ	*th* of *thin*.
ž	*z* of *azure*.	ð	*th* of *then*.
		j	*y* of *yet*.

Other symbols have the same value as in ordinary script.

PART I

NOTE.—Names marked with an asterisk are not found on the modern map.

Abberwick (Eglingham). 1169 Pipe *Alburwic* ; 1278 Ass. *Alberwick, Alburckwick* ; 1291 Ipm. *Aburwick,* 1333 *Abberwyke,* 1346 *Alburwyke* ; 1428 F.A. *Awberwyke* ; 1586 Raine *Awberwick* ; 1610 Speed *Averwick* ; 1663 Rental *Alberwick* ; 1689 Ingram *Abberwick.*

" The *wīc* (Part II) of *Alubeorht* (m.) or *Aloburh* " (f.). Cf. L.V.D. *Alubercht, Al(u)burg.* Phonology, §§ 39, 24. We should have expected a modern form *Awb(e)rick.* *Abb-* in 1333 is probably an error for *Alb-.*

Abshiels (Stanton). 1286 Plac. For. *Abscheles.* " Abba's *scheles* " (Part II). Cf. Abload, Glouc. (Baddeley, p. 2).

Acklington (Warkworth). 1176 Pipe *Eclinton* ; 1186 *Aclinton* ; *c.* 1250 T.N. *Aclington* ; 1663 Rental *Acklington* ; 1695 Lesbury *Ecklington.*

O.E. *Aecceling(a)tun*=farm of Aeccel or of his sons. Searle gives *Aecci* and *Acca* ; from these might be formed a diminutive *Aeccel.* Cf. *æclesmor* (K.C.D. 570), *æcelesbeorh* (B.C.S. 902), Goth. *Accila* (Schönfeld) and O.H.G. *Eccila, Echila* (f.) (Förstemann), O.Sw. *Aklunge* (Hellquist). Phonology, § 2.

Acomb [jekəm] (Bywell St Peter). 1268 Ipm. *Akum;* 1414 N. vi. 119 *Acomb.* (St John Lee) 1296 S.R. *(Ak)um.*

O.E. *(æt þǣm) ācum*=(at the) oaks. Cf. Acomb, Yorks. (Moorman, p. 2). Final *b* is due to the influence of the numerous words in *mb* in which *b* is silent, more especially to the common place-name suffix *-comb.* Phonology, §§ 14, 17.

Acton (Blanchland). 1269 N. vi. 313 *Akedene* ; 1663 Rental *Acton* alias *Acden.*

A

"Oak-valley" *v. denu* (Part II). Phonology, §§ 14, 21;
App. A, § 1.

(Felton) *c.* 1250 B.M. *Aketon*; 1255 Ass. *id.*; 1313 R.P.D.
Ayketon.
O.E. *Aca(n)-tūn*=Aca's *tūn* (Part II). Cf. B.C.S. 1289
and Aketon, Yorks. (Moorman, p. 6). Björkman (Z.E.N.,
p. 12) suggests O.N. *Aki*, but this seems unnecessary in face
of the well-established English name. Alternatively the
name may be O.E. *āc-tūn*=oak-farm. *Ayketon* is due to
association with names showing *Ayk* from O.N. *eik*, "oak,"
as in Aikton, Cumb. (Sedgefield, p. 2).

Adderstone [eðəsən] (Bamburgh). 1233 Pipe *Edredeston*,
1234 *Edreston*; 1242 Cl. *Hethereston*; 1288 Ipm. *Edderston*;
1346 F.A. *Hetherston*, 1428 *Ederston*; 1663 Rental *Etherston*;
1785 N.C.D. *Adderston*; 1833 Map *Edderstone*.
"Eadred's *tūn*." Cf. Atherstone by Tamworth,[1] Warw.
(Duignan, p. 16), earlier *Edredestone, Aderestone*, Addersey,
Som. K.C.D. 73 *Eadredeseie.* Phonology, §§ 29, 53.

Agarshill Fell, Agars Hill (Whitfield). 1278 Ass.
Algerseles. "The *scheles* of *Alger.*" *Alger* is from O.E.
Ealdgar or *Aelfgar.* Phonology, §§ 39, 53; App. A, § 7.

Aislaby (Egglescliffe). 1228 F.P.D. *Askelbi*; 1311 R P.D.
Aselackeby, 1314 *Aslagby*, 1344 *Aslakby*; 1382 Hatf
Aslayby; 1570 Eccl. *Aisleyby.*
"The *by* (*v. býr.*, Part II) of *Áslakr.*" The 1228 form
may be an alternative name derived from O.N. *Áskell*, but
more probably it is due to sporadic metathesis. Both are
well-known Norse names (Björkman, N.P. pp. 16-20). Cf.
Aislaby, Yorks, and Aslackby, Lincs. [eizəlbi]. Phonology,
§ 51.

Akeld (Kirknewton). 1169 Pipe *Achelda*, 1176 *Hakelda*;
1246 Ipm. *Akekeld*[2]; 1255 Ass. *Akil(d)*; *c.* 1320 Sc. *Ak(h)ille*;
1428 F.A. *Akyld*; 1694 Edling. *Akell*; 1733 Norham *Yakeld*.

---

[1] In Atherstone the variant vowel is due to the twofold development
of O.E. *Ead-* to M.E. *Ad-* or *Ed-*. This will not explain an *Ad-* developed
so late as in *Adderstone*. Here it may be due to the influence of Nthb.
edder, S. Eng. *adder.*

[2] This would point to O.N. *eik*=oak, but the *k* is probably a scribal
error.

O.W.Sc. *á*, river+*kelda*, well or spring. *Keld* is used locally of a marshy place (Heslop, *s.v.*), and the whole name is descriptive of the position of Akeld on the edge of the Till valley. Phonology, §§ 14, 17, 56; App. A, § 9. There seems now to be no trace of the old pronunciation with loss of final *d*.

*Akenside (Elsdon). 1332 Cl. *Akenside*; 1663 Rental *id.* " Oaken side " (*side*, Part II), i.e. hill grown over with oaks. Cf. Birkenside *infra*. Phonology, § 14.

Aldin Grange (Broom). *c.* 1170 Finch. *Aldingrig*; 1267 F.P.D. *Alderigg, Aldingrig*, 1539 *Aldyngryge*; 1637 Camd. *Aldernedge*.

O.E. *Ealding*(*a*)*hrycg*=ridge of Ealda or his sons. Phonology, § 27. App. A., § 12 .

Aldworth (Mitford). *c.* 1120 Brkb. *Aldewurth*.

O.E. *se ealda weorþ*=the old *weorþ* (Part II), or *Ealdan weorþ*=Ealda's *weorþ*. Cf. B.C.S. 358 *to ealdan wyrðe*, where we probably have the adjective.

Allen, R. 1275 H. 2. 3. 443 *Alwent.*

This river name is explained by Holder (s.n. *Alventium*). He suggests that it is from **Albentio*, a derivative of **albanto* or **albento*, "shining white," a participial form descriptive of the river itself, and connected with the adj. stem *albo-*. Cf. Alwin and Alwent *infra*. Phonology, §§ 49, 56.

Allendale. 1226 B.B.H. *Alwentedale*; *c.* 1250 T.N. *Alwendale, Alwennerdale*; 1273 R.H. *Alwennerdall, Alwendale*; 1663 Rental, *Allendaile*.

"Alwent-dale" (*v.* Allen *supra* and *dalr*, Part II). The spellings with *er* are difficult. Possibly they are due to confusion with Ennerdale, Cumb. (*v.* Lindkvist, p. 41). Phonology, § 49.

Allendale Town. 1245 Gray *Alewenton.*
" Farm on the Alwent or Allen," *v. supra.*

Allensford (Shotley). 1382 Hatf. *Aleynforth*; 1580 Halm. *Allonsford.*

Allenshiel (Hunstanworth). 1304 Cl. *Aleynsheles.* " Aleyn's ford and shiels." *Alayn* is a common M.E. name. Cf. Elliscales, Lancs., earlier *Alaynscheles.* Wyld (p. 118) takes it to be from O.E. *Aeþelwine.* The *shiels* were named from Alan the Marshal, their one-time owner (Hatfield, p. 124). The neighbouring ford may have been named from the same man.

Allerdean (Ancroft). 1108 F.P.D. *Elredene*; *c.* 1250 T.N. *Alvereden*; 1228 F.P.D. *Alredene*, 1539 *Allerdene.*
" Aelfhere's dene," cf. the history of several of the Yorkshire *Allertons* (Moorman, p. 7). Phonology, §§ 1, 50.

Allerhope Burn (Kidland). *a.* 1240 Newm. *Alrehopeburn*; 1536 Arch. 3. 8. 20 *Alrope.*

***Allerside** (Shotley). 1261 Ipm. *Alarseth*, 1262 *Allerseth*, *Alleriset*; 1454 Pat. *Allerside.*

Allerwash (Warden). 1205 Pipe *Alrewas*, *Allerwas*; 1323 Ch. *Allerwasch.*

The *hop*, *sǣte* and *wǣsc* (Part II) overgrown with alders. *Aller* is the common Nthb. form of *alder* (Heslop *s.v.*). For the last name cf. Alrewas, Staffs. (Duignan, p. 3). App. A § 8.

Allery Burn (Chatton). 1292 Ass. *Alriburn.*
" Aldery-burn," i.e. grown over with alders, or O.E. *alra burna* =burn of the alders. Cf. *alra broc*, B.C.S. 361.

Alne, R. [eil, jel]. 2nd c. Ptolemy *Ἄλαυνος*, *c.* 720 Bede *Alne*; 1539 Tate ii. 23 *Water of Ale.*
For this name cf. Alne, Warw. (B.C.S. 157 *Aeluuinnae*) and Ellen, Cumb., earlier *Alne*, *Alin*, *Alen* (Sedgefield, p. 47). Duignan (pp. 10, 11) connects this river name with that of the *Allen* or *Alwen* in Flint, and the French rivernames *Allain*, *Aline*, *Allaine*. These go back to some Celtic adj. related to Gaelic *aluin*, *alainne*, *ailne*, " fair, handsome," Welsh *alain*, *alwyn*, with the same sense, also " bright, clear, lucid." Cf. Ayle *infra*. Phonology, §§ 17, 56.

Alnham [jeldəm]. 1228 F.P.D. *Alneham*; 1304 Orig. *Aneham*; 1507 D.S.T. *Aylnam*; 1663 Rental *Ailnham*; 1680 Mitford *Aledome*; 1712 Ingram *Yeldam.*

Alnmouth [1] [jelməθ]. 1205 R.C. *Anyemue*; 1230 Pat.

---

[1] It has been suggested that this is Bede's " juxta fluvium Alne in loco qui dicitur *Adtuifyrdi* (i.e. at Twyford) quod significat ad duplex vadum." There are two fords across the river here, though the name Twyford has not survived here or elsewhere on the Alne.
Before the place was called Alnmouth, it seems also to have been known as *burgus de Sancto Walerico*. Wm. the Lion granted Wm. de Vescy

Alnemuth ; *c.* 1250 T.N. *Auneimuwe* ; 1255 Ass. *Allemue,*
Alnemue ; 1314 R.P.D. *Alemuth.*
  Alnwick [anik]. *c.* 1160 Ric. Hex. *Alnawic* ; 1213 Pat.
Aunewyk ; 1268 Ass. *Annewyk* ; 1434 Pat. *Alnewyk* ;
1496 N.C.W. *Awnewik* ; 1585 Tate i 273 *Anwik.*
  " Homestead by (*v. hām* Part 11), mouth of, *wīc* by
the Alne." The forms of Alnmouth and Alnham show a
twofold phonetic development, (1) [ɑln] > [ɑuln] > [ɑˑln] >
[eiln] which ultimately prevailed, (2) [ɑln] > [ɑuln] >
[ɑˑn]. Cf. Calne, Wilts. [kɑˑn], Jespersen, 10.452. If the
second had survived, the modern pronunciation would have
been [einəm] or [jenəm]. Note the denasalisation of *n* to
d in Alnham. In Alnwick later shortening of [ɑˑn] to [an]
has taken place. *Ann* in 1268 is probably merely an error
of transcription for *aun*. Phonology, §§ 17, 49. For *-mue*
v. Zachrisson, pp. 93 f.
  Alwent (Gainford). 1238 Cl. *Alowent* ; 1306 R.P.D.
Alwent ; 1732 Gainf. *Alwen.*
  Cf. Allen, R., *supra.* The place must have been so called
from the stream, now Alwent Beck, on which it stands.
For such names from Celtic river-names *v.* Bradley in
Essays and Studies by Members of the English Association,
vol. i. p. 10. Phonology, § 56.
  Alwin, R. (Alwinton). 1228 Newm. *Al(e)went, v.* Allen
supra.
  Alwinton [aləntən]. *c.* 1240 Newm. *Alwenton* ; 1346
F.A. *Alnowenton (sic)* ; 1539 Arch. 3. 4. 116 *Alanton.*
  " Farm on the Alwin." Phonology, § 49.
  Amble. 1203 R.C. *Ambell* ; 1212 Perc. *Ambbill* ; 1292
Q.W. *Anebelle* ; 1296 S.R. *Ambel* ; *c.* 1250 T.N. *Ambell* ;
1347 Perc. *Anebill.*
  Probably Celtic. Cf. Kemble, Glouc., earlier *Kenebelle*
and Kimble, Bucks., D.B. *Chenebella.* Phonology, § 51.
  Amerston (Elwick). 1243 Finch. *Aymundeston, Amund-*
iston ; 1320 Cl. *Aymundeston.*
  " Farm of *Eymundr,*" a name of Norse origin. Cf.
Amotherby, Yorks., earlier *Aymunderby.* Phonology, § 53.

in 1152, the right to have a court at St Waleric, " qui vocatur *Neubiginge*,"
i.e. the new town carved out of Lesbury Parish. (N. II. 439, 469–70.)

Ancroft (Islandshire). *c.* 1180 D.S.T. *Ancroft*; 1228 F.P.D. *Anecroft*.

O.E. *se āna croft*=the single or lonely *croft* (Part II.). Cf. Onehouse, Suff. (Skeat, p. 124), and Onecote, Staffs. (Duignan, p. 111). Phonology, §§ 14, 21.

Andrews House (Tanfield) 1430.33 *Androwehous*. So called from its one-time owner.

Angerton (Hartburn). 1186 Pipe *Angerton*. The first element in this name and in Ingram *infra* may be identical with Ongar, earlier *A*(*u*)*ngre*, Essex. *v. Essays and Studies*, u.s., vol. iv. p. 56, where the present writer suggests that all alike contain a lost English cognate of O.H.G. *angar*=grass-land, as opposed to forest or arable land.

Alternatively the first element may be the O.W.Sc. personal name *Ásgeirr*, with Latinised form *An*(*s*)*garus*, which Björkman (Z.E.N. p. 15) finds in Angerby and Angerton, Lancs.

Anick[1] [einik] (St John Lee). *c.* 1160 Ric. Hex. *Aeilnewic*; 1225 Gray *Einewic*; 1226 B.B.H. *Ainewik*; 1296 Ipm. *Anewyke*; 1479 B.B.H. *Aynewyk*, 1536 *Anyk*. "Aeþelwine's *wīc.*" O.E. *Aeþel-* > Late O.E. and E.M.E. *Aegel-*. *v.* Zachrisson, p. 101.

Anton Field (Aldin Grange). 1438 Acct. *Antonfeld*. Cf. Anton Hill, Nthb. Possibly the first element is the name *Anthony*.

Apperley (Bywell St Peter). 1261 Ipm. *Appeltreley*; 1428 F.A. *Appirley*. (ib. St Andrew) 1359 Pat. *Apirley*. "Apple-tree-*lēah* (Part II) or clearing." Cf. K.C.D. 538 *apaldreleage*. Phonology, § 53.

Ardley (Hexhamshire). 1228 Gray, *Herdeley*; 1287 B.B.H. *Erdeley*. "Earda's *lēah* (Part II) or clearing," *Earda* being a shortened form of such a name as O.E. *Eardwulf*. Cf. Earsdon *infra*.

Ashington (Bothal). 1170 Pipe *Essende*, 1199 *Esinden*; 1255 Ass. *Essenden*; 1428 F.A. *Esshenden*; 1487 Ipm. *Eshenden*; 1637 Camd. *Assinton*; 1663 Rental, *Ashington*.

---

[1] In Brkb. Chart there is mention of an *Aynewik* in Cowpen (*c.* 1154–89). This probably goes back to O.N. *ein*=one+*wīc* (Part II). Cf. Ancroft, *supra* and Aintree, Lancs. (Lindkvist, p. 43).

O.E. *Aescinga-denu* =valley of Aesc or of his son(s), *Aesc* being a short form of one of the numerous O.E. names in *Aesc-*. Phonology, § 1, App. A. § 1.

Auckland, Bishop, North, and West. *c.* 1050 H.S.C. *Alclit*; 1085 D.S.T. *Alcleat*; 1104-8 S.D. *Alclit*; 1143-52 F.P.D. *Alclet*; *c.* 1190 Godr. *Alcleat, Alclent*; *c.* 1180 F.P.D. *Alklet, c.* 1200 *Aclent, Auclent, Alklint*; 1202 Pipe *Auclint*, 1213 *Aclent*; 1219 F.P.D. *Auclent*; 1226 Pat. *Acclent*, 1227 *A(u)clent*; 1228 F.P.D. *Auclent, Acclent*; 1214-33 *Auclent*; 1237 Cl. *Akeland, A(u)clent*; 1238 Pat. *Aclent*, 1240 *Acland*; 1248 D.S.T. *Aukland*; 1274 Cl. *Aucland*; 1283 Pat. *Alkeland*; 1283 Pap. *Aukeland*. B.B., A. *Alcland, Aclet, Alclet*; B., C. *Auckland, Aukeland*.

There can be little doubt that this name is of Celtic origin, and that the wide diversity of forms is due to attempts to anglicise the name. Lindkvist tried to show (*Namn og Bygd*, vol. i. pp. 67-74) that the original form was O.N. *auk-land* =additional land taken into cultivation, and that the other forms can be explained as perversions of that, made when *alk* had come phonetically to be the equivalent of *auk*. The present writer (ib. pp. 149-51) showed that by a fuller gleaning of forms, such a theory became untenable. *Auc-* forms are found 150 years before *Alc-* ones. The development may have been the other way about, viz., that the Celtic name has been modified, in part at least, under the influence of a Norse word. Phonology, §§ 39, 55.

*Aunchester, Anterchesters. 1367 Pat. *Antrechestre*; 1379 Ipm. *Antrichestre*; 1542 Bord. Surv. *Anterchester*; 1584 Bord. *Aunchester*.

A name of Romano-Celtic origin (*v. ceaster*, Part II). The first part of the name may be associated with the Celtic name *Antros* (an island, now Médoc), *Antrum* or *Antricinum* (an island in the Loire), *Antrum* (river) and *Antrum* (now Antre, Franche-Comté), given by Holder, col. 162. Bates (*Border Holds*, p. 32 n.) says that the name was later corrupted to *Turn Chesters*, and is so marked on old maps of Nthb. Phonology, § 5.

Axwell Park (Ryton). 1344.31 *Aksheles*; 1361.35 *Axsels*; 1382 Hatf. *Asshels*; 1386.32 *Axsheles*; 1396.35

Axelsheles ; 1411 Arch. 2. 24. 118 *Axelfeld* ; 1416.33 *Axschelles.*

" Oak-*sheles*," i.e. by the oak(s). *ksh* > *ks* giving *axel*(s), and then in the 1396 form, *sheles* is once more added. Phonology, § 14, 21. App. A, § 7.

Aycliffe [jakli]. Type I : *c.* 1050 H.S.C. *Heaclif* ; 1109 D.S.T. *Heaclif*(*f*). Type II : 1085 D.S.T. *Aclea* ; *c.* 1125 F.P.D. *id* ; *c.* 1160 Ric. Hex. *Aclech* ; 1203 R.C. *Acle* ; 1312 F.P.D. *Akleye, Akelei, Ackelay, Akeley* ; 1335 Ch. *Acleia* ; 1343 Bury *Acley* ; 1507 D.S.T. *Acle* ; 1539 F.P.D. *Acley* ; 1680 Houghton **Yakely**. Type III : 1378.32 *Aclyf* ; 1381 Pat. *Aclif* ; 1391 D.S.T. *Aklyff*, 1400 *Aclif* ; 1402 F.P.D. *Akclyff* ; 1576 N.C.W. *Accliffe* ; 1731 Bp. Wearm. *Ackliff*.

In addition to these forms it may be that the synods of *Aclea* (A.S.C. 782 and 789 E.) were held at Aycliffe,[1] as also the synod of *Hacleah* in 805 (B.C.S. 322, Haddan and Stubbs, iii. 558).

This is a name which offers great difficulties, and one cannot be certain of their solution. It would seem that Type III cannot be related to Type I in spite of the similarity of suffix. Aycliffe is frequently referred to, and it is impossible to suppose that Types I and III are the same with a gap of 260 years in their history, quite apart from the difficulty of initial *h*, and that O.E. *Hea-* should give *He-* (cf. Healey and Heaton *infra*). Probably the place under Type I is not *Aycliffe* at all. Type II has its exact parallel in Ockley, Surrey, from *Aclēa* (A.S.C. 851 Ā) =oak-clearing, the form in Ric. Hexh. representing the nominative *Āclēah* (cf. Part II). It is possible that Type III has developed from this nominative form. Final *h*, pronounced as [χ], and later as [f] (cf. *saugh* > [saf] in Nthb.) may have led to a pronunciation with final [klef], and subsequent confusion with the common word *cliff*. The modern pronunciation may be derived directly from Type II, or it may be that it is from Type III with loss of final *f*, for which there are other local parallels. Phonology, §§ 14, 17, 21, 56.

Aycliffe, School. Type I : B.B. *Sculacle* (B., C.) *Sculacley* ;

---

[1] Haddan and Stubbs (III. .439 n.) believe this to be true only of the Council of 789.

1351.31 *Scolakley*; 1382 Hatf. *Skulacley*; 1440 D.S.T. *Sculacley*. Type II : 1410.33 *Scolakliff* ; 1410.35 *Skolaclyf*. " Skuli's Aycliffe," so called from the Viking chieftain Scula (O.W.Sc. *Skúli*), to whom, together with one Onlaf-beald, King Rægenwald gave the patrimony of St Cuthbert *c*. 920 (H.S.C., § 23).[1] The same name is found in Scoulton and Sculthorpe, Norf., D.B. *Sculetona, Sculetorpa*.

Aydon (Corbridge). 1225 Ass. *Ayden* ; 1279 Ipm. *id* ; *c*. 1250 T.N. (*H*)*ayden* ; 1298 B.B.H. *Hayden* ; 1305 Ch. *Eyden* ; 1322 Ipm. *Hayden* ; 1346 F.A. *Haydon*, 1428 *Aydon*. (Alnwick), 1325 Perc. *Haydene* ; 1346 F.A. *Ayden, Haydon*.

The suffix is O.E. *denu*, " dene." If the *h* is original (cf. Ilderton *infra*) the first element is O.E. (Angl.) *hēg*, " hay." If *h* is inorganic, it may be O.E. (Angl.) *ēg*, " island, peninsula." This name is not impossible, as applied to the Corbridge Aydon, for Aydon Castle is partly encircled by the windings of the Cor Burn. This stream used to be called the Ay Burn, but that is probably a back-formation from *Aydene*. Phonology, § 35. App. A, § 1.

Aydon Shiels (Hexhamshire). 1341 B.M. *Aldenscheles* [2] ; 1362 Ipm. *Aldenschole*.

" Ealdwine's *scheles*." Phonology, § 39.

Ayle, R. (Kirkhaugh). 1258 H. 2. 3. 59 n. 1 *Alne*. Cf. Alne *supra* and Ale, Roxburghshire (Johnston, p. 9), earlier *Alne*. Phonology, § 56.

***Backstonerigg** (Kirkheaton). 1322 Inq. a.q.d. *Backe-stanrigg*. " Ridge where *backstones*, i.e. flat stones for baking cakes may be found." Cf. Heslop *s.v.*, *Bakstanside* in Bamburgh (Pat. 1358), *Baxstansyde* in Sandhoe (B.B.H. 1479), *lez Bakstanes* in Heugh (*ib.*), and Baxterwood, *infra*. Phonology, § 21.

Backworth (Bywell St Peter).[3] 1271 Ipm. *Backewrth*.

---

[1] Surtees (III. 314) says that the name is derived from a school which was once established here by the Prior and Convent of Durham, but no confirmation of this statement has been found.

[2] There was another *Aldensheles* near Alwinton (Pat. 1317, Ipm. 1334, R.C. 1341, Ipm. 1391), which cannot now be identified.

[3] The hamlet is now known as Letchouses, but stands on Backworth Letch.

(Earsdon) 1203 R.C. *Buxwurtha, Bucwortha* (*sic*) ; 1271 Ch. *Bachiswrd, Bacwrth.*

" Bacc(a)'s *weorþ* " (Part II).

Bagraw (Hexham). 1385 N. iv. 11 n. 6 *Bagraw* ; 1663 Rental, *Baggaraw.*

Probably " Bacga's row " (*rāw* Part II). There are place-names Baggarah and Baggrow, Cumb., but Sedgefield (p. 9) gives no early forms. Phonology, § 16.

Bamburgh [bambri]. 10th *c.* A.S.C. *Bebbanburh, Bæb-banburh* ; 1097 Colding. *Bebbanburch* ; 1129 Pipe *Baenburg,* 1165 *Baemburc* ; *c.* 1170 Jord. *Bane(s)burc* ; *c.* 1160 Ric. Hex. *Bahanburch* ; 1182 Pipe *Baenburc* ; 1199 R.C. *Bamburg* ; 1280 Ch. *Baumburg* ; 1284 De Banco, *Bamburne* ; 1311 R.P.D. *Baunburgh* ; 1332 Ch. *Beaumburc* ; 1353 F.P.D. *Baumburgh* ; 1430 Pat. *Bamburgh* ; 1516 N. i. 150 *Bawmbourgh* ; 1575 *ib.* 152 *Bambrough* ; 1602 *ib.* 158 *Balmbrough* ; 1663 Rental, *id.* ; 1705 N. i. 170, *Balmburgh.*

Bede (III. 10) speaks of this place as " urbs regia quae a regina quondam *Bebba* cognominatur." This Bebba was the queen of Aeþelfriþ of Bernicia. The alternative form *Bæbba* gave rise to the form *Bæbbanburh,* from which the later forms develop through *Babnburh, Banburh, Bamburh.* Phonology, §§ 53, 57. For *aum, aun,* ib. § 5. Later, *au* > [ɑ·] as in the 17th and 18th *c.* spellings in *alm* and [ɑ·] > [a] as in Nthb. [igzampl], [tʃam(b)ər]. For *Banesburc, v.* Zachrisson, p. 119. App. A., § 10.

Barford (Winston) B.B. *Bereford* ; 1436 Acct. *Barforth.*

A common place-name. Cf. Barford, Oxon., Norf., Northt., Warw., Wilts (2) and Barforth, Yorks, all of which have D.B. *Bereford.* Alexander (*s.n.*) suggests derivation from an O.E. name *Bera,* but it seems unlikely that eight fords should happen to be owned by a man bearing a very doubtful O.E. name. Offa signed a charter (*v.* B.C.S. 264) at a place called *Aetberanforda.* In B.C.S. 627 we find *to bæran forda,* in B.C.S. 446 *bere ford,* the latter being in a comparatively late copy. One might suggest that these contained O.E. *bær* = *bare,* used of any unsheltered place, but

---

[1] Bamburgh was in pre-English times called *Dinguaoroy.* Nennius, ed. Stevenson, § 63.

the early development of *e* makes this very unlikely. Ekblom (*s.n.* Barford, Wilts) suggests O.E. *bere-ford*, ford by the barley (field), but such a compound is not very probable or convincing. Phonology, §§ 8, 30.

Barhaugh (Knaresdale) [bɑrəf]. 1279 Iter. *Berhalu, Berehalche*; 1566 F.F. *Berehawgh* alias *Barrow in Tynedale*. O.E. *bere-healh* =barley-haugh (*healh* Part II). Hodgson (2. 3. 67) says that "the rich and sunny haughs of the place are still adapted to the growth of the grain." Phonology, § 8.

Barley Hill (Shotley). 1225 Coram *Birlawe*, 1230 *Berlauwe*; *c.* 1250 N. vi. 250 n. 7 *Beirallawe*. O.E. *bere-hlāw* =barley-hill (*hlāw*, Part II). Cf. Bearl *infra*. The modern *hill* is pleonastic. Possibly *Beirallawe* stands for an alternative *bere-hyll* (cf. Bearl)+pleonastic *lawe* =hill. App. A, § 2.

Barlow (Ryton). B.B. *Berleia*; 1380 R.P.D. *Berley*. "Barley-field" (*lēah*, Part II). App. A, § 2. Cf. Barlow, Salop. D.B. *Berlie*. Phonology, § 8.

Barmoor (Lowick) [bɛəmuər]. 1231 Cl. *Beiremor*, 1232 *Beigermore, Beygermore*; 1289 Ch. *Bayremore*; 1346 F.A. *Bayrmore*; 1539 F.P.D. *Barmour, Barmore*; 1542 H. 3. 2. 190 *Byermore*. "Beaghere's *mōr*" (*v.* Part II). Searle does not record this name but gives several similar names in *Bēag-*. *Bēaghere* would give M.E. *Beӡer, Beyer, Bayr*. Cf. also Byermoor *infra* and Bairstow, Yorks, earlier *Bayrestowe*, for which Goodall (p. 65) offers an unlikely explanation.

Barmpton (Haughton-le-Skerne). *c.* 1110 F.P.D. *Bermentun, c.* 1150 *Bermestuna*; 1203 R.C. *Bermeston*; 1430 F.P.D. *Bermpton*; 1539 *Barmtone*; 1633 Comm. *Barmton*. "Farm of *Be(o)rm(a)*." This name is not recorded by Searle but is found in Barming, Kent (D.B. *Bermelinge*, F.A. *Barmlinge*) in the dimin. form *Bermel*, and in a place called *Bermintune* in Hampshire D.B. (V.C.H. I. 511). It is probably a pet-form for *Beornmær, Beornmod*, or *Beornmund*. Phonology, §§ 8, 55.

Barmston (Washington). 1361.45 *Berneston*; 1400 Acct. *id.*; 1471.35 *Bermeston le Ford*; 1596 Wills *Barmston*.

" Beorn's farm." Cf. Barmston, Yorks., earlier *Berneston*, " le Ford," because by a ford on the Wear. For *le v.* Chester-le-Street *infra*. Phonology, § 52.

Barnard Castle [ba·ni kasl]. 1197 Pipe *Castellum Bernardi*; 1312 R.P.D. *Chastel Bernard*; 1486 Pat. *Barney Castell*.

The castle was built by Bernard Baliol. Phonology, §§ 8, 53.

Barneystead (Simonburn). 1373 Ipm. *Bernerstede*, 1415 *Barnarstede*; 1649 Comps. *Barnarsteed*; 1663 Rental *Barnett Steed*.

" Beornhard or Bernard's stead " (*stede*, Part II). For Barnett, cf. Garretlee *infra* and Barnard Gate, Oxon., Pron. Barnett Yat (Alexander, p. 49). Phonology, §§ 8, 53.

Barns (Knaresdale). 1325 Ipm. *le Bernes*.

"Barns," cf. Barnes, Surrey.

Barrasford (Chollerton). *c.* 1250 T.N. *Barwisford*; 1255 Ass. *Barewesford*; 1292 Ass. *Barwisforth*; 1298 B.B.H. *Barweford*; 1324 Ipm. *Bar(o)wesford*; 1479 B.B.H. *Barousford, Barassford*.

Apparently O.E. *bearwesford* = ford of or by the *bearu* or grove, though we should have expected *bearuford* = grove-ford. Cf. the common name *Woodford*. Phonology, §§ 49, 30.

Barrow Law (Kidland). 1304 Pat. *Brerylawe*, 1307 *Brerilawe*.

"Briary-hill." M.E. *brere*, "briar" becomes Nthb. [briər]. If this identification is correct the name was changed later.

Barton (Whittingham). 1199 Pipe *Barton*; 1253 Ipm. *id.*

Barton is a very common place-name and usually goes back to O.E. *bere-tun* = barley-farm, later " the demesne lands of a manor let out to tenants but retained for the lord's own use." In these names O.E. *beretun* > M.E. *berton* > Mod. Eng. *barton* (Phonology, § 8) but this change from *e* to *a* did not take place in the 12th century, and either we must take *Barton* here to be from some Anglian form *bær-tūn* (cf. Orm's *barrliȝ* < O.E. *bær-līc*) or explain it in some entirely different way, e.g. (*se*) *bara tūn* = the bare farm or *bār-tūn* = boar-enclosure.

Battleshield (Kidland). *c.* 1225 Newm. *logia quondam Willelmi Bataile.*

Dixon (*Upper Coquetdale*, p. 29 n. 6) notes that the above reference shows that the shiel was so named from its former owner and not from some raiding foray, *v.* Brunton *infra* and cf. Battails in Bradwell, Ess., so called because granted to Amauri Battaile (F.F. 1207).

Bavington (Kirk Whelpington). 1255 Ass. *Babinton*; *c.* 1250 T.N. *Babington*; 1257 Ch. *Babbinton*; 1479 B.B.H. *Babyn(g)ton*; 1610 Speed *Bauinton*; 1677 St John Lee *Babington.*

O.E. *Babbing(a)tūn* = farm of Babba or of his sons. Cf. *babbingthorn*, B.C.S. 1289. There is also a Frisian name *Baba*, cf. Winkĺer (p. 22), who gives a patronymic *Babinga* and a place-name *Babinga-sete*. Phonology, § 24.

*Baxterwood [1] (Durham). 1199 Finch. *Bakestaneford*; *c.* 1300 D.S.T. *Bácstanford*; 1472 Acct. *Baxstanford.*

"Ford from or near which *backstones* are taken." [2] Cf. Backstonerigg *supra*. App. A, § 4.

Baydales (Darlington). *c.* 1190 Godr. *Badele*; B.B. *Bathela*; 1340 R.P.D. *Bathel-spitel*; 1382 Hatf. *Bathley*; 1784 Coniscl. *Badelbeck.*

If this identification is correct, this is the same name as Bale, Norf., D.B. *Bathele* and Bathley, Notts., F.A. *Batheleye*, i.e. Baða's clearing. *Baða* is probably a pet form of one of the numerous O.E. names in *Beadu-*; *-spitel* because there was once a hospital here, and *-beck* from a neighbouring stream. Phonology, § 42; App. A, § 7.

Beadnell (Bamburgh) [bi·dlən]. 1160 Pipe *Bedehal*, 1176 *Bedenhala*, 1253 *Bedenhall*; 1273 R.H. *Beednal*; 1753 Lesbury *Beadlin.*

"Bedwine's *healh*" (Part II). The name is found in K.C.D. Phonology, § 49. At first sight one would take this to be identical with Bednall, Staffs., or *Beadanhalan* (B.C.S. 936), the first element being gen. sg. of O.E. *Beada*, *Bǣda* or *Bēda*, but it is impossible to believe that the

---

[1] Found in old maps, near Aldin Grange.
[2] Acct. Rolls mention a quarry here.

suffix -*an* could thus have survived in Nthb. (Introd. p. 27). For the metathesis, cf. Kidland *infra*.

Beal (Kyloe). 1228 F.P.D. *Beyl*; 1248 Sc. *Behulle*; 1340 R.P.D. *Behill*; 1387 Raine *Beil*; 1539 F.P.D. *Beyll*.
"Bee-hill," i.e. where they often swarm. Cf. *byohyll* (B.C.S. 1027) and *beodun* (ib. 797). Middendorf (p. 13) also gives *beo-cumb* and *-leah* = bee-valley and field. *-hull* is a Southern form. Phonology, § 36.

Beamish (Chester-le-Street).[1] 1288 N. ix. 251 *Bewmys*; 1388.45 *Beawmys*; 1449.34 *Bewmys*; 1480.35 *Beamyssh*; 1487.36 *Beaumyssh*.
O.E. *beau-mis* = well-placed. Cf. Surtees (2.222). " B. stands in the deep wooded valley of the Team . . . richly cloathed with luxuriant forest trees." Phonology, § 20.

Beanley (Eglingham). *c.* 1150 Perc. *Benelegam*; 1663 Rental *Beanley*.
"Bean-field." Cf. *bean-leah* (B.C.S. 763), *bean-æcer* and *-stede* (Middendorf, p. 12).

Bearl (Bywell St Andrew). 1239 Ipm. *Berehill*, 1249 *Berhull*; 1346 F.A. *Berill*, 1428 *Berhill*; 1624 Arch. 2. 1. 139 *Bearle*.
O.E. *bere-hyll* = barley-hill. Cf. Ryle *infra*. Phonology, § 36.

Bear Park (Broom) [bi·ər]. 1267 Ch. *Beaurepeyr*; 1311 F.P.D. *Beurepair*, *Bellus Redditus*; 1398 Accts. *Berepark*; 1429.33 *Berpark*; 1456.34 *Beurepark*.
O.E. *beau-repaire* = beautiful retreat. The place was used as a *refugium* or country-seat by the monks of Durham. Cf. Beurepair(e) in Headcorn, Kent, and near Bramley, Hants., also Belper, Derbys. (Walker, p. 58). Phonology, § 20; App. A, § 12.

Beaufront (St John Lee). 1356 B.M. *Beaufroun*; 1479 B.B.H. *Beuanfront*, *Beaufront*; 1638 Freeh. *Befront*; 1610 Speed, *Bewfront*; 1750 Map *id*.

[1] Sawtry Beaumes or Beams, Hunts., was so called from its owner, Walter de Beumes. Skeat (p. 338) takes *Beumes* to be O.Fr. *beau mes*, L.L. *bellus mansus*, but the persistent *-mys* is against this derivation for Beamish. There is also a Manor of Beams, earlier *Beaumees*, in Shinfield, Wilts., probably named from the same family, whose name is often spelled *Belme(i)s*.

"Fine brow," from its position facing south across the valley of the Tyne. In Horsley's time (18th c.) it was pronounced [bi·vrən] (N. iv. 202 n.). Phonology, §§ 20, 56.

Beaumont (Chollerton). 1232 Ch. *Beaumont*; 1296 S.R. (De) *Bello Monte, Beumound, Bemound*; 1298 B.B.H. *Beumond*, 1479 *Beaumond*; 1622 N. iv. 259 *Beamont*.

Beaumont Hill (Coatham Mundeville). 1382 Hatf. *Beaumond, Bewmond*; *c.* 1570 Eccl. *Beamon(t)-hill*; 1582 N.C.W. *Beamond Hill*; 1637 Camden *Beamond*.

"Fine-hill." Phonology, § 20; -*mont* and -*mond* are variant A.N. forms from Lat. *montem*. The modern name is pleonastic.

Bebside (Horton). 1203 R.C. *Bibeshet*; 1271 Ch. *id.*; 1292 Q.W. *Bepeset*; 1296 S.R. *Bebisset*; 1388 Ipm. *Bebset*; 1428 F.A. *id.*; 1638 Freeh. *Bebside*.

The first element is the O.E. name *Bibba* or *Bebba*. If stress is laid on the *sh* in the two earliest spellings, the suffix is O.E. *scēat* (Part II), as in Bagshot, Surr., earlier *Baggeshete*, with later change from *sh* to *s* under the influence of the more common suffix -*set* from O.E. *sǣte* (Part II). Or, if the *h* is an error, the suffix is that word itself. Phonology, §§ 7, 10; App. A, § 8.

Beckley (Tanfield). 1344.31 *Bekkeley*.

"Becca's clearing." Cf. *beccan ford* (B.C.S. 309) and Beckley, Suss., earlier *Beccanlea* (Roberts, p. 15).

Bedburn (Witton-le-Wear). 1313 R.P.D. *Bedburne*; 1314 *Bedeburn*.

The burn of *Bǣda, Bēda*, or *Bēada*.

Bedlington. *c.* 1050 H.S.C. *Bedlingtun*; 1085 D.S.T. *Bethlingtun*; 1104-8 S.D. *Betlingtun*; *c.* 1150 D.S.T. *Bellingtona*; *c.* 1170 Reg. Dun. *Bethligtone, Betligtun*; *c.* 1175 Hist. Reg. *Betlingetun*; 1203 F.P.D. *Bellingeton*, 1228 *Bellington*; 1291 Tax. *Bedelinton*; 1315 R.P.D. *Bedelington*; 1335 Ch. *Bellington*; 1507 D.S.T. *Bedlyngton*.·

O.E. *Bēdeling(a)tūn* = farm of *Bēdel* or of his sons. *Bēdel* is a diminutive of *Bēda*. (Cf. *Beccel, Bosel, Mannel*). Spellings with *ll* are probably due to an assimilation never fully established [1] (Phonology, § 51), those with *tl* are due

[1] In Billing, Northt., earlier *Bethlinge*, it was carried out.

to A.N. influence (Zachrisson, p. 43 n.), those with *thl* are due to the common interchange of *ðl* and *dl* in certain Anglian words. Cf. Budle *infra*.

Beechburn (Auckland) [bitʃbəˑn]. 1304 Cl. *Bycheburn*; 1388 D.S.T. *Bicheborne*; 1637 Camd. *Bichborne*; 1768 Map *Bitchburn*.

"Bicca's stream." The ordnance form is fast ousting the original one, "from motives of delicacy."

Belasis (Billingham). 1305 R.P.D. *Belasis*; 1446 D.S.T. *Belassis*; 1539 F.P.D. *Bellces*. **Bellasis** (Durham). 1312 F.P.D. *Belasis*. (Stannington) 1267 Ipm. *Beuasis*, 1270 *Beuasys*; 1278 Ass. *Beleassis*; 1377 Ipm. *Belasyse*.

"Beautiful seat." Cf. Bellasis, Norf., Bellasize, Yorks. E.R., Belsize nr. Peterborough and Belsize in Hampstead.

Beldon Burn (Blanchland). *a.* 1214 Dugdale vi. 2. 886 *Beldene*; 1608 N. vi. 355 *Beldoune*.

Belford. 1249 Ipm. *Beleford*; 1255 Ass., 1258 and 1290 Pipe *id.*; 1300 Pat. and 1301 Cl. *Belleford*; 1313 Ipm. *Beleford*, 1314 *Belford*, 1323 *Belforth*; 1460 H. 3. 1. 27 *Belfurthe*; 1550 H. 3. 2. 207 *Belforth*; 1610 Speed *Belford*.

"Dene or valley and ford of *Beola* or *Bella*." Cf. Belstead Suff. (Skeat, p. 86), and *bellan ford* B.C.S. 454. Phonology, § 30; App. A, § 1.

Bellingham [belindžəm]. *c.* 1170 Reg. Dun. *Bainlingham*; 1278 Ass. *Bellingham*; 1279 Iter. *Belingjam*, *Belingeham*; 1332 B.B.H. *Belyncham*; 1386 Newm. *Bellingham*; 1524 Raine *Belling(e)ham*; 1542 Bord. Surv. *Bellyngeam*.

Apart from the spelling in Reg. Dun. we should take the name to be O.E. *Beolinga-* or *Bellinga-hām* =homestead of the sons of *Beola* or *Bella*. Cf. *Belleghem*, W. Flanders, for which Winkler (p. 30) gives earlier *Bellinghem*. If, however, the spelling in Reg. Dun. is correct and not a scribal blunder we must connect it with O.N. *Beinir*, O.Norw. *Beini*, or perhaps with M.E. *Beyn*, which Björkman (N.P. p. 25) takes to be a nickname from O.N. *beinn*, ready. *Bainel* would be a diminutive of it, and *Bell-* would show assimilation and shortening of the vowel. All this, however, is very doubtful. Phonology, §§ 51, 34.

Bellister (Haltwhistle). 1279 Iter. *Belester*; 1305 Ipm. *Belestre*; 1355 Orig. *Belecestre*; 1405 Ipm. *Belistre*; 1663 Rental *Bellister*.

" Bella's *ceaster* " (Part II). -*cester* here, as in Craster *infra*, is due to A.N. influence (Zachrisson, pp. 18-21). For the reduction of the suffix, cf. Gloucester, Leicester and Craster itself.

Bell Shiel (Rochester). 1330 Fine *Belleshope*; 1370 Cl. *le Belles*; 1375 Ipm. *Belleshopa*; 1376 Cl. *the Belles*; 1663 Rental *Bell Sheele*.

Bell's *hop* (Part II), cf. Belsay *infra*. *Belles* alone means " Bell's " in the same way that we speak of " Smith's," meaning " Smith's farm or house."

Belsay (Bolam). 1162 Pipe *Bilesho*, 1170 *Belesho*; 1166 R.B.E. *Bellesso*; 1203 R.C. *Billesho*; *c.* 1250 T.N. *Belsou*; 1255 Ass. *Beleshowe, Belleshou*; 1270 Ch. *Beleshou*; 1296 S.R. *Belsow*; 1318 Inq. aqd. *Belshowe*; 1346 F.A. *Belsham* (*sic*); 1433 Pat. *Belsowe*; 1542 Bord. Surv. *Belso*; 1638 Freeh. *Belshaugh*; 1663 Rental *Belsey*.

The *hōh* (Part II) or heel of ground of *Bell*, strong form of *Bella*. App. A, § 7.

Benfieldside (Lanchester). 1297 Pap. *Benfeldside*; 1307 R.P.D. *Benfelside*.

" Bean-field side or hill."

Benridge (Ponteland). *c.* 1240 Newm. *Benrig*; 1322 Ipm. *Benerig*, 1408 *Benriche*; 1593 N.C.W. *Benrych*; 1663 Rental *Benridg*.

O.E. *bēan-hrycg* = bean-ridge. Phonology, §§ 21, 27, 58.

Bensham (Gateshead-on-Tyne). 1241-9 Allen *Benchelm*; 1529 Anc. D. *Bencham*.

A difficult name. The suffix may be the word *helm*, discussed under Helm *infra*, and found also in *Denshelm* (F.P.D. *c.* 1270). If so, it refers to the hill on which Bensham stands. The first element may be *Be(o)rnic*, a diminutive of *Be(o)rn*, or a name derived from *Bernicia*, the Celtic name of the old Northumbrian province (Redin, p. 150). Cf. Bensham, Surr., earlier *Benchesham*. App. A, § 8.

Benton, Long and Little. *c.* 1190 Godr. *Bentun*.

O.E. *beonet-tūn* = farm on the " bents," or long, coarse

B

grass, cf. the common Bentley, or *bēan-tūn* = bean farm. Phonology, § 21.

Benwell (Newcastle-on-Tyne). Type I: *c.* 1050 H.S.C. *Bynnewalle.* Type II: 1251 Ch.; 1255 Ass. *Benewell;* 1261 Ipm.; 1346 F.A. *Benwell;* 1448 Pat. *Bennewell.*

Type I is difficult, but the suggestion may be hazarded that the original name of Benwell was " *binnan wealle,*" i.e. within wall. This aptly describes its position on the site of the Roman settlement of Condercum, immediately south of the Wall. For place-names of this type cf. Twining, Glouc., B.C.S. 350 *bituinæum,* i.e. between (lit. by two) rivers, and B.C.S. 344 " in loco qui dicitur *binnan ea* . . . inter duos rivos gremiales fluminis," and St Mary Bynnewerk at Stamford, i.e. within the *werk* or castle. Type II is probably an attempt to explain the earlier name by associating it with the more familiar *-well* or spring. If the name is really new, the first element would be derived from O.E. *Beonna* (m.), or *Beonnu* (f.)[1]

Berrington (Kyloe). 1278 Ass. *Beringdon;* 1342 Colding. *Beryngdon;* 1370 Sc. *Beryngton;* 1610 Speed *Barrington.*

O.E. *Bæringadūn* = hill of *Bǣre* (Angl. *Bēre*), or of his sons. Cf. Berrington, Worc. (Duignan, p. 8). Phonology, §§ 8, 22; App. A, § 1.

Berwick Hill (Ponteland). *c.* 1250 T.N. *Berewic;* 1428 F.A. *Berewic super montem;* 1595 Bord. *Barricke of the hill.*

O.E. *bere-wīc* = " barley dwelling " primarily, but later, like *barton,* used to denote demesne farm. Phonology, § 8, and cf. Barwick, Norf., D.B. *Berewica.*

Bewclay (nr. Grottington). Type I: *c.* 1250 Gray *Boclive;* 1296 S.R. *Bokelef;* *c.* 1356 B.M. *Boclif;* 1479 B.B.H. *Boclyve;* 1547 Hexh. Surv. *Buckcliffe.* Type II: 1296 S.R., 1298 Arch. 3.2.2. *Bokeley;* 1382 Pat. *Bucle;* 1663 Rental *Bukeley.*

A difficult name. The second element may be O.E. *clif* = cliff, oblique case *clife* > *clive,* cf. Cleeve, Glouc. This would suit the outstanding position of Bewclay. The first may be O.E. *bōc* = beech, hence " beech-hill," but one would

---

[1] Type I might be taken, in its first element, as a variant of Type II. (Phonology, § 7.)

hardly expect beeches to grow in so exposed a position.
For the shortening of the vowel, implied in *Buckcliffe*, cf.
buck-mast and *-wheat*, which are derivatives of O.E. *bōc*.
Type II shows loss of final *f*, for which there are several
local parallels (Phonology, § 56). The spelling *Bewc-*
represents Nthb. [bjuk] from O.E. *bōc*. Phonology, § 18 ;
App. A, § 7.

Bewdley (Stanhope). 1382 Hatf. *Bewdley*.
Possibly the same as Bewdley, Worc. (Duignan, p. 19),
earlier *Beaulieu, Bewdeley*, from Fr. *beau lieu* = beautiful
place, cf. Bewley *infra*. The *d* is unexplained.

Bewick (Eglingham) [bju·ik]. *c.* 1136 D.S.T. *Beuuiche* ;
1166 Pipe *Bowich*, 1200 *Bewich* ; 1203 R.C. *Bowic*.
O.E. *bēo-wīc* = bee-dwelling or farm. Cf. Bewick, Yorks.,
D.B. *Biuinch (sic)*, B.M. *Bewick*, and Beal *supra*. The
farm must have been famous for its bees when honey and
beeswax were more highly prized than now. The forms
with *Bo-* point to O.E. *beó-* with rising stress, instead of the
more usual *béo-* with falling stress.

Bewley (Billingham). 1197 Pipe *Beulaco; p.* 1336
D.S.T. *Bealou, Bellus locus, c.* 1360 *Belu*, 1446 *Beaulieu* ;
1539 F.P.D. *Bewley*.

v. Bewdley *supra*, and cf. Beaulieu, Hants. [bju·li].

Bickerton (Rothbury). 1245 Brkb. *Bykerton* ; *c.* 1247
Newm. *Bikerton, Bykertone* ; 1346 F.A. *Bikerton*, 1428
Bekerton.

Bickerton, Yorks., is explained by Moorman (p. 25) as
derived from O.N. *bekkjar*, gen. sg. of *bekkr*, " a stream,"
and the whole name taken to mean " enclosure by the water,"
with raising of *e* to *i* before *k* (cf. Phonology, § 7). Wyld
(p. 67) similarly explains the *Bicker-* of Bickerstaff, Lancs.,
with a good deal of support from unchanged forms with *e*.
The same element is clearly found in Beckering, Lincs.,
and Beckermet, Cumb. (Lindkvist, pp. 5, 6), with only one
form in each case with *i* for *e*. The difficulties in thus
explaining the Nthb. name are (1) the almost uniform occurrence of *i*- forms, (2) the otherwise unparalleled use of O.N.
bekkr, " beck," in this county. (2) could only be got over
if we imagined the name as a whole to have been taken

straight from that of some Scandinavian or Anglo-Scandinavian farm. In coming to any conclusion we should note that there are Bickertons in Cheshire and Herefordshire (D.B. *Bicretone* [1] and *Bicretune*), counties where Scandinavian influence is rare,[2] a Bixton, Norf., 1316 F.A. *Bykerston* and Bycardike, Notts., earlier *Bikeresdic, Bikerisdik.* The last-named place is explained by Mutschmann (p. 29) as a corruption (with double gen. suffix) of O.N. *bekkjardik* = dike of the stream. Such a derivation is very doubtful in the entire absence of *e-* forms, and it is certainly more natural to take the first element here and in Bixton as a personal name. It may be O.N. *bikarr* = bowl, goblet, which was used as a nickname (Fritzner *s.v.*). The dialectal *bicker* for *beaker* may be derived from this word,[3] and we might render these names " Beaker's farm and dike." The absence of any *s* in the forms of the Nthb., Heref., and Cheshire names makes such a derivation unlikely for them, and two possibilities remain :—(1) that the first element, at least in the Nthb. name, is Byker (*v. infra*). The meaning would then be "farm by the marsh," Phonology, § 22, (2) that it is M.E. *bicker* (of uncertain origin), meaning "strife, quarrel," and that the names refer to a question of disputed ownership, as in Threapwood *infra*.

Biddick (Houghton-le-Spring). 1190 Godr. *Bidich,* B.B. *South Bedic* ; 1268 F.P.D. *Bedyk, Byddyke* ; 1339.31 *Bidykwaterville* ; 1382 Hatf. *Bedyk.* (Washington) B.B. *Bedyk Ulkilli* ; 1382 Hatf. *Bedyk, Bydik* ; 1442.34 *Bedic by Wessington* ; 1603 Houghton *Beddicke,* 1611 *Bidwick.*

The second element is apparently O.E. *dīc* = ditch, dyke (Part II). The first may be *Bēda,* O.E. *Bēda(n)-dic* > M.E. *Beddik* (Phonology, § 21) > *Biddik* (ib. § 7), cf. Biddenden, Kent, earlier *Bedyngdenne.*[4] *Waterville,* because of its position in a bend of the Wear, *Ulkelli* from its owner, *Ulkell* or *Ulfketill* (O.W.Sc.), perhaps the same from whom Ouston *infra* took its name.

---

[1] Later *Bikerton, Bykerton.*   [2] Apart, of course, from the Wirral.

[3] Björkman (*Scand. Loan Words,* p. 211) is very doubtful on this point.

[4] If the original vowel was *i* and not *e,* we might compare *Bydictun* (B.C.S. 390), where Ceolwulf of Mercia signed a charter. This name is equally difficult.

Biddlestone (Alwinton) [bitlstən]. 1181 Newm. *Bitlesden*; *c.* 1250 T.N. *Bidlisden*; 1268 Ass. *Bydlisdene*; 1307 Ipm. *Bydellesden*; 1313 Perc. *Bideliston*; 1314 Ipm. *Bydelesden*, 1324 *Bedilsden*, *Bitelsden*; 1346 F.A. *Betlesdon*, 1428 *Bedelesdon*; 1486 Ipm. *Bedilsden*; 1542 Bord. Surv. *Byttylsden*; 1638 Freeh. *Bittleston*; 1663 Rental *id.*; 1755 Wallis II, 509 *Bittlesdon*.

The first element is the gen. sg. of a personal name. It is difficult to be certain of the name, because of the fluctuation between *t* and *d* forms in M.E. The preponderance of evidence is in favour of a name in *d*, which might be either **Bidel* or **Bydel*, diminutives of *Bida* or *Byda* (Searle). *t* for *d* would be an example of A.N. confusion of *t* and *d* (cf. Zachrisson, p. 43 n.), which ultimately affected the pronunciation of the name, cf. Battlesden, Beds., D.B. *Badelestone*, *c.* 1200 *Badelesdone*, 1428 F.A. *Battlesden*.

If the original consonant was *t*, the name would be *Bitel*, a diminutive of O.E. *Bita*, a shortened form of such names as *Bit-beald*, *Bit-beorht* (Searle), though these names are not found in O.E. This name would seem to be found in O.E. place-names—*bytlescumb* (K.C.D. 408) and *bytlesmor* (ib. 470). *d* would then be explained as due to voicing of *t* before following *l* (E.D.G., § 283), cf. Lidlington, Beds., earlier *Litincletone*, *Litlington*, Biddlesden, Bucks., earlier *Bettlesden*, *Bittlesden*. App. A, § 1, 7. Hence "valley of Bidel, Bydel, or Bitel."

Biggin (Hamsteels). 1490.35 *Biging nigh Hampstels.* "Building," *v. bygging*, Part II.

Bildershaw (West Auckland). Type I, 1312 R.P.D. *Byllershaugat.* Type II, 1432.35 *Billyngshawe.*

"The *sceaga* (Part II) or wood of *Bilheard* or **Bilhere.* The latter is not actually found, but is a likely name, cf. O.H.G. *Bilihari* (Heintze *s.v. Bil-*). Type II is either a blunder or an attempt to alter the name to the common type, with a patronymic as the first element.[1] Phonology, § 55.

Billingham-on-Tees. *c.* 1080 D.S.T. *Bellingaham*, 1125 *Billingeham*, *c.* 1150 *Billingaham*; 1203 R.C. *Billingeham*;

---

[1] It is just possible that we have here the occupative surname, *Biller*, "a maker of bills or axes" (Weekley, p. 114).

1335 Ch. *Belingeham*; 1430 F.P.D. *Billyngham*, 1539 *Byllinghame*.

O.E. *Billinga-hām* = homestead of the sons of *Bill(a)*, cf. Billingford, Norf.; Billingborough, Lincs. ; Billingbrook, Worc. Phonology § 10.

Billingside (Lanchester). 1297 Pap. *Billingside*.

Billy Mill (Tynemouth). 1320 N. 8.316 *Molendinum de Billing*.

" Billing's hill and mill."

Billy Row (Brancepeth). 1334.31 *Billey*, 1425.45 *Billyraw*.

Bilton (Lesbury). 1288 Ipm. *Bilton*.

" Clearing and farm of Billa," *Billa* (D.B.) being a shortened form of compound names such as *Bilfrith, Bilgils*.

For Row *v. rāw*, Part II.

Binchester (Auckland). *c.* 1050 H.S.C. *Bynceastre* ; 1104-8 S.D. *Bincestre* ; 1341 R.P.D. *Binchestre*.

Binchester stands on the site of the Roman station of *Vinovia* (Ptolemy Οὐιννούϊον), and the first element *Bin-* probably represents that name. For *v > b* cf. on the Continent, Besançon (*Vesontio*), Bolsena (*Volsinii*), Dietrich von Bern (*Verona*), and, in Britain, Richard of Cirencester's *Benonis* for *Venonis* (i.e. High Cross, Leic.) in the Itinerary. For the second element *v. ceaster*, Part II.

Bingfield (St John Lee). 1180 Pipe *Bingefeld* ; 1290 Abbr. *Bingefeud* ; 1298 B.B.H. *Byngefeld*.

Cf. Bingham, Norf., D.B. *Bingheham*, Bingley, Yorks., D.B. *Binghelai*. Moorman has provided the solution of these names when he quotes D.B. *Bingelie* for Billingley, Yorks. All alike show the patronymic *Billing* in compressed form. Bingfield is therefore O.E. *Billingafeld* = field of Billa's sons. For *-feud* v. Zachrisson, p. 146.

Bingfield Comb. 1479 B.B.H. *le Grene-came*. " Green ridge," *v. camb*, Part II. Phonology, § 4.

Birchope (Charlton). 1325 Ipm. *Byrchensop* ; 1330 Fine *Birchenshop* ; 1373 Pat. *Brechenshop*.

This name is difficult, and the identification is not quite certain. If correct it may be suggested that it is " Beorhtwine's *hop* " (Part II). O.E. *Beorhtwine* has metathesised forms in *Briht-*, *Breht-* (cf. Mod. Eng. *Bright-wen*), and is found in D.B. as *Brictuin, Brichwinus.* Names in *Beorht-* were a great puzzle to A.N. scribes and speakers, as in Brightlingsea, Ess. [brikəlsi), earlier *Brihtlingese,* Bricklehampton, Glouc., earlier *Brihthelmetun,* and the history of Birchope may be *Beorhtwineshop* >M.E. *Brihten-sop, Berhtensop > Brechensop, Byrchinshop* (where *ch* = k), and *Birkensope > Birch(ens)ope* under the influence of St. Eng. *birch* as against North Eng. *birk.*

Birkenside (Shotley). 1262 Ipm. *Byrkinside*; 1454 Pat. *Brekenside*; 1705 Shotley *Breckenside.*

"Hill overgrown with *birks* or birches." Cf. Akenside *supra.* Phonology, § 25, 54.

Birling (Warkworth). 1186 Pipe *Berlinga*; *c.* 1210 Newm. *Byrlyngs*; 1248 Ipm. *Birling*; 1346 F.A. *Berlyng,* 1428 *Birling.*

Cf. Birling, Kent, earlier *Baerlingas* (B.C.S. 183), with *ae* for the more usual *e* (Bülbring, § 92 n. 1), M.E. *Berlinges, Birlinges,* plur. of a patronymic from **Berel(a)*, a dimin. of **Bera.* These names are not found in O.E., but cf. O.H.G. *Berilo,* Mod. H.G. *Berle,* and the patronymic *Bierling* (Heintze *s.v. ber-*). It is found also in *Barlinghem, Ber-linghen* (Artois), *Bierlingen* (Würtemberg), *v.* Taylor, p. 107, Winkler, p. 32, and in Barlings, Ess. and Lincs., probably also in Birling, Suss., and Birlingham, Worc. Duignan (p. 20) derives the latter from O.E. *byrle,* "cupbearer"; and Roberts (p. 24) explains the Sussex place-name, rather hesitatingly, in the same fashion. Such an explanation would not fit the forms of Birling, Kent.

Birtley[1] (Chester-le-Street). B.B. *Britleia*; 1344 R.P.D. *Birteley.* (Chollerton) [baˑtli] 1229 Pat. *Birtleye*; 1255 Ass. *Brutteleg*; 1346 F.A. *Britelay.*

O.E. *beorhtan lēage* (dat.) = bright clearing. Phonology, § 54.

Bishopley (Stanhope). 1307 R.P.D. *Biscopley.*

Bishopton. 1104-8 S.D. *Biscoptun.*

"Field and farm of the Bishop of Durham."

Bishopwearmouth, *v.* Wearmouth, Bishop.

---

[1] There was also an unidentified Birtley in Auckland (1401.33, *Bretlay*).

Bitchfield (Stamfordham). 1242 Cl. *Bechefeud*; 1268 Ipm. *Bechefeld*, 1421 *Bichfeld*; 1542 Bord. Surv. *Bechefeld*; 1628 Arch. 1.3.95, *Bitchfeild*.

O.E. *bēce-feld* = beech-field (*feld*, Part II). Phonology, §§ 21, 7.

Black Blakehope (Troughend). *c.* 1230 H. 2.1.16 n. *Blachope*; 1663 Rental *Black-blakeup*.

M.E. *blake-hop* = pale coloured "hope," from O.E. *blāc* = pale, livid.

Black Bog (Billingside). 1382 Hatf. *le Bog*. **Blackburn** (Lanchester). 1313 R.P.D. *Blakburn*. **Black Dene** (Stanhope). 1382 Hatf. *Blakden*. **Black Hall** (Harperley). 1371.32 *le Blakhall*. ***Blacklaw** (Simonburn). 1348 Cl. *Blaclawe*. **Black Lough** (Edlingham). *c.* 1200 Newm. *Blakemere*. **Blackwell** (Darlington) B.B. *Blakwella*.

" Black," from the colour of the soil, materials, or waters.[1] Cf. *a black water* = one from the moors (*Compl. Angler*).

***Blackmiddingmoor** (Bamburgh). 1333 Fine *Blacmyddingmore*; 1360 Pat. *Blakmyddingmore*.

" Black-midden-swamp" (*mōr*, Part II). The second element in this forbidding place-name is M.E. *middyng* = dung-heap, a word of Scandinavian origin.

Blagdon (Stannington). 1255 Ass. *Blakeden*; 1346 F.A. *Blakden*; 1443 Ipm. *Blakdon*; 1628 Freeh. *Blagdon*.

" Black-dene." Hodgson (2.2.317) says it had its name from a " dark woody dene or dingle, the water of which runs into the Blyth a little below Bellasis bridge." Phonology, § 51, and cf. Blagdon, Som., so named from Black Down above it. App. A, § 1.

Blakeston (Norton-on-Tees). *c.* 1100 D.S.T. *Bleikestuna*; 1100-35 F.P.D. *Bleichestona*; 1203 R.C. *Blekestone*; 1335 Ch. *Blakeston*; 1345 R.P.D. *Blaykeston*; 1539 F.P.D. *Blaxtone*.

" Bleik's farm," *Bleik* being a Norse nickname from O.W.Sc. *bleikr*, " pale." Lindkvist (p. 25) notes the name *Alanus Bleik* in a 13th c. document. Cf. Kahle, p. 70, and Jónsson, p. 209. Later the name was spelled as if from the cognate O.E. *blāc*, North. M.E. *blake*. Phonology, § 21.

---

[1] Possibly some of these may contain O.E. *blāc*, "pale, livid," Dial. *blake*, with shortening of the vowel before the consonant group.

Blanchland. 1165 Chron. de Mailros, *Blanchelande* ; 1242 Pat. *Blanca Landa* ; 1270 Ch. *Alba Landa* ; B.B. *Blauncheland*.

The abbey of Blanchand was probably named after, though not affiliated to, the abbey of Blanchelande in the diocese of Coutances, near Cherbourg, founded as a Premonstratensian house in 1154. In the Norman name, *lande* has the sense of " untilled ground " ; *v.* Ducange, *s.v. landa*, planities inculta et vepribus obsita. The abbey was situated among the uncultivated moors still called " les Landes de Lessay " from the neighbouring abbey and village of Lessay (Exaquium). Similarly there was a priory of " Landa" or Laund (Austin canons) in East Leicestershire, now wrongly called Laund *Abbey*, and an abbey at Byland (=Bella Landa) in Yorkshire, both names derived from their site. *Blanche* no doubt refers to the white habit of the canons, just as the abbey of Whitland, also known as *Blanchland* or *Alba Landa*, in Caermarthenshire, is doubtless so called from the white habit of the Cistercian Order. Froissart, in his chronicles (ed. Kervyn de Lettenhove, ii. 160), refers to Blanchland as " une blanche abbaye qui était tout arse, que on clammoit au temps le roi Artus le Blanche Lande," but his account of the age of the name need hardly be taken seriously (A. H. T.). Phonology, § 5.

Blaydon (Ryton). 1340 R.P.D. *Bladon*.

The first element is possibly North. M.E. *bla* (< O.N. *blá*), " bluish-grey, livid," applied to the colour of the soil of the hill (*dūn*, Part II), but the paucity of early forms makes it difficult to be certain. Cf. Wyld, *Lancs. Placenames*, p. 294, and Bladon, Oxon. (Alexander, p. 56).

Bleaklaw (Chatton). 1251 Pipe *Blakelawe*.

The first element may go back to an inflected form *blace* or *blacan* of O.E. *blæc*, " black," or it may be from M.E. *blāke* (O.E. *blāc*), " pale." The two are completely confused in North. M.E. (N.E.D. *s.v. blake*).[1] In modern times *blake* has been replaced by *bleak*, a Scandinavian loan-word from the Norse cognate of O.E. *blāc*.

---

[1] The confusion finds its echo in the Chatton Registers, where we have *Blakelaw* (1729), *Bla(c)klaws* (1735), *Blakelaws* (1745).

Blenkinsopp (Haltwhistle). 1177 Pipe *Blencheneshopa*; 1255 Ass. *Blenkeneshop*; *c*. 1250 T.N. *Blenkeinshop, Blekenishop, Blencanishop*; 1428 F.A. *Blankensop*, 1346 *Blenkaneshope.*

"*Blenkin's hope*" (Part II). The name *Blenkin* is common but its origin is obscure. Weekley (p. 88) suggests that it is from the place-name Blencarn, Cumb. This receives slight support from the *Blencan* form given above, but it is very doubtful if we should get a personal name from a place-name as early as 1177. The alternative is to take it as a dimin. in -*chen* or -*kin* of Low German origin, though no such name is given by Winkler.

Blyth, R. 1257 Ch. *Blye*; 1267 Brkb. *Blythe.*

This river-name is found in a list of Northamptonshire boundaries (B.C.S. 792) *þær bliðe utscyt, andlang bliðan* and also in Notts., Staffs., Suff. and Warw. This can hardly be the common O.E. *bliðe*="merry, pleasant," for no river-names of this type are known. It must be Celtic. For loss of *th* in *Blye* cf. Zachrisson, p. 82 f.

Blyth (Town). (*a*) 1236 Newm. *Blithemuthe*; (*b*) 1208 Abbr. *Snoc de Bliemue*; 1386 Cl. *le Blithsnoke*; 1423 Abbr. *Blythesnuke*; 1550 Waterf. *Blythesenooke.*

The modern name has lost its second element. The "snook" is part of the town. We read (N. ix. 349) that "the term has been applied both to the promontory on the north side of the river and to the tongue of land on the south, but more properly belongs to the latter." For *snook* in place-names *v. Essays and Studies, u.s.* Vol. iv. pp. 67-8. It means "sharp-pointed projection" and is found also in Snook Bank *infra* and the Snook of Holy Island.

Bockenfield (Felton). 1206 N. vii. 353 *Bokenfeld*; 1244 Brkb. *id.*; 1307 Ch. *Bockinfeld*, 1340 *Bokenfeld*; 1346 F.A. *Bokinfeld.*

"Beechen-field," i.e. grown over with beech-trees. Cf. Akenside and Birkenside *supra*. The regular O.E. adj. would be *bēcen*=beechen, but an unmutated form *bōcen* may also have existed (cf. *ācen* and *ǣcen*). Similarly Bochidene, Warw., earlier *Bokindene* (Duignan, p. 31) and Bockingfold, Kent, earlier *Bokenefeld*. Phonology, § 32.

Bolam (Nthb.). *c.* 1155 B.M. *Bolum*; 1167 Pipe *Boolon*; 1166 R.B.E. *Boolun*; 1270 Ch. *Bolum*; 1324 Ipm. *Bolom*; 1339 Bury *Bolum*; 1507 D.S.T. *Bolom.* (Gainford) 1316 R.P.D. *Bolom.*

" Homestead by the rounded hill " from *Bolham* (*v.* App. A, § 6). The element *bol* in English place-names is fully discussed in *Essays and Studies, u.s.* Vol. iv. pp. 59-60. Since that was written the following passage has been noted in Hodgson (2.1.331): " I think that Bolam has its name from being situated as it is on a *bol* or high swell of land, as Bol-don in the county of Durham has from a rounded hill under which it is situated."

Boldon [boudən]. 1153-95 F.P.D. *Boldun*; 1312 Ch. *Boldon*; 1637 Camd. *Bowdon.*

" Rounded hill," [1] *v.* Bolam *supra.* Phonology, § 39.

Bollihope (Frosterley). 1382 Hatf. *Bolyopshele*; 1377 Ipm. *Bolyhopsheles.*

Eggleston (p. 55) gives an earlier form *Bothelingopp* from a charter of Bishop Bek. This form has not been traced, but if correct would point to a patronymic **Bodeling* from **Bodel*, a dimin. from O.E. *Bod(d)a.* Hence " Bodeling's hope." Apart from this, one might equally well take the first element as *Bolling* from *Bolla.* Phonology, § 51.

Bolton (Edlingham). 1200 Pipe *Bolton*; 1226 Ch. *Bodelton*, 1227 *Boulton*; 1227 Gray *Boelton*; 1229 Pat. *Boulton*; *c.* 1250 T.N. *Bowilton*; 1697 N. vii. 218 *Bowton* alias *Boulton.*

There are nine Yorkshire Boltons, eight of which show D.B. forms *Bodeltone, Bodeltune*, two in Lancs., earlier *Bodeltone, Bothelton*, one in Cumb., earlier *Bochelton (sic)*, one in Haddingtonshire, earlier *Boteltun, Botheltun* (Johnston, p. 44), and there can be little doubt that Moorman is correct when (pp. 28-9) he explains the first element

---

[1] It is just possible that the first element in this name is a personal name. F.P.D. (p. 10 n.) has an isolated form *Bollesdon*, as if from a personal name *Boll*, but the form may be due to confusion with Bowsden *infra*, often mentioned in F.P.D. documents. It receives perhaps some slight confirmation in the existence of a *Bolleburn* stream in the neighbourhood (F.P.D. p. 10).

as another form of O.E. *botl* or *bold*, " building " (cf. Budle *infra*), and takes the whole to mean " enclosure of land with some sort of building on it." Wyld (pp. 72-3) is doubtful of its history, but neither he nor Sedgefield (p. 18) offers alternatives which fully satisfy the phonological requirements. Phonology, § 44.

Bolts Law (Stanhope). 13th c. R.P.D. *Boltislawe*. " Bolt's hill." Cf. Boltby, Yorks. D.B. *Boltebi*. O.W.Sc. *Boltr* is common as a nickname (Fritzner, *s.v.*, Björkman Z.E.N. p. 26).

Bothal [bɔtl]. 1233 Pipe *Bothalle*; *c.* 1250 T.N. *Bothal*; 1255 Ass. *Bot(t)ehale*; 1270 Ch. *Bothala*; 1346 F.A. *Bottal*, 1428 *Bottell*; 1507 D.S.T. *Bothall*.

O.E. *(æt) Bōtan hēale* i.e. (at) the *healh* (Part II) of *Bōta* (m.) or *Bōte* (f.). Phonology, § 21 ; App. A, § 6.

Boulmer (Long Houghton) [bu·mə]. 1161 Pipe *Bulemer*, R.B.E. *Bolimer*; 1296 S.R. *Bulmer*; 1579 Bord. *Bowmer*; 1663 Rental *Boomer*.

O.E. *bulan-mere*=bull's mere or Bull's. *bul(l)a* is not found in independent use in O.E., but the evidence of place-names makes it fairly clear that it was already in use and also employed as a nickname (*v.* M.L.R. vol. xiv. p. 236). The "mere" must be a sea-pool, referring to the shallow lagoons found on the sea-shore here (Tomlinson, p. 420). Phonology, § 39.

Bowmont Water. *c.* 1050 H.S.C. *Bolbenda*; 1292 Ass. *Bolbent*; 1580 Bord. *Bowbaynt*; *c.* 1590 Map. *Bowbent*.

Corrupt in its modern form and clearly of pre-English origin. Phonology, § 39.

Bowsden (Lowick) [bauzən]. 1228 F.P.D. *Bollesdene*; 1239 Ipm. *Bollisdun*, 1250 *Bollisdon*; 1335 Ch. *Bolesdon*; 1337 F.P.D. *Bollesden*; 1428 F.A. *Bollesdon*; 1539 F.P.D. *Bolsden*; 1579 Bord. *Bowsdenn*.

" Boll's valley or hill." App. A, § 1. Cf. Bolsover, Notts., earlier *Bollesouere* and, with weak form *Bolla*, *Bollanea* (B.C.S. 144). Phonology, § 39.

Bracks Farm (Auckland). 1382 Hatf. *le Brak*, *v.* Brakkes *infra*.

Bradbury Isle (Sedgefield). 1104-8 S.D. *Brydbyrig* (*sic*) ;

B.B. *Bradbire* ; 1344 R.P.D. *Bradbery* ; 1490.36 *Le Ile near Bradbery.*

Bradford (Bamburgh). 1267 Ipm. *Bradeford* ; 1460 II. 3.1.28 *Bradfurth.* (Bolam) 1271 Ipm. *Bradeford,* 1377 *Bradferthe.*

Bradley (Haltwhistle). 1279 Iter. *Bradeley.* (Medomsley) 1340 Ipm. *Bradley.* (Ryton) 1382 Hatf. *id.* (Wolsingham) B.B. *Bradleia.*

" Broad *burh* (Part II), ford and clearing." Phonology, §§ 21, 30.

Brafferton (Aycliffe). 1091 F.P.D. *Bradfortuna* ; B.B. *Bradfertona* (B., C. *Brafferton*).

"Farm by the broad ford." Cf. Brafferton, Yorks., D.B. *Bradfortune,* Bretforton, Worc., D.B. *Bratfortune* and Swinnerton, Staffs., earlier *Swinforton.* Phonology, §§ 21, 51.

Brainshaugh (Acklington). 1104-8 S.D. *Bregesne* ; *n.d.* F.P.D. *Brainesleie* ; 1438 Acct. *Braynley* ; 1480 N. v. 483 *Branssehalgh* ; 1534 N. v. 485 *Braineshaugh* ; 1663 Rental *Brainshaugh* ; 1676 Warkw. *Bransehaugh.*

Cf. Bransford, Worc. (Duignan, p. 25), earlier *Bregnesford.* This may point to an O.E. name *Bregn or the first element may be gen. sg. of O.E. *Bregwine,* in syncopated form. Hence " haugh or clearing of Bregn or Bregwine." The form in S.D. is corrupt, and is also an example of a place being known by its owner's name (cf. Harle *infra*) with no suffix added.

Brakkes (Heighington). 1382 Hatf. *les Brakkes.*

Pl. of either (1) M.E. *brak,* " strip of uncultivated ground between two plots of land " (Jamieson), a derivation of the vb. *breken,* " to break," or (2) M.E. *brak*=bracken, apparently a shortened form of that word itself.

Brancepeth. 1085 D.S.T. *Brentespethe* ; 1155 F.P.D. *Brandespethe* ; *a.* 1196 Finch. *Brenspad* ; 1254 D.S.T. *Branspath* ; 1311 R.P.D. *Braun(de)spath,* 1312 *Brancepath,* 1340 *Brauncepath* ; 1408 D.S.T. *id.* ; 1796 Sherb. *Brawnspeth.*

O.E. *Brandes-pæð*= Brand's path (*pæð* Part II). *Brand* is a name of Scandinavian origin. For the form cf. Brauncewell, Lincs., and Braunston, Leic., Northts.,

Rutl., earlier *Branteswell, Brandeston.* Phonology, §§ 53, 5, 1.

Brandon (Brancepeth). *c.* 1190 Godr. *Braindune (sic), Brandun*; 1217 Pap. *Brandun*; *n.d.* Finch *Bramdun.* (Eglingham) Type I: *c.* 1150 Perc. *Bremdona*; 1292 Q.W. *Bremedon.* Type II: 1247 Sc. *Bromdun*; *c.* 1250 T.N. *id.*; 1308 Ipm. *Bromdon.* Type III: 1255 Ass. *Bramdon*; 1346 F.A. *Brampdon*; 1350 Cl. *Brandon*; 1428 F.A. *id.*; 1480 Ipm. *Braundon.*

Branton (Eglingham). Type I: *c.* 1135 Perc. *Bremetonam*; 1247 Sc. *Bremtone*; *c.* 1250 T.N. *id.* Type II: 1308 Ipm. *Brombton.* Type III: 1334 Perc. *Brampton*; 1350 Cl., 1450 Pat. *id.*; 1480 Ipm. *Braunton*; 1498 H. 3.2.127 *Branton.*

The different types have arisen through confusion between plant-names of similar form. Type III shows dialectal *brame*=briar or bramble. Cf. Bramham, Yorks. (Moorman, p. 32). Type II shows O.E. *brōm*=broom. Type I is due perhaps to the influence of O.E. *brēmel*= bramble. The Durham Brandon belongs definitely to Type III. The second elements are *tūn* and *dūn* respectively Phonology, §§ 51, 55, 5.

Branxton (Glendale). 1249 Ipm. *Brankeston*; 1346 F.A. *Branxston*; 1343 Bury *Branxtone.*

Cf. Branscombe, Dev., B.C.S. 553 *Branecescumbe,* Branxholm, Roxburgh, earlier *Brancheshelm.* These three point to an O.E. name *Brannoc,* perhaps dimin. of *Brand.* Cf. O.H.G. *Brandico,* M.H.G. *Brancke* (Heintze). The name *Brand* in England is usually taken to be of Norse origin, but it may be noted that, as early as 1046, we find Bransbury, Hants., as *Brandesburh,* while Branston, Staffs., is *Brantestun* in a charter (B.C.S. 978) dated 956. In neither place is it very likely that we have to do with a Scandinavian name.

Breaks, The (Windlestone). 1420 Acct. *les Brakes.* " The thickets," apparently the plural of *brake,* " a clump of bushes, brushwood, or briers " (N.E.D.).

Breamish, R. *c.* 1050 H.S.C. *Bromic*; 1532 Raine *Bremish*; 1610 Speed *Bromish*; 1637 Camd. *Bramish*;

1645 Map. *Bromish*; 1755 Wallis II, 149 *Bramish*; 1833 Map *Bremish*.

A river-name of Celtic origin.

Brenkley (Ponteland). 1177 Pipe *Brinchelawa*; 1271 Ch. *Brinkelawe*; 1298 B.B.H. *Brinkelagh*; 1346 F.A. *Brenklawe*; 1479 B.B.H. *id.*; 1628 Freeh. *Brinkley*, 1638 *Brenkley*.

The second element is O.E. *hlāw*, "hill." The first is the personal name *Brynca*, L.V.D. The common word *brink* does not suit the character of the hill. Phonology, § 10; App. A, § 2.

Brierdene (Earsdon) [bri·ədən]. 1295 Ty. *Brerden*; 1596 N. ix. 96 n. 2 *Breerden*.

Brierton al. **Brearton** (Stranton). 1315 R.P.D. *Brereton*.

"Brier-valley and farm." [bri·ər] is the correct Nthb. form of *briar*.

Brinkburn (Coquetdale). Type I: *c.* 1120 Brkb. *Brinkeburne*; 1216-27 Newm. *id.*; 1313 R.P.D. *Brenkeburn*; 1507 D.S.T. *id.*; 1542 Bord. Surv. *Brenkborne*; 1663 Rental *id.*; 1728 N. vii. 492 *Brenckburn*. Type II: *c.* 1175 Joh. Hex. *Brincaburch*. Type III: 1104-8 S.D. *Brincewelæ*.

"Brynca's burn, *burh* (Part II) or spring," or "the burn, *burh* or spring beneath the brink or hill." Cf. Brenkley *supra*. Phonology, § 10; App. A, § 10.

Broadstruthers Burn (Cheviot). 1255 Sc. *Bradstoir*.

"Burn through the broad *strother*" (Part II). *Brādstrother* >[bradstruðə] >[bradstuə] >bradstə]. Phonology, §§ 14, 21. Cf. Anstruther, pron. [anstə]. The present form is entirely artificial.

Broadwood (Wolsingham). 1153-95 B.B. *Bradewode*; B.B. *Bradwode*; 1382 Hatf. *Bradeworth*.

"Broad-wood." The normal development would be *Bradwood*. Cf. Bradley *supra*. *Broadwood* is due to St. Eng. and the independent *broad*. App. A, § 3.

Brockley Whins (Hedworth). 1382 Halm. *Brockleys*.

Brockley Hall (Rothbury). 1309 Ipm. *Brockleygehirst*.

"Badger-haunted clearing," O.E. *brocc*, "badger," and *hyrst*, "wood."

Brockwell (Winlaton). 1398.35 *Brokwelstrother.*
O.E. *brōc-* or *brocc-wielle*=brook- or badger-spring.
Phonology, § 21. *v. strother*, Part II.
Broom (Durham). 1153-95 Finch. *Brom.* **Broom-
haugh** (Bywell St Andrew). 1262 Ipm. *Bromehalwe*; 1268
Ass. *Bromhalgh*; 1346 F.A. *Bromhalf.* **Broomhope** (Chol-
lerton). *c.* 1250 T.N. *Bromhop.* **Broomley** (Bywell St
Peter). 1255 Ass. *Brumleg*; 1268 Ipm. *Bromley*, 1425
Brumilee. **Broomshiels** (Lanchester). 1297 Pap. *Bromsteles
(sic)*; 1382 Hatf. *Bromeschels.*
All named from the broom (O.E. *brōm*) growing there.
Cf. Broom, Worc., Broomhall, Salop, earlier *Bromhale,*
Bromley, Kent (B.C.S. 506 *bromleag*). Phonology, § 18.
Broomy Holm (Chester-le-Street). 1326.45 *Bromywhome*;
1384.32 *Bromyngholm*; 1382 Hatf. *Bromemyngholme (sic).*
Probably, " the *holm* " (Part II) by the broom-covered
ing. (Introd. p. xxvii.).
Brotherlee (Stanhope). 1457.35 *Brotherleshele.*
Brotherwick [brɔdrik]. 1251 Ipm. *Brothirwike*; 1273
R.H. *Broyerwyk (sic)*; 1275 Ipm. *Brothirwyk*; 1663 Rental
Brotherick; 1734 Warkw. *Broderick.*
" Brother's clearing and dwelling." *Brother* is a
Scandinavian name by origin (O.W.Sc. *Brōðir*). Cf.
Brotherton, Yorks. (Moorman, p. 37), and Brothertoft,
Lincs. (Lindkvist, p. 214). Phonology, §§ 41, 49.
Browney, R. *c.* 1125 F.P.D. *aqua de Brun*; *c.* 1170
Finch. *Brune flumen*; 1479 B.B.H. *Broune.*
Apparently " brown " from the colour of its waters,
but how *-ey* came to be added is a mystery. Is it a survival
of forms with O.E. *ēa*=river ?
Brownridge (Chatton). 1330 Ass. *Brunrige.* **Brown-
side** (Evenwood). 1312 R.P.D. *Brounsyde.* Self-ex-
planatory.
Broxfield (Embleton). 1256 Ass. *Brokesfeud*; 1307 Ch.
Brockesfeld.
" Field of the brock or badger " or " belonging to Brock."
Bruntoft (Elwick). 1304 Cl. *Bruntoft*; 1389 Pat.
Burnetoft.
The second element is Scand. *toft*=clearing (Part II).

The first is either O.E. *burna*=stream, with metathesis (cf. Brunton *infra*) or the cognate O.W.Sc. *brunnr*=spring or fountain (Lindkvist, p. 214).

Brunton (Embleton). *c.* 1250 T.N. *Burneton Bataill*; 1377 Ipm. *Burneston.* (Gosforth) *c.* 1250 T.N. *Burneton.*

" Burn-farm," with metathesis of *r.* Phonology, § 54. *Burneston* shows pseudo-genitival *s. Bataill*, from its connexion with the family of that name (cf. Battle Shield *supra*).

Buckton (Norham). *c.* 1250 T.N. *Buketon*; 1344 R.P.D. *Bukton*; 1560 Raine *Buckton.*

" Farm of Bu(c)ca " or " goat-farm." Cf. Buckden, Hunts., which Skeat (p. 324) derives from O.E. *buccandenu*, taking *bucca* to mean " he-goat " or a man named *Buck.*

Budle (Bamburgh) [bʌdl]. Type I: 1165 Pipe *Bolda*, 1177 *id.* Type II: 1196 Pipe *Bodle*; *c.* 1250 T.N. *Bodhill*; 1288 Ipm. *Bodell*; 1346 F.A. *Bodil*, 1428 *Budill*; 1538 Must. *Buddill.* Type III: 1205 Pipe *Bodlum*; 1319 Ipm. *Bodlom*, 1328 *Bodlum.* Type IV: 1314 Ipm. *Botel.*

The Teutonic type *buthlo*=building, gives rise to four different forms: I, *boðl*, as in the first element in Bolton *supra*; II, *bodl* (< *boðl*), cf. M.E. *fiðele* > *fiddle* (Jespersen, 7.42); III, *bold* (< *bolð* < *boðl*) with metathesis and subsequent change of open ð to stopped *d* (cf. *seðel* and *seld*, *spādl* and *spāld*, Bülbring, §§ 444, 452). This is a distinctively Mercian type, cf. Newbold, Derbys. IV, *botl*< *bodl* with unvoicing of *d* to *t* (cf. *spādl* and *spātl*, *seðl* and *setl*). Cf. Wallbottle *infra.* These forms explain Types I, II, and IV. Type III represents dat. pl. (*æt þǣm*) *boðlum* =(at the) buildings. For the phonetic development to [bʌdl] cf. Sc. [bʌdi] and [bʌdm] for *body* and *bottom.*

Bulbeck Common (Slaley). 1236 Pat. *Bolebec.*

The common formed part of the ancient barony of Bolbec or Bulbeck, so called from " Bolbec, a village near the mouth of the Seine, the cradle of the race of the Norman knight upon whom Henry I conferred one of the baronies created out of the wide lands that once belonged to the official earldom of Northumberland " (N. vi. 221). N.Fr. *Bolbec* is from O.W.Sc. *bolla-bekkr*, i.e. beck or stream of *Bolli* or Bull.

c

Burdon (Bishopwearmouth). *c.* 1050 H.S.C. *Byrdene*; 1390 Finch. *Byrden*; 1433 D.S.T. *Birdene*.

"Byre-valley," i.e. with a byre or cow-shelter in it; App. A, § 1.

(Haughton-le-Skerne) 1109 D.S.T. *Burdune*; 1335 Ch. *Burdon*. Cf. Burradon *infra* and Burton, Glouc., earlier *Burgtune* (Baddeley, p. 28).

Burn Hall (Durham). *c.* 1225 F.P.D. *Brune*; 1330 D.S.T. *Burn*.

O.E. *burna*, "stream," with occasional metathesis.

Burnhope (Lanchester). 1307 R.P.D. *Brunhop*; 1372 Acct. *Brunhopschel*; 1382 Hatf. *Burnhop*.

"The hope by the burn," *hop* Part II. cf. Bruntoft *supra*.

Burnigill (Brancepeth). *c.* 1190 Godr. *Brunninghil*; 1261 F.P.D. *Burnyngyll*, 1268 *Brunynghille*, *Brunninghille*; 1313 R.P.D. *Bruni(n)ghill*, *Burnynghill*; 1343 R.P.D. *Burnyngill*.

O.E. *Brūning(a)-hyll*=hill of Brown or of his sons, *Brūning* being a patronymic from O.E. *Brūn*. Phonology, § 54; App. A, § 7.

Burradon (Alwinton) [bɔ·rdn]. *c.* 1200 Sc. *Burhedon*; *c.* 1250 T.N. *Burweton*; 1313 Perc. *Boroghdon*, 1323 *Burghdon*; 1324 Ipm. *Borouden*; 1628 Freeh. *Burrowdon*. (Earsdon) *c.* 1150 Perc. *Burgdon*; *a.* 1162 N. ix. 44 n. 2 *Burgdunie*; 1346 F.A. *Boroudon*; 1638 Freeh. *Burroden*; 1662 Arch. 2. 24. 122 *Burradon*.

O.E. *burh-dūn*=fort-hill, possibly from some early stronghold which crowned the hill, cf. Burdon *supra*.

Burton (Bamburgh). 1257 Ch. *Burton*; 1346 F.A. *id.*, *Bourton*. A very common English place-name, going back to O.E. *burh-tūn*=fortified enclosure.

Bushblades (Lanchester). 1312 R.P.D. *Burseblades*; B.B. *Bursebred*; 1382 Hatf. *Buresblades*; 1669 Lanch. *Busblaids*, 1717 *Bushblaids*.

The elements are probably Nthb. *birse*, "bristle" (O.E. *byrst*, M.Sc. *brust*), and the common word *blade*. Cf. *Bursyland* in Stanhope (Hatf. Surv.) and *Burbladthwayt*, Yorks. (Lindkvist, p. 106). Hence, "place where the bristly blades grow." Possibly there may have been an actual plant-name *burseblade*. For *-bred* *v.* Zachrisson, pp. 120-3.

Buston (Warkworth). 1166 Pipe *Buttesdon, Butteston,* 1186 *Buttesdun* ; 1248 Ipm. *Butlesdon* ; *c.* 1250 T.N. *Budlisden, Butlesdon* ; 1255 Ass. *Botleston,* 1278 *Boteleston* ; 1293 Perc. *Botliston* ; 1307 Ipm. *Botilston* ; 1346 F.A. *Butelston, Bot(el)eston* ; 1428 F.A. *Buston.* O.E. *Buteles-dūn* or *-tūn*=Butel's hill or farm. For *t* >*d*, cf. Biddleston *supra.* The change may have been assisted by the analogy of the variant forms *bodl* and *botl* of O.E. *botl*=building (cf. Budle *supra*). For the assimilation in the Pipe Roll forms cf. Bottesford, Leic. (D.B. *Bottesford*) and Lincs. (D.B. *Budlesford*), Lincs. Surv. *Botlesford*.[1] Phonology, §§ 50, 53 ; App. A, § 1.

Buteland (Birtley). 1255 Ass. *Buteland* ; 1265 Sc. *Boteland* ; 1296 S.R. *Botland* ; 1324 Ipm. *Botelond* ; 1628 Freeh. *Buteland* ; 1663 Rental *Beutlands.* " Bota's land," cf. Bothal *supra.* Phonology, § 18.

Butsfield (Lanchester). 1312 R.P.D. *Botesfeld* ; 1334.45 *Butlesfeld* ; 1382 Hatf. *Butesfeld.* " Butel's field," cf. *butlesleage,* B.C.S. 279 A. It is a dimin. of *But(t)a,* which gave rise to Buttington, Glouc. (A.S.C. *Buttingtun*). Phonology, § 53.

Butterby (Durham). 1242 D. Ass. *Beutrone (sic)* ; 1352.31 *Beautrove* ; 1355 Acct., 1381.45 *id.,* 1429.33 *Bowtreve,* 1491.36 *Beautroby,* 1500 *Beatreby* ; 1592 Wills *Butterbye.* Fr. *beau trouvé*=well-found, a name probably bestowed by the earliest Normans, " who discovered and appropriated the beautiful sequestered spot hid in the bosom of woods and waters " (S. iv. 109). Phonology, § 20. The later changes show assimilation to a common Scandinavian type of place-name.

Butterknowle (Lynesack). 1313 R.P.D. *Boterknoll.*
Butter Law (Newburn) 1251 Ipm. *Bottirlawe* ; *c.* 1250 T.N. *Buterlawe* ; 1309 Ipm. *Botirlawe* ; 1428 F.A. *Butterlawe.*
Butterwick (Sedgefield) 1131 F.P.D. *Boterwyk* ; B.B. *Buterwyk* ; 1314 R.P.D. *Buttrewik* ; 1337 R.P.D. *Boterwyk.* " Butter-knoll, hill and dwelling," referring to ground

---

[1] It is tempting to connect all these names with O.W.Sc. *Buðli,* but that name does not seem to have been used in historical times ; *v.* Lind. *s.n.*

and farms with rich pasturage. The same element is probably found in Butterleigh, Dev., Butterwick (2) and Butterworth, Yorks., Butterley and Birley, Heref., Bitterley, Salop., Butterhill, Staffs. (Duignan, p. 30), and Butterwick, Lincs. (2). Similarly in Norway we have *smørbøl* (N.G. III. 282)=butter-dwelling, and *Smørstad* (N.G. iv. 215), and Jakobsen (p. 188) notes that in the Shetlands some place-names are formed with *smjǫr* as their first element, denoting fertility, e.g. Smerrin < *smjǫr-vin*. It should be noted, however, that there are some place-names in which *Butter*- must rather represent a personal name, e.g. Butterford, Dev., and Buttermere, Wilts. and Cumb., and it may do so in some of the names given above. In Cumberland this might be taken to be from *Buter=O.W.Sc. *Butr* (D.B. *Buterus*), with rare preservation of inflexional *r*, but this seems unlikely in Dev., and still more in Wilts., as the form *Butermere* is found in a charter of 863. We should hardly expect a Scandinavian settler to be well established here by this date, cf. Sedgefield and Ekblom, *s.n.*

Byermoor (Whickham). B.B. *Becchermore*; 1385.45 *Byrmore*; 1656 Ryton, *The baremore*.

The same as Barmoor *supra*. The variant forms are due to the great diversity of development of O.E. *ēag* in M.E. and Mod. Eng. Cf. E.D.G., § 185.

Byers Green (Auckland). 1345 Pat. *the Byres*. **Byerside** (Medomsley). 1382 Hatf. *Bires*; 1421.35 *Biressyde*. **Byers** (Lambley). 1239 B.B.H. *Byres*.

Pl. of O.E. *bȳre*=cow-byre.

Byker (Newcastle-on-Tyne). 1249 Pipe *Byker*; 1286 Ipm. *Biker(r)*; 1298 Ch. *Biker*; 1428 F.A. *Byker*; 1490 Pat. *Bycarfelde*.

M.E. *bi-ker(r)*=neighbouring upon a marsh (*kjarr*, Part II). Such place-names in *By*- are fairly common. Cf. Byfleet, Surrey (B.C.S. 39 *biflete*); Bygrave, Herts. (K.C.D. *biggrafan*); Byfield, Northants.; Bythorne, Hunts.; Biford, Glouc.; Bywood, Dev.; Byworth, Suss.; and Bywell *infra*. The length of vowel makes it impossible to connect the name with O.W.Sc. *bekkr*, pl. *bekkir*, streams. That is probably found in Bicker, Lincs. (D.B. *Bicker*, T.N. *Biker*).

Bywell-on-Tyne. 1104-8 S.D. *Biguell*; 1174 D.S.T. *Biwell*; 1346 F.A. *Bywell*.
" By the spring," cf. Byker *supra*. *Big-* is a common early spelling of *bī*, cf. *biggrafan* quoted under Byker.

Caistron (Rothbury). 1184 Pipe *Kersten*, 1240 *Kesterne*; 1244 Ch. *Kersthirn*; 1256 Ass. *Crestern, Casterne*; 1290 Ch. *Kestern*, 1307 *Kerstern*; 1428 F.A. *Kestryn*; 1538 Must. *Krestron*; 1663 Rental *Kaistrin*.
The second element is O.E. *þyrne*, " thorn-bush " (Part II). Cf. Casterne, Staffs. (Duignan, p. 33), earlier *Cætes-thyrne* and Chawston, Beds. (Skeat, p. 56), earlier *Calvesterne*. The first is *carse* (M.E. *kers*), in common use in Scots dialect and place-names (e.g. Carse o' Gowrie), meaning " fen, low wet land, low alluvial land on the banks of a river " (N.E.D.). Hence " thorn-bushes in low marshy land." Cf. Carsthorne, Kirkcudbright, given by Johnston (p. 67). *sþ* > *st* as in *nostrils* < *nosepirles*. Phonology, § 54.

Callaly (Whittingham). Type I: 1160 Pipe *Calualea*; 1177 Pipe *Caluwelei*; 1244 Brkb. *Calweley*; 1247 Ch. *Calveley*; c. 1250 T.N. *Caluley*; 1425 Ipm. *Calele*; 1428 F.A. *Calole*. Type II: 1210-2 R.B.E. *Calverlega*; 1273 R.H. *Calverley*.
O.E. *calwa(n)leage* (dat.) = bare clearing. Cf. *on calwan hyll* (B.C.S. 1108), Cow Honeybourne, Glouc., earlier *Calughhonyburn*, Callaughton, Salop., earlier *Caleweton*. The adj. is descriptive of some barren, infertile stretch of country. Type II may be a mere blunder due to the influence of such names as Callerton *infra*, or it may point to an alternative name with *Calver-* from O.E. *cealfra*, gen. pl. of *cealf*, " calf," as its first element, cf. Calverley, Yorks., and Callerton *infra*.

Callerton (Ponteland). 1100-35 Ty. *Calverduna*; c. 1250 T.N. *Calverdon*; 1228 Pipe *Caluerton*; 1292 Q.W. *Calverton*; 1346 F.A. *Calverdon*; 1350 Cl. *Callerdon*; 1428 F.A. *Callerton*. **Black Callerton** (Newburn). c. 1250 T.N. *Blackalverdon*; 1311 Ipm. *Black Callirdon*. **High Callerton.** 1296 S.R. *Calverden de Valence*; 1428 F.A. *Callerton Valkens*.
O.E. *cealfra-dūn*=hill of the calves. Phonology, § 51;

App. A, § 1. "Black," probably from the soil, also known as Callerton Delaval, because it formed a part of the barony of Delaval (T.N.). High Callerton was once held by a member of the great house of Valence.

Cambo (Hartburn) [kamə]. 1230 Sc. *Camho*; 1240 Pat. *Kamho*; 1253 Pat. *Cambhou, Cambhogh*; 1255 Ass. *Camhou*; 1258 Pat. *Cambhogh*; 1277 Ch. *Cambhou*; 1278-81 Perc. *Kambou*; 1346 F.A. *Cambow*; 1583 Bord. *Cammo*; 1715 Arch. 3. 13. 8 *Camma*.

A difficult name. The second element is *hōh* (Part II), but there is no *camb* (Part II) or "kame" here. Possibly the first element is *cam* (N. Cy. *camb*), used of slate. Slate is quarried near here, and the name may be "heel of land where slate (*cam*) is quarried."

Cambois (Bedlington) [kaməs]. *c.* 1150 F.P.D. *Kambus*, 1203 *Cambus*; 1236 Newm. *Kamhus, Camhous*, 1246 *Cambhus*; 1255 Ass. *Camhus*; 1335 Ch. *Cammus*; 1344 R.P.D. *Cambhus*; 1359 Pat. *Cambowes*; B.B. *Camhus* (B. *Camboise*, C. *Cambous*); 1363 Ipm. *Cambois*; 1551 N. ix. 224 *Cammosse*; 1637 Camd. *Cammus*.

This name may be Celtic and connected with Gael. and Ir. *camus*, "bay, creek." Thus Adamnan (*Vita S. Columbae*, ed. Fowler, p. 63 n. 1) speaks of "locus qui Scotice vocitatur Cambas," and Camus, Camas, and Cambus are common place-names in Scotland and Ireland. Cf. Gillies, *Place-Names of Argyll* (p. 13), Hogan s.n. *Camas* al. *camus*, Cambus, Cambos, and Cambas in Stirling (Johnston, p. 60), and Camus, Joyce II, p. 398. Cambois stands on a broadly curving bay. The chief difficulty in this interpretation is that the *-hus* forms could only be explained by very early etymologising on the part of the scribes. The spelling *-bois* is due to the influence of A.F. *bois*, "a wood." Cf. Warboys, Herts., Theydon Bois, Ess. and Boisfield,[1] Co. Durham, the last held by the family of *De Bosco* or *Bois*.

Capheaton (Kirk Whelpington). 1274 Swinb. *Magna Heaton*; 1428 F.A. *id.*; 1454 Pat. *Cappitheton*; 1465 Ipm. *Capitheton*; 1538 Must. *Captheton*; 1536 N. vii. 468 *Capheton*.

---

[1] Now merged in Pespool.

v. Heaton *infra.* Originally distinguished by the epithet *Great*, it was qualified in the 15th cent. by the Lat. *caput*=head or chief, now reduced to *Cap-*, cf. *cap-castle* (N.E.D.)=chief village of a district.

Carham-on-Tweed. *c.* 1050 H.S.C. *Carrum*; 1104-8 S.D. *id.*; *c.* 1250 T.N. *Karh'm*; 1251 Ch. *Karram*, 1252 *Karrum*; 1255 Ass. *Karham.*

O.E. *carr-hām*=homestead by the rock, or (*æt þæm*) *carrum*=(at the) rocks, *carr* being an O.E. word of Celtic origin. Richard of Hexham (*Chronicles of Stephen*, Rolls Series, Vol. 3, p. 145) speaks of " *Carrum*, quod ab Anglis *Werch* (i.e. Wark *infra*) dicitur," suggesting that Wark was an English name trying to oust an earlier Celtic one. App. A, § 6.

Carlbury (Coniscliffe). 1271 Ch. *Carlesburi*; 1313 R.P.D. *Carlebiry.*

" Carl's stronghold." *Carl* is either the common word *carle*, the Scand. equivalent of English *churl* (cf. Charlbury, Oxon.), or, more probably, that word used as a personal name. Searle gives many examples. Cf. Björkman N.P. pp. 77-8.

Carlton (Redmarshall). *c.* 1050 H.S.C. *Carltun*; 1109 R.P.D. *Carlentune*; *c.* 1190 Godr. *Karletun*; 1307 R.P.D. *Carleton.*

O.W.Sc. *karlatún*=farm of the carls. A very common place-name in Scandinavian England corresponding to the equally common native Charlton. *Carlen-* is probably from a pseudo-weak gen. pl. *carlena*. Cf. Carlton-upon-Trent, Notts D.B. *Carlentune* (Mutschmann, p. 31).

Carp Shield (Muggleswick). 1339 Acct. *Garpschele*; 1380 Acct. *Cappeschel*; 1387 *id.*; 1469 D.S.T. *Carpshele.*

Possibly " shiel of *Garpr*," an O.W.Sc. name, or it may be from some name allied to that which presumably lies behind Carperby, Yorks., for which V.C.H. (*North Riding* I. 207) gives earlier *Chirprebi*, 14th c. *Kerperby.*

Carraw (Newbrough). 12th c. B.B.H. *Charrau*; 1279 Iter. *Karrawe*; 1280 Ch. *Cadrere*; 1296 S.R. *id.*; 1298 B.B.H. *Carrawer*; 1354 Pat. *Carraure*; 1479 B.B.H. *Carraw.*

Clearly not an English name. The first element may be the stem *cadro-* (Holder *s.v.*) found in Welsh *cader*, "chair," O. Bret. *cadr*, "beautiful." For the suffix we may perhaps compare Stranraer, earlier *Stranrever*, *Stranraver*, which Johnston (p. 275) takes to be from Gael. *sron reamhar*, "thick point." Cf. also Knockrower (Joyce I. 20)< *cnocreamhar*, Canrawer and Carrigrour (ib.). Phonology, § 8.

Carrick (Elsdon). *n.d.* Swinb. *Kairwych* ; 1324 Ipm. *Carwyk*, 1331 *Cairewik*, *Kayrwik* ; 1344 Pat. *Carewyk* ; 1586 Raine *Caricke* ; 1628 Freeh. *Cairwick*, *Carrick*.

A name of hybrid origin. The first element is cognate with Welsh *caer*, "fort." Cf. *Cair Ebrauc*, *Caer Efrawg*, the Welsh name for York city. The second is O.E. *wīc*, "dwelling," hence "dwelling by the fort," perhaps some old earthwork.

Carriteth (Simonburn). 1325 Ipm. *le Caryte*, 1328 *le Karite* ; 1330 Cl. *le Carite* ; 1597 Bord. *Caryteth* ; 1663 Rental *Carrieteeth*.

This would seem to be O.N.F. *carite(dh)*, M.E. *carited*, *caritet*, *cariteth* < Lat. *caritatem*. This word is found in its Central French form as a place-name in *La Charité-sur-Loire*, Nièvre Dept. (Cl. 1245 *le Karyte*). Godefroy takes this to be (*Dict. de L'Anc. langue française*, II. 73-4), O.F. *charité*, *c(h)areti*=établissement charitable. Carriteth would then mean "land used for some charitable or religious purpose." If this is the origin of the name we must believe that there were two M.E. forms of the name, one *Car(r)iteth* which has chanced to survive only in late documents, the other *carite* which was the one more commonly used.

Carrycoats (Thockrington). *a.* 1245 Newm. *Carricot* ; 1542 Bord. Surv. *Carre Cottes* ; 1663 Rental *Carye Coats*.

Bates (*Border Holds*, p. 46 n.) suggests that this is from the Celtic *Caer-y-coed*, i.e. stronghold in the wood. This may be correct but, to judge from Carrick *supra*, we should have expected early forms in *Cair-*.

Cartington (Rothbury). 1233 Pipe *Kertindon* ; 1297 Ipm. *Kertinton* ; *c.* 1250 T.N. *Kertindun* ; 1314 Ipm. *Cartyngdon* ; 1346 F.A. *C(h)arty(n)gton*, 1428 *Cartyngton*.

" Kiartan's hill," *Kiartan* being a common Scandinavian
name, ultimately of Celtic origin. App. A, § 1.
Cassop (Byers Green). B.B. *Cazehope* (B. *Cassehopp,*
C. *Cassop*) ; 1382 Hatf. *Casshop* ; 1339 R.P.D. *Cassop.*
O.E. *Casan-hop*=Casa's *hop* (Part II). Cf. *to casan
þorne* (B.C.S. 1005) and Casewick, Lincs., D.B. *Casuic.*
Castle Eden. *v.* Eden, Castle.
Catch Burn (Morpeth). 1278 Ass. *Cacheborn, Gache-
born* ; 1296 S.R. *Chaceburn* ; 1317 Pat., 1363 Cl. *Cache-
burne* ; 1663 Rental *Catchburne.*
"Cæcca's burn." Cf. *cæccan wel* (B.C.S. 865).
Catcherside (Kirk Whelpington). 1270 Swinb., 1296
S.R., 1324 Ipm. *Calcherside* ; 1401 Ipm. *Calchersyde* ;
1595 F.F. *Cachersyde* ; 1650 Map *Catchaside.*
"Cold-cheer-hill" (M.E. *caldchere-side*). Phonology,
§§ 3, 53.
Catcleugh (Elsdon). 1279 Iter. *Cattechlow.* **Catlaw
Hall** (Hutton Henry). *n.d.* Finch. *Kattelawe.* **Catraw**
(Stannington). 1479 B.B.H. *Catrawe.* **Catton** (Allendale).
c. 1225 B.B.H. *Cattedene,* 1298 *id.* ; 1343 Pat. *Catton* ;
1547 Hexh. Surv. *Cadden* ; 1610 Speed *Caddon* ; 1637
Camd. *id.*
" The clough (*clōh*, Part II), hill, row (*rāw*, Part II), and
valley belonging to *Catta*," cf. *cattan-eg,* B.C.S. 1176 or,
" haunted by the wild animal of that name " (O.E. *catt* m.,
catte, f.). Cf. *on catedenes heafdan,* B.C.S. 216.
Catterick Moss (Stanhope). 1311 F.P.D. *Katerick-
saltere* ; 1382 Hatf. *Catryk.*
Clearly pre-English. Cf. Catterick, Yorks., Ptolemy
κατουρακτόνιον, Anton. Itin. *Cataractone,* and for the
element -*altere* cf. Holder's *Altrum,* now Antre, and Ottercops *infra.*
Causey Hall (Tanfield). 1277 Pat. *Kaltysete, Kaldesete* ;
1399.45 *Cawce,* 1450.34 *Caweset.*
" Cold farm " (*sǣte,* Part II). Phonology, § 39. Its
development has been influenced by association with
cawse=causeway, as in
Causey Park (Hebron). 1221 Brkb. *capella de Calceto* ;
c. 1250 T.N. *La Chauce* ; 1324 Ipm. *Le Cauce* ; 1346 F.A.

La Chauce; 1455 Ipm. *le Cawse*; 1491 Newm. *Calcekyrke*; 1517 Arch. 2. 24. 118 *Cawsee Park*; 1663 Rental *Cawsey Park*.

The causey or causeway referred to is "an ancient paved way along the eastern boundary (of the park) on the line of the present North Road" (H. 2. 2. 131). The *capella* or *kyrke* was a chapel which once stood within its precincts. For forms *v.* N.E.D. *s.v.*

Caw Burn (Haltwhistle). 1669 Pipe *Caweden*. "Cawa's valley." Cf. *Caua* in L.V.D.

Cawledge Park (Alnwick) [kaliʃ]. Type I: 1241 Perc. *Caweleg, a.* 1252 *Cauleche,* 1270 *Cauleth, c.* 1280 *Caulathe*, 1352 *Cauleg*; 1479 N. ii. 453 *Caulage*; 1764 N.C.D. *Calledge*; 1663 Rental *Callis Park*. Type II: *c.* 1190 Godr. *Claubec*; *c.* 1280 Perc. *Claubache*.

Type I shows the personal name *Caua* as first element. The second is *letch* (M.E. *leche, lache*), in common use in Nthb. to denote "a long narrow swamp in which water moves slowly among rushes and grass" (Heslop, *s.v.*). *th* is a common error of transcription for *ch*. For *-age*, *-edge*, cf. Debach, Suff., locally pronounced *Debbidge* (Skeat, p. 5), Burbage, Wilts. and Leic., which contain the same element. Phonology, § 58.

Type II seems to have as its second element M.E. *bache*, a stream (Part II). The first part cannot be explained.

Charlaw Moor (Langley). 1232 Ch. *Cherlawe*; 1382 Hatf. *Charlawe*.

O.E. *Ceorran-hlāw*=Ceorra's hill. Cf. Charsfield, Suff. (Skeat, p. 25).

Charlton (Bellingham). 1279 Iter. *Charletona*. (Ellingham) 1166 Pipe *Cherletona*.

O.E. *ceorla-tūn*=farm of the ceorls or freemen.

Chatterley (North Bedburn). 1428.33 *Chaterley*; 1464.35 *id.*

Cf. Chatterley, Staffs. (13th c. *Chadderley*) and Chadderton, Lancs., for which Sephton (pp .164-5) gives early forms, *Chaderton, Chatherton, Chat(t)erton*. These may possibly contain an O.E. name *Cæd-here*. Cf. *Cædbæd, Cædbeald*, and *Cædwalla* in Searle. The Durham name is not found

before the 15th cent., and *Chater* there may be the same
as *Chaytor*, a personal name from Fr. (*a*)*cheteur* (Weekley,
p. 120).

Chattlehope Burn (Elsdon) [tʃatləp]. *c.* 1320 B.M. *Chetil-
hopp*; 1317 Ipm. *Shetilhop*; 1610 Speed *Chetlop*; 1663
Rental *Chattlehope*; 1716 Elsdon *Chetlup*.

O.E. *cietel-hop*=kettle-shaped hope or, possibly, "belonging to *Cietel*" (later English *Chettle*). Cf. a similar
use of O.N. *ketill*. *cietel* is found in O.E. place-names
as a descriptive element, cf. *cytelwyll* (B.C.S. 610) and
cytelflod (ib. 682), referring to the bubbling up of the water
in the spring or stream. For *sh v.* Zachrisson, pp. 156-7.

Chatton [ʃatən]. 1177 Pipe *Chetton*; 1253 Ch. *Chatton*;
1255 Ass. *Chetona, Chatton*; *c.* 1250 T.N. *Chatton*; 1307
Ch. *Chatton*; 1323 Ipm. *Chattoun*; 1342 Bury *Chetton*;
1663 Rental *Chatton*.

"Farm of Cetta or Ceatta." Cf. *cettantreo* (B.C.S. 210)
and *ceattanbroc* (K.C.D. 636), and Chettisham and Chattis-
ham, Cambs. (Skeat, pp. 21, 50). Phonology, § 26.

Cheeseburn Grange (Stamfordham). 1286 Ch. *Cheseburgh*;
1292 Q.W., 1479 B.B.H. *id.*; *c.* 1536 B.B.H. *Chesborne*.

"The *burh* (Part II) famous for its cheeses." Cf. Cheswick *infra*, Chiswick, Midd., Cheswardine, Salop, and
(probably) Cheesden, Lancs. The vowel should be short.
Phonology, § 21, App. A, § 10.

Chesterhope (Redesdale). 1298 B.B.H. *Chestrehop*; 1628
Freeh. *Chestrop*; 1663 Rental *Chesterup*. **Chesterwood**
(Haydon). *c.* 1150 H. 2. 3. 383 *Chest'wada*, 1364 Ipm.
Chesterword.

"Hope and enclosure (*weorþ*, Part II) by the *chester* or
fort." App. A, § 3.

Chester-le-Street. 1104-8 S.D. *Cun(e)cacestre*; *c.* 1160
Ric. Hex. *Kunkacestra, Cestra*; 1400 D.S.T. *Cestria in Strata*.

This may be O.E. *Cuneca(n)-ceaster*, i.e. Cuneca's fort,
from the one-time owner of the site of the Romano-British
settlement, cf. Consett *infra*, and *cunecanford* (B.C.S.
610), but it is probable that the first element is Celtic.
Chester has been identified with the *Congavata* of the
Notitia Dignitatum, and there is a Cong Burn (*v. infra*)

flowing into the Wear near Chester. *Cuneca* is therefore a possible Anglian corruption of some misunderstood Celtic name. Chester has also been identified with Bede's *in Cuneningum* (v. 12), but it is difficult to connect this with either *Congavata* or *Cunecacestre*. Later the first element was dropped (cf. *Chester* itself) and then Chester was distinguished from other Chesters as *in Strata*, i.e. on the Roman Road from Darlington to Newcastle. The *le* is not the definite article but the O.F. preposition *lès*, near, as in *Plessis-lès-Tours* in France.

Chesters (Humshaugh). 1104-8 S.D. *Scytlescester juxta murum* ; *c.* 1160 Ric. Hex. *Cithlescester, Scydescester.*

If this identification is correct the modern name should be *Shittlechester*, and the site must once have been owned by one *Scytel* (cf. Shitlington *infra*).

Cheswick (Islandshire) [tʃizik]. 1228 F.P.D. *Chesewic* ; 1639 N.C.D. *Chesswick.*

O.E. *ciese-wīc* = cheese-dwelling. Cf. Cheeseburn *supra*. Phonology, §§ 21, 7, 19.

Cheveley (Warkworth). 1299 Ipm. *Chiveleye* ; 1341 Bury *Cheveleye* ; 1597 Bord. *Cheveley.* **Chevington** (ib.) [tʃivəntən]. *c.* 1050 H.S.C. *Cebbingtun* ; 1230 Pat. *Chivinton* ; *c.* 1250 T.N. *Chini'gton* (*sic*) ; 1268 Ipm. *Chyvington*, 1335 *Chevyngton* ; 1428 F.A. *id.* ; 1430 Pat. *Chyvyngton* ; 1724 Warkw. *Chiventon.*

"Clearing of *Cifa* (*Ceofa*)," "farm of the same or of his sons." Cf. Chieveley, Berks., B.C.S. 1055 *Cifanlea* (Stenton, p. 47) and Cheveley, Chesh., earlier *ceofanlea* (B.C.S. 1041). Cf. Chevington, Worc., earlier *Civincgtune* (Duignan, p. 30) and Chivington, Suff. (Skeat, p. 96). The form in H.S.C. may go back to O.E. *Ceobba*, which may be interpreted as a short form of *Ceolbeald, Ceolbeorht*, or as a derivative of *Ceofa*, with gemination of *f* (=*bb*), (Redin, p. 88), or it may be due to the influence of the neighbouring Choppington *infra*, or even be that place itself.

Cheviot [tʃiv(i)ət]. 1181 Pipe *Chiuiet* ; 1239 Ipm. *Chyviot* ; 1244 Ch. *Chyvietismores* ; 1597 Bord. *Chiveot.*

Clearly pre-English. Cf. Chevet, Yorks., early *Ceuet* (D.B.), *Chevet, Chyvet* (Goodall, p. 100).

Chibburn (Widdrington). 1228 Pipe *Chibrnemue*; 1292 Ass. *Chilburne*; 1404 Ipm. *Chibburne*; 1574 F.F. *Chilbourne*.

" Cilla's stream." Phonology, § 51. For *-mue v.* Zachrisson, p. 93.

Chillingham [ʃiliŋəm]. 1186 Pipe *Cheulingeham*; 1231 Cl. *Chevelingham*; 1291 Ch. *Chevingleham* (*sic*); 1346 F.A. *Chevelyngham*; 1348 H. 3. 2. 119 *Chillyngham*; 1470 Ipm. *Chelingham*; 1507 D.S.T. *Chillyngham*.

" Homestead of Ceofel or of his sons." This name is not found in O.E., but cf. Chilswell, Berks., earlier *Cheveleswell*. It is a dimin. of *Ceofa*. Phonology, §§ 51, 7.

Chilton (Merrington). 1091 F.P.D. *Ciltona*; 1195 Pipe *Chilton*.

O.E. *Cillan-* or *cilda-tūn*, i.e. farm of Cilla or of the young men. Skeat takes the latter to be the history of Chilton, Berks., D.B. *Cilletone* (p. 93).

Chipchase (Chollerton). 1229 Pat. *Chipches*; 1255 Ass. *Chipeches*; 1298 B.B.H. *Chipchesse*; 1298 Arch. 3. 2. 3 *Chipchace*; 1346 F.A. *Chipchesse*; 1542 Bord. Surv. *Chypchase*.

" Chip's chase," *Chip* being from O.E. *Cippa*. Cf. Chippenham, Wilts., A.S.C. *Cippanhamm* and Chipstable, Som., D.B. *Cipestaple*. For the suffix, cf. Scots. *chess* (N.E.D.) and local [tʃes] for " chase." N.E.D. gives no example before 1440.

Chirdon (Greystead) [dʒɔrdən]. 1255 Ch. *Chirden*; 1279 Iter. *id.*; 1325 Ipm. *Chirdene*; 1610 Speed *Chirden*; 1663 Rental *Chirdon*.

Chirton (Tynemouth). 1203 R.C. *Chertona*; 1255 Ass. *Chirton*; 1271 Ch. *Chertun, Cherton*; *c.* 1250 T.N., 1346, 1428 F.A. *Chirton*.

O.E. *Ceorra(n)-denu* and *-tūn*=Ceorra's valley and farm. Cf. Churton, Chesh., earlier *Chirton*. App. A, § 1.

*****Chirland.**[1] 1178 Pipe *Childerlund*, 1167 *Chirlund*; *c.* 1250 T.N. *Chirland*; 1273 R.H. *id.*

If the Pipe Roll forms were correct the second element

---

[1] Identified by Dixon (*Upper Coquetdale*, p. 302) with Chirnells Moor, S.E. of Cartington.

would be O.W.Sc. *-lundr*, " a grove," with later substitution of *land* as in Toseland, Hunts., and Timberland, Snelland, Lincs., but such a suffix is very unlikely in Nthb., and the Pipe Roll forms are probably mistakes for *-land* (Part II). The first element is O.E. *cildra*, gen. pl. of *cild*=child. Cf. Childerley, Cambs. (Skeat, p. 66) and Chilton *supra*. In such names the word is probably used in its technical sense as applied to a young noble awaiting knighthood. Cf. Childs Wickham, Glouc.

Chollerton. 1154-95 Swinb. *Choluerton*; 1229 Pat. *Colerton*; 1232 Ch. *Chelreton*; 1241-6 Newm. *Chollerton*; 1257 Swinb. *Choluerton*; 1265 Sc. *Cholverton*; 1273 R.H. *Cholvirton*; 1278 Abbr. *Colverton*; 1298 B.B.H. *Cholverton*; c. 1250 T.N. *Chelverton*; 1316 R.P.D. *Cholverton*; 1346 F.A. *Chollerton*.

O.E. *Ceolferðes-tūn*=Ceolferth's farm. For M.E. *Chelver-* and *Cholver-* cf. Learchild *infra*. The early forms forbid our connecting this place with the *Cilurnum* of the *Notitia Dignitatum* (M'Clure, p. 115). No early forms for the neighbouring Chollerford have been found. Phonology, § 50.

Choppington (Bedlington). Type I: 1181 Pipe *Chabiton*; 1325 Fine *Chabinton*; B.B. *Chabyngton*; 1381.32 *id.* Type II: 1310 Pat. *Chapynton*, 1359 *Chapyngton*; 1363 Ipm. *Chapington*; 1563 Raine *id.* Type III: 1358 Cl. *Chepynton*; 1621 Arch. 2. 1. 24 *Cheapington*; 1682 Arch. 2. 24. 122 *Cheppington*.

Type I is O.E. *Ceabbing(a)tūn*=farm of Ceabba or his sons. Type II, if it is not due to an otherwise unparalleled development of medial *b* to *p*, suggests the name *Ceapa* instead of *Ceabba*. Type III may be due to the analogy of the common *cheaping*=market, found in Chipping Norton, Glouc., and Chipping Ongar, Essex. For *chop-* v. Chopwell.

Chopwell (Ryton). 1153-9 Newm. *Cheppwell*; 1278 Ass. *Cheppewell*; 1313 R.P.D. *Chapwell*; 1316 Pat. *Chepwelle*; 1416 J. and W. *Chapwell*.

" Ceappa's well." Phonology, § 1. For *a* and *o* cf. Choppington *supra*. There is a good deal of evidence for such rounding of *a* to *o* before a following labial, as shown

by the following forms :—Sopley, Hunts. D.B. *Sopelie*,
Ch. *Sappeleia*, Copley, Chesh. D.B. *Capelis*, Scopwick,
Lincs. D.B. *Scapwic*, later *Scaupewic, Scopwick*, Chobham,
Surr., earlier *Chabbeham*. Some places, on the other hand,
show *a* for earlier *o*, e.g. Clapham, Beds. D.B. *Clopeham*,
Shabbington, Beds. D.B. *Sobintone*, Grappenhall, Chesh.
D.B. *Gropenhale*, Clapton, Northts. D.B. *Clotone*, Surv.
Cloptone. These may be due to the influence of the names
already dealt with,

Clarewood (Corbridge). 1247 Ch. *Clavrewurth*, 1296
Clavreworth; 1428 F.A. *Claverworthe*; 1453 Pat. *Cla(ve)r-
worth*; 1538 Must. *Clarewod*.

O.E. *clæfre-weorþ*=clover-enclosure. Cf. Claverley, Salop,
pronounced *Clarely*, Clarborough, Notts., earlier *Claure-
burg*, Claverdon, Staffs., earlier *Cla(ve)rdon* (Duignan, p. 43),
and O.E. *næfre* > ne'er [nɛ·ə]. App. A, § 3.

Claxton (Greatham). 1091 F.P.D. *Clachestona*; 1312
R.P.D. *Claxton*.

" Klakk's farm," *Klakkr* being a common Danish name,
found also in Claxton, Norf., and Clawson, Leic. (D.B.
Clachestane).

Cleadon (Whitburn). 1280 Ch. *Clyvedon*; 1307 R.P.D.
Clivedon; B.B. *Clevedona* (B. *Clyvedon*).

O.E. *cleofa*, M.E. *cleve*=steeply sloping hill+*dūn*. Cf.
Clevedon, Som. D.B. *Clivedone*; Cleveland, Yorks.; Cleeve,
Glouc. (Baddeley, p. 44).

Cleatlam (Gainford). *c.* 1050 H.S.C. *Cletlinga*; 1271
F.P.D. *Cletlum*; 1313 R.P.D. *Cletlame*; 1446 D.S.T.
Cletlam; 1607 S. 4. 33 *Cleatlam*; *c.* 1740 Map
Cletlam; 1646 Staindrop *Cleat(e)nam*.

This difficult name is probably a compound of *Cletley*
and the common suffix -*ham* (cf. Riddlehamhope *infra*).
Cletley would be a hybrid formation from O.W.Sc. *klettr*,
" rock, cliff," Dan. *klint* found in *Clints infra*. Hence
" clearing on or by the cliff." Surtees (iv. 33) says that the
village stands on a high exposed brow, and we may note
that Cleatham, Lincs., D.B. *Cletham*, similarly stands on
ground rising steeply from the Lincs. flats. *Cletlinga* in
H.S.C. is probably for O.E. *Cletlingas*=dwellers at Cletley.

Cf. B.C.S. 506 *bromleaginga*=dwellers at Bromley, Kent.
Phonology, § 36.

Clennell (Alwinton). 1181 Newm. *Clenil*; 1255 Ass.
Chenhull (sic); 1346 F.A. *Clenhill*, 1428 *Clenell*.
"Clean hill," i.e. free from weeds or barren. Cf. Clan-
field, Oxon. (Alexander, p. 80) and Clandon, Surr., B.C.S.
697 *Clendone*. Phonology, §§ 21, 36.

Clifton (Stannington). *c.* 1250 T.N. *Clifton*.
"Hill-farm," a very common name.

Clints Wood (Stanhope). 1382 Hatf. *les Clyntes*.
"The rocks." A Scand. loan-word (N.E.D. *s.v.*).

Close House (Heddon-on-the-Wall). 1414 Inq. a.q.d.
le Cloos.
O.Fr. *clos* (< Lat. *clausum*)=enclosure.

Coanwood (Haltwhistle). 1279 Iter. *Collanwode*; 1373
H. 3. 2. 33 *id.*; 1575 F.F. *Counwood*; 1610 Speed *Conewood*.
"Collan's wood."[1] *Collan* may be the *Collanus* who
was once provost of Hexhamshire (*Hexh. Priory*, I, p. viii.).

Coastley (Hexham). *c.* 1250 Gray *Cotisley*; 1279 Iter.
Cocheley; 1280 Wickw. *Cocelay*; 1295 S.R. *Coceley*; 1324
N. iv. 10 *Cosselay*; 1385 N. iv. 11 *Coscele*; 1479 B.B.H.
Cocelye; 1538 Must. *Cosle*; 1547 Hexh. Surv. *Costeley*;
1682 Arch. 2. 1. 107 *Coastley*.
"Cocc's clearing." Cf. *coccanburh*=Cockbury, Glouc.
(B.C.S. 246). The forms show traces of both strong and
weak gen. forms, *cocces* and *cocca(n)*. The latter gives
Cocheley, where *ch=k* (Zachrisson, p. 36). *Cotisley* is pro-
bably an error of transcription for *Cocisley*. Later *Cocces*-
may have undergone the same metathesis which we find in
Fewston, Yorks., earlier *Foscetun* < *Foxatun* (Moorman,
p. 72). This would give *Coscele* and *Cosselay* or *Cocelay*.
Later *t* developed between *s* and *l* (cf. Eslington *infra*)
giving *Costle*.

Coatham Mundeville (Haughton-le-Skerne). *c.* 1200
D.S.T. *Cotum Super Scyren*; 1313 R.P.D. *Cotum*, 1344
Cotum Maundevill; 1446 D.S.T. *Cotom*. **Coatham Stob**
(Long Newton). 1379 S. 3. 218 *Cotom*.
O.E. (*æt þǣm*) *cotum*=(at the) cotes. Cf. Coton, Cambs.

[1] Note also *Collanland* in Stanhope (Hatf. Surv.).

Mundeville because once held by the family of Amundevylle
(D.S.T. lx.) who derived their name from *Emondeville* or
Amundavilla in Normandy, i.e. the " vill " of *Ámundr*, its
Norman settler (Jakobsen in *Danske Studier*, 1911, p. 68.)[1]
Sometimes it was distinguished as " on the Skerne." *Stob*
was probably so called from some prominent stubbed
tree. It was also known as Coatham Conyers, from its
onetime owner.

Coatsay Moor (Heighington). 1446 D.S.T. *Cotes*; 1539
F.P.D. *Cottes super moram.* App. A, § 6.
" Cotes on moor " > " Cotes a' moor " > Coatsay Moor.

Cocken (Chester-le-Street). 1138-40 Finch. *Coken*; *c.*
1150 F.P.D. *Cochena*, 1185 *Koken*, 1203 *Cochen*.
Pre-English.

Cockerton (Darlington). *c.* 1050 H.S.C. *Cocertun*; 1304
Cl. *Cokerton*.

It stands on the Cocker Beck, but this river-name may be
a back-formation. There is, however, a river Cocker, Cumb.
(Sedgefield, p. 36), earlier *Cocur*, and in Lancs. Sephton (pp.
79, 133) takes this to be of Celtic origin and identical with the
Kocker, a tributary of the Neckar. There is also a Cocker
Beck, Notts. (Ch. *Cokerbec*). From these river-names are
derived Cockermouth, Cumb., and Cockersand and Cocker-
ham, Lancs. There is also a Cockerington, Lincs., Surv.
Cockringtuna and Coker, Som., D.B. *Cocre.* In neither
of these places has any trace of a river *Co(c)ker* been
found.

Cockfield (nr. Barnard Castle). 1314 R.P.D. *Cokefeld*;
1507 D.S.T. *Cokfeld*. **Cocklaw** (St John Lee). 1479
B.B.H. *Coklaw*; 1652 Comps. *Cockley*. **Cockle Park**
(Hebron). 1314 Ipm. *Cockhill*; 1517 Arch. 2. 24. 118
Cokyll Park; 1628 Arch. 1. 3. 94 *Cockle Park*.

All named from the bird or from a man so-named.
Phonology, § 36.

Cocklaw (Adderstone). 1296 S.R. *Creklawe, Crokelawe.*

---

[1] The Durham Assize Roll gives this place as villata de *Aedmundesville*.
This form seems to be an unauthorised anglicising of the French name.
It receives no support from the forms found in *Calendar of Documents
relating to France.*

D

If the identification is correct the modern form is corrupt and the old name is identical with Kirkley *infra*.

Coe Burn (Edlingham). 1295 N. vii. 104 *Coveburn*. The first element is probably North Eng. *cove*=hollow or recess in a rock, cave, cavern or den. In the same document we have mention of *Meldircoveslade, Meldercove, Ebscove*, as though there was more than one *cove* in the neighbourhood. Cf. Cove, Hants., D.B. *Coue* and Suff. D.B. *Coua*. Phonology, § 46.

Coldcoats (Ponteland). *c.* 1250 T.N. *Caldecotes*. **Coldcotes** (Simonburn). 1279 Iter. *Kaldecotes*. **Coldlaw Burn** (Cheviot). 1255 Sc. *Caldelauburne*. **Coldstrother** (Kirkheaton). 1232 Ipm. *Caldestrother*. **Coldtown** (Corsenside). 1331 Ipm. *Caldton*; 1618 Redesd. *Caldtowne*. **Coldwell** (Bavington). 1324 Ipm. *Caldewell*; 1663 Rental *Coldwell*. (Stannington) *c.* 1226 Perc. *Caldewele*. **Coldmartin** (Chatton). 1195 Pipe *Calemerton*; 1255 Ass. *Caldemerton*; 1288 Ipm. *Caldemarton*; 1346 F.A. *Cal(d)merton*; 1574-96 Bord. *Caldmartyn*; 1663 Rental *Cold Martin*; 1715 Chatton *Caldmartine*.

The first element calls for no note except that, on the map at least, St. Eng. *cold* has replaced Northern [kad], [kauld], [ka·d]. For the suffixes *v.* Part II. Coldstrother now forms part of Kirkheaton, and the name went out of use in the 16th cent. *Martin* is a common form for *Marton*. Cf. Martin (twice), Lincs. and Notts., Martin Hussingtree, Worc. (Duignan, p. 109). Mutschmann (p. 90) points out that an unstressed vowel after a dental, especially before another dental, is often pronounced *ĭ* in English dialects. *Marton* < *merton*=O.E. *mǣre-tun*, " boundary-farm," or *mere-tūn* =farm by the mere or pool.

Colepike Hall (Lanchester). 1350.31 *Colpighill*; 1382 Hatf. *id.*, 1456.35 *Colpikhill*; 1654 Lanch. *Colpihill*, 1670 *Coepichell, Cowpeighell*.

The modern form is clearly corrupt. *pighill, pickhill* and *pickle* are North. dialect forms of the old word *pightle*, " a corner of land, small field or enclosure " (N.E.D. *s.v.* and Goodall, pp. 227-8). Cf. *le Pighill* in Benfieldside, *Pighill* in Stockton and *Pyghel bank* in Newton Cap (Hatf. Survey). *Col-*, probably from some surface coal-working here. App. A, § 7.

CONSETT   51

Collierley (Lanchester). 1297 Pap. *Colyesley*; 1378
Pat. *Colyerlye*.
"Collier's field." Cf. *Colyerland* (Bishopley and Ryton)
in Hatf. Surv.
Colpitts (Slaley). 1255 Ass. *Colpittes*; 1296 S.R. *Col-
pottes*; 1663 Rental *Colepits*.
"Coal-pits," from some old workings. Cf. *Colpittes* in
Finch, Cart. *c.* 1270. *-pottes* is not impossible, for *-pot* is
in common dialectal use for a deep hole, the shaft of a
mine.
Colwell (Chollerton) [kɔləl]. 1255 Ass. *Colewell*; 1318
Ipm. *Colwell*; 1323 Inq. a.q.d. *Collwell*; 1326 Ipm. *Colle-
well*; 1479 B.B.H. *Col(le)well(e)*; 1663 Rental *Collell*.
O.E. *cōle wielle*=cool spring. Cf. Colwall, Heref., D.B.
Colewelle. Phonology, §§ 21, 49.
Combfield House (Muggleswick). 1446 D.S.T. *Camhouse.*
"House on the ridge," *v. camb*, Part II.
Cong Burn (Chester-le-Street). 1382 Hatf. *Clonglech*;
1423 S. 2. 368 *Conkburn.*
Clonglech is probably miswritten for *Conglech*, the *l* of
the second element being anticipated in the first. The
burn unites with Twizell Byrn just by Chester-le-Street,
and it is possible that the first element in the original form
of the latter name (*v. supra*) is this river-name. For *-lech*
v. Part II.
Coniscliffe [kʌnsklif, kʌnzli]. Type I: A.S.C. *Cininges-
clif*; *c.* 1050 H.S.C. *Cincgesclife*. Type II: 1203 R.C.
Cunesclive; 1271 Ch. *id.*; 1298 Pat. *Conescliue*; 1314
R.P.D. *Conysclyf*; 1336 Ipm. *Consclyf*; 1507 D.S.T.
Cunyngsclif; 1637 Camd. *Cunsley*; 1665 Coniscl. *id.*
"King's cliff." Type I is O.E. Type II has been
modified under the influence of O.W.Sc. *konungr*, "king,"
found in Conisborough, Yorks., Coniston, Conishead, Lancs.,
Conisholme, Lincs., D.B. *Coningisholm*. For [kʌn] cf.
conduit and *v.* Horn, § 64. Phonology, § 56; App. A, § 7.
Consett (Lanchester). 1297 Pap. *Conkesheued*; 1312
R.P.D. *Couckeheved (sic)*; B.B. *Conekesheued*; 1443.34
Counsett; 1479.35 *Conset, Consyd, Consed*; 1577 Barnes
Consyde; 1580 Wills *Consett*; 1687 Ebch. *Conside.*

"Cunec's headland" (*hēafod*, Part II). Cf. *cunecanford*,
B.C.S. 610 and the name *Chunico* given by Förstemann
v. Chester-le-Street *supra*. Phonology, §§ 36, 53; App. A.
§§ 7, 12.

Copeland House (Auckland). 1104-8 S.D. *Copland*;
1313 R.P.D. *Coupland*, 1340 *Coupeland*. **Coupland** (Kirknewton). *c.* 1250 T.N. *Coupland*; 1255 Ass. *Couplaund*;
1663 Rental *Copeland*.
This name is explained by Lindkvist (pp. 145-6). It
is from O.W.Sc. *kaupa-land*, purchase-land=*kaupa-jǫrð*,
opposed in a way to *óðals-jǫrð*, an allodial estate. *kaup*, a
bargain=O.E. *cēap*. Cf. Copeland, Cumb., and the Copeland
Islands off Belfast Lough, which Bugge wrongly explains as
from O.N. *kaupmanna eyjar*=merchants' islands (*Norges
Historie*, Vol. I, p. 297).

Copley (Auckland). 1315 R.P.D. *Koppeleyker.*
"Coppa's clearing," cf. *Coppanford* (K.C.D. 699), or
possibly "clearing on the hill-top," from M.E. *coppe*=hill.
Baddeley (p. 49) takes this to be the history of Copley,
Glouc., earlier *Coppeleye.*

Coppy Crook (Auckland). 1409.35 *Copecrokes*; 1420.35
Copicroche.
M.E. *coppid-croke*(s)=the crook(s) (*krók*, Part II) with
the copped or pollarded trees. Cf. Copthall, Ess., earlier
Copyd Hall and *Copid Hall*, Berks. (B.M. ii. 410). For
coppy- cf. Copythorne, Hants. V.C.H. gives no early forms
for this, but elsewhere we have *copped thorn* (B.C.S. 740)
and *Coppid-thorne* (Hants. V.C.H. v. 218).

Coquet, R. [koukit]. *c.* 1050 H.S.C. *Cocwuda*; 1104-8
S.D. *Coqued*; 1200 R.C. *Coket.*
A pre-English name.

Corbridge-on-Tyne. Type I: *c.* 1050 H.S.C. *Corebricg*;
c. 1154 S.D. *et Corabrige*; 1157 Pipe *Corebrigge*; *c.* 1160
Ric. Hex. *Corabrigham*; 1203 R.C. *Corbrigg*, 1204 *Corig-
brige*, 1205 *Corebrig*, 1212 *Corbrug*; 1217 Pat. *Corebrigg*;
1507 D.S.T. *Corbrige.* Type II: *c.* 1110 Hexh. Pr. Suppl.
ix. *Colebruge*; 1135-7 N. x. 45 *Coleb'*; 1158 Pipe *Colebi'*,
1169 *Cholebrige*; 1198 N. viii. 67 *Colebrug*; 1203 R.C.
Collubrug; 1273 R.H. *Colbrige.*

The first problem in this difficult name is the relationship of Types I and II,[1] and in any attempt to solve it we must bear in mind that there is a similar problem in the relation of the name of the Roman settlement on its western side— *Corstopitum* of the Antonine Itinerary—to the medieval and early modern names for the site of that settlement, viz., *Colchester*, for which we have (N. x. 47 n. 5) early forms *Colchester* (1356, 1549), *Colchestre* (1394), *Colecester* (Leland), *Colecestre* (Camden). Zachrisson solves this problem (pp. 120-2) when he shows how with Anglo-Norman scribes *r——r >l——r* by a dissimilatory process (cf. *Schorpshire* and *Salopescira*) and *l——r >r——r* by a less common assimilatory process. Thus *Corebrigge* might become *Colebrigge* and *vice versa*. The first alternative is the more probable because (1) it is the more common process, (2) otherwise we must believe the identity of initial syllable between *Corstopitum* and *Corebrigge* to be a mere coincidence.

Heslop (Arch. ii. 8. 95) attempted to solve the problem of *Colchester* by explaining the first element as Lat. *Colonia*, but this is declared impossible on historical grounds (N. x. 49). Leland tried to solve it by the suggestive *Colus flu* in the margin of his MS., suggesting apparently that this was the early name of the Cor Burn. There is no authority for such a form, and its existence would lead us to the difficulty that we should have to believe that *Colchester* had no connection at all with the ancient name *Corstopitum*. Rather it may be suggested that forms in *Cor-* did exist side by side with those in *Col-* in the Middle Ages, but have not chanced to survive. If this is so the double forms could be explained in the same way as *Corbrige* and *Colbrige*.

What, then, is the meaning of this element *Cor-* in *Corstopitum, Corbridge* and *Colchester*? Maclure (p. 155 n. 1) takes the name *Corstopitum* to be identical with Corsept (on the Loire), which is supposed to go back to the

---

[1] Leland (*Itinerary* v. 112) was aware of the problem. He writes : " As far as I can perceive by the Boke of the Life of S. Oswin the Martyr, *Colebrige* is always put there for Corbridge. (Cf. *Vita S. Oswini*, Surtees Soc., vol. 8, p. 83.)

tribal name *Coriosopites* (cf. also Corseul, Côtes-du-Nord, from the *Coriosolites*), but W. H. Stevenson has shown (N. x. 9) that this is phonologically impossible, *Corio-* would yield O.E. *Cyre-*. As to the suffix, Professor Chadwick suggests to me that the form in the *Antonine Itinerary* may be corrupt, and that the true suffix should be *ritum* "a ford." Cf. O.Cy. *rit* gl. *vadum*, Welsh *rhyd*, and such names as *Augusto-ritum* (Holder s.v. *rĭtŭ-*). There is a well-known ford across the Tyne just under *Corstopitum*, and the place may have been called from it. The first English name may have been *Corst-ford*, and in any case *Cor(st)-bridge* was so called from the *bridge* which took its place.[1]

Cornforth (Bp. Middleham), B.B. *Corneford*. **Cornhill-on-Tweed** [kɔ·nəl]. *c.* 1180 D.S.T. *Cornehale*; *c.* 1250 T.N. *id.*; 1228 F.P.D. *Cornhale*; B.B. *Cornehall*; 1335 Ch. *Cornehale*; 1539 F.P.D. *Cornell*; 1558 V.N. *id.* **Cornsay** (Lanchester). 1154-95 B.B. *Cornesho*; B.B. *Cornshowe*; 1312 Pat. *Cornesough, Cornesowe*; 1547 Lanch. *Cornsew.*

There are many English place-names in *Corn-* and their etymology is by no means clear. The common word *corn* is not found as an element in O.E. place-names, but in a Worc. charter we have *corna* in *corna-broc, -wudu* and *-liþ*, for which no satisfactory explanation has been offered. Later we have Corley, Warw., D.B. *Cornelie*, Cornworthy, Dev. D.B. *Corneorda*; Curworthy, Dev. (D.B. *id.*), Cornhill in Perivale, Midd., Ch. *Cornhull*, Cornard, Suff. T.N. *Cornerth*, Cornley, Notts. B.M. *Cornelay*, which from the nature of the compounds might contain the common word *corn*, though the medial *e* at times makes this doubtful. We cannot, on the other hand, have this word in Cornwood, Dev. D.B. *Cornehuda*, Cornwell, Ox. B.C.S. 222 *cornwelle*, and it is very doubtful if we have it in Cornbury, Oxf. D.B. *Corneberie*; Cornbrough, Yorks. B.M. *Cornebrug*, Corneybury, Herts. D.B. *Cornei*. Corndean, Glouc., may contain *corn*, but Baddeley (p. 49) thinks the first element is *corne*, found elsewhere in Gloucestershire as the name of a water-way.

---

[1] Cor Burn is pretty certainly a back-formation from Corbridge. The bridge is not over the Cor at all. No old forms are known.

In these latter names we pretty certainly have to do with some Celtic element of unknown meaning (cf. *Corn*wall), or possibly with an otherwise unknown personal name *Corna*. That some such personal name did exist is clear from Cornsay, which shows the gen. of its strong form *Corn*. Cornforth is from the personal name or contains the unsolved *corn*. Cornhill is "corn-haugh" (*healh*, Part II). Cf. Tomlinson (p. 544), who describes it as in the midst of rich cornlands, and note Barhaugh *supra*. Cornsay is "Corn's *hōh* (Part II) of land." Phonology, §§ 30, 36; App. A, §§ 6, 9.

Corsenside (Redesdale). *c.* 1250 T.N. *Cressenset*; 1291 Tax. *Crossenset*; 1306 R.P.D. *Crossansete*; 1507 D.S.T. *Crossynsyde*; 1586 Raine *Corsenside*; 1722 Ponteland *Crosenside*.

The suffix is *sǣte* (Part II). The first element is perhaps the Gaelic name *Crossan* found in *Kerke(by) Crossan*, Cumb. (Ekwall, p. 28). App. A. § 8.

Cottingwood (Morpeth). 1257 Ch. *Cotingwud*. "Wood of Cot(t)a or of his sons."

Cottonshope (Elsdon). *c.* 1230 H. 2. 1. 16 *Cotteneshopp*; 1278 Ass. *Cotnesop*; 1324 Ipm. *Cotynghopp*; 1331 Ipm. *Cotynshope*; 1618 Redesd. *Cottenshope*.

"Cot(t)ens hope." Cf. *Cotten*, L.V.D. and *Cotenesfeld*, B.C.S. 472.

Coundon (Auckland). 1197 Pipe *Cundun*; 1313 R.P.D. *Cundon*; B.B. *Conduna* (B., C. *Coundon*); 1365 Halm. *Coundon*.

Duignan takes the first element here and in Coundon, Warw., D.B. *Condone, Condelme* (p. 47) to be the Gaulish *cond*, found in Fr. Condé and Condat, "confluence of streams," and suggests that in each case it was brought over by Roman legionaries from Gaul. This is highly doubtful and, in any case, does not suit the position of the Durham Coundon. The name might possibly be O.E. *cūna-dūn* = hill of the cows, cf. *cunden*, B.C.S. 343 = Cowden, Kent, or it may contain the name *Cund(a)* found in *Cundes-leage, -broc* and *-fen* (B.C.S. 890), and *cunding æceras* ib. 1282, and in *Cunda*, the name of a bishop in

B.C.S. 416. Forssner (p. 57) connects this with *Cundwalh,*
Cundigern in L.V.D., and takes the first part of these names
to be Celtic.

Coupland, *v.* Copeland *supra.*

Cowden (Chollerton). *c.* 1250 T.N. *Colden.*
O.E. *cōle denu* = cool valley. Phonology, § 39.

Cowgate (Newcastle-on-Tyne). 1290 De Banco *Cougate.*
"Cow-going or walk," (*v.* Part II), used technically of a
pasture over which a cow may range, right of pasturage for a
cow in common land " (Heslop *s.v.*). Cf. Cowgate, Edinburgh.

Cowpen (Horton). 1153-95 Brkb. *Cupum*; *c.* 1190
Newm. *id.,* 1250 *Copoun*; 1271 Ch. *Copun*; 1295 Ty.
Cupun; 1346 F.A. *Copon*; 1428 F.A. *Coupowne*; 1560
V.N. *Coopon.* **Cowpen Bewley** (Billingham). *c.* 1150 R.C.
Cupum; 1335 Ch. *Cupum in Werehale*; 1446 D.S.T.
Coupon; 1539 F.D.P. *Cowpon.* **Cowpen Marsh** (ib.).
c. 1330 Acct. *Coponmersk.*

The name is clearly a dative plural, and the suggestion
has been made (*Essays and Studies,* u.s., vol. iv. p. 61)
that it is from O.W.Sc. *kúpa* = cup or bowl, used also of
a cup-like depression or valley, referring perhaps to old
saltpans found in both places, or that it is associated with
Sw. dialectal *kupa* = a small cottage or household. Hence
"at the hollows" or "at the cottages." *Bewley,* because
part of the manor of that name (*v. supra*). For *Werhale*
v. Introd. § 1. St. Eng. *marsh* has replaced dialectal *marsk*
due to Scand. influence. Cf. Marske by the Sea, Yorks.

Coxhoe (Kelloe). 1277 Finch. *Cockishow*; 1304 Cl.
Cokeshou; 1344 R.P.D. *Coxhowe*; 1639 Redm. *Coksey.*
"Cocc's *hōh* or heel of land." Cf. Cockfield *supra.*

Cragshiel (Simonburn). 1291 Ipm. *le cragscriel* (*sic.*);
1663 Rental *Craggsheel.*
"Shiel by the crag."

Cramlington. *c.* 1130 F.P.D. *Cramlingtuna, c.* 1150
Cramlingatuna, Cramilintona, 1203 *Crameligton*; 1270
Ch. *Cramlington*; 1292 Ass. *Cramelton.*
"Farm of the sons of Cramel." *Cramel* is not elsewhere
known. It would seem to be a dimin. of a name **Cram.*
Cf. Dan. dialectal *kram* = narrow, tight, harsh, severe,

(Falk and Torp. *s.v. kram*), perhaps used as a nickname. Phonology, § 59.

Cranerow (Hamsterley). 1382 Hatf. *Cranrawe.*
" Crane's row," *Crane* being used as a surname. Cf. Tranwell *infra.*

Craster (Embleton) [kreistə]. 1244 Ipm. *Craucestre*; 1346 F.A. *Crau(u)cestre*, 1428 *Crauecestre*; 1460 H. 3. 1. 30 *Craister*; 1538 Must. *Crawstor*; 1550 H. 3. 2. 207 *Craster*; 1663 Rental *Craister.*

Crawcrook (Ryton). 1242 D. Ass. *Krakruke.* **Crawley** (Eglingham) [krala]. 1225 Pipe *Crawelawe*; 1460 H. 3. 1. 28 *Krawlawe*; 1498 H. 3. 2. 127 *Crawley*; 1628 Freeh. *Crawlaw*; 1663 Rental *Crawley*; 1670 Egling. *Cralla*, 1685 *Cralaye*, 1697 *Craly.*
" The chester," *ceaster* Part II, crook and clearing of a man named Crow, or frequented by the bird." For later developments of Craster and Crawley cf. Nthb. dialectal [kra], [kra·] for *crow.*

Cresswell (Woodhorn). Type I: 1234 Cl. *Kereswell*; 1255 Ass. *Kercewell.* Type II: 1255 Ass. *Cressewell, Grescewell*; 1450 Ipm. *Cresswell.* Type III: 1450 Ipm. *Carswell*; 1637 Camd. *id.* Type IV: 1265 Ipm. *Crassewell*; 1346 F.A. *Crasswell.*
" Cress-spring." Hodgson (2. 2. 199) says that the place " has its name from a spring of fresh water at the east end of the village, the strand of which is grown up with water-cress." Type I shows O.E. *cerse*, M.E. *cerse, kerse*; Type II O.E. *cresse*, the unmetathesised form from Teut. **krasjo*; Type III is from Type I (Phonology, § 54); Type IV from Type III with fresh metathesis. Cf. Nthb. [kras], [kars], [ka·rs] for *cress* (E.D.G. p. 391).
This very common place-name is found as Cresswell in Derbys., Beds., Som. (5), Staffs., Kerswell (thrice), Dev., Carswell, Glouc., Beds., Coarswell, Dev., Caswell, Northt.

Crimden Beck House (Monk Hesleden). 1270 Ch. *Crumeden.*
" The *crum* or crooked valley." Cf. *crumdæl*, B.C.S. 356. Phonology, § 13.

Cronkley (Shotley). 1268 Ipm. *Crombeclyve*; 1296

S.R. *Crumclef*; 1298 Ipm. *Crumcliffe, Crommeclive*; 1306 N. vi. 208 *Crounclef*; 1663 Rental *Cronkley*. O.E. *crumbe-clif* = crooked cliff. Cf. Crunkley Gill, Yorks. D.B. *Crumbeclive* and Cronkley Scar, nr. High Force. O.E. *crumb* > Mod. Dial. *crum* and *crom*, the latter representing a spelling pronunciation as in *Cromwell* and *Crompton*, usually pronounced with [ɔ]. Phonology, §§ 52, 56; App. A, § 7.

Crook (Brancepeth). *c.* 1270 F.P.D. *Cruketona*; B.B. *Cruktona* (B., C. *Croketon*); 1304 Cl. *Crok.* **Crook Burn** (Haltwhistle). 1479 B.B.H. *Crokeburne.* **Crookdean** (Kirkwhelpington). 1324 Ipm. *Crokeden*, 1424 *Croketon*; 1663 Rental *Krookden.* **Crookham** (Ford). 1244 Ch. *Crucum*; 1304 Ch. *Crukum*, 1340 *Crocum*; 1428 F.A. *Crokome*; 1542 Bord. Surv. *Croukham.* **Crookhouse** (Howtel). 1323 Ipm. *le Croukes.* **Crooks** (Thirlwall). 1479 B.B.H. *le Crowkes.*

All these names alike contain the common word *crook* (*krók*, Part II). In Crook Burn, Crookdean, Crookham and Crookhouse it probably refers to the windings of a stream or valley. Crookham looks like a dat. pl. (cf. Acomb *supra*) but it is difficult to believe that a Scand. loan-word would be thus inflected, and we must take -*um* to be an early weakening of the suffix -*ham*. Crookhouse is apparently a corruption of pl. *crook-es*, i.e. the windings of Bowmont Water (cf. Harbourhouse *infra*). In Crook and Crooks it probably means an odd nook or corner of ground of crooked shape. It is common as a field-name, e.g. *crukes* in Preston (N. ii. 319), *les Croukes* in Bamburgh (N. i. 131). Note also Crookes, Yorks. Crook originally had the common suffix -*ton*. Cf. *Króktún* in Iceland (Jónsson, p. 469).

Crooked Oak (Shotley). 1318 Inq. a. q. d. *Crokedhake*, 1378 *Crokedake*; 1663 Rental *Crookoak.*

Cf. Crookdake, Cumb., earlier *le Crokedaik.* Phonology, § 14, 38.

Crowsfield (Bedburn). 1491.36 *Crawfeld.*

Probably owned by the same *Robert Crawe*, or one of his family, who gave his name to *Crawescroft* in Bedburn (Hatf. Surv.).

Croxdale (Durham). *c.* 1190 Godr. *Crokestail*; 1214

D.S.T. *Croxtayl*; 1335 Ch. *Crokesteil*; *c.* 1570 Eccl. *Croxdaill*; 1580 Halm. *Croxdall.*

" Crook's (O.W.Sc. *Krókr*) *tail* of land." *tail* is used in Mid. Scots. to mean " a piece or slip of irregularly bounded land, jutting out from a larger piece." App. A, § 8. For the first element cf. Croxteth, Lancs, Croxton Norf., Leic., Lincs., and Croxby, Lincs.

Cullercoats (Tynemouth). *c.* 1600 N. 8. 281 *Culvercoats*; 1693 N. viii. 283 *Cullercoats.*

" Dove-cotes." *culver* (O.E. *culfre*) is an old name for the wood-pigeon, and *culver-house* is still used in some parts of England for pigeon-house. Phonology, § 50.

Cushat Law (Kidland). *n.d.* Newm. *Cousthotelaw (sic)*; 1536 Arch. 3. 8. 20 *Cowshotlaw.*

O.E. *cūscote-hlāw* = cushat or wood-pigeon hill. Cf. *cuscetes haga*, K.C.D. 987. *t* for *c* in the first form is a common error.

Dally Castle (Simonburn). 1279 Iter. *Daley*; 1610 Speed *Dala Cast*; 1663 Rental *Dallie Castle.*

The first element may be O.E. *dāl* = part, used in the compound *dāl-land*, " common-land, hence " common clearing," or it may be O.E. *dæl*, " valley," which Skeat finds in Dalham, Suff. (p. 50) B.C.S. 612 *Dælham*. This may also be found in Dawley, Midd. D.B. *Dallega*. Dawley, Salop, D.B. *Dalelee* and Dalley or Delley, Dev., D.B. *Dalilea* may have the same history but more probably contain the O.E. personal name *Dealla.*

Dalton (Hexhamshire). 1271 Ch. *Dalton*. (Stamfordham) 1268 Ipm. *Dalton*, 1436 *Dawton*. **Dalton Piercy** (Hart). 1637 Camd. *Dawton.*

These may have the same history as Dalton-le-Dale *infra*, but more probably go back to O.E. *dæl-tūn* = valley-farm. *Piercy* because once in the possession of the Percy family, from whom it passed in 1370 to the Nevilles (S. 3. 98).

Dalton-le-Dale. *c.* 900 Bede, Hist. Abb. *Daltun, Daldun*[1]; *c.* 1050 H.S.C. *Daltun*; 1314 R.P.D. *Dalton-in-Valle*; 1584 Houghton *Datton*, 1604 *Dawton*; 1637 Camd. *Dawton.*

---

[1] *Daltun* and *Daldun* are variant MS. readings. It is just possible that Dawdon is intended. In any case the names have been confused.

Dawdon (Dalton-le-Dale). *c.* 1050 H.S.C. *Daldene*; 1230
F.P.D. *Daudene.*

The history of these names must be taken together, for
it is almost certain that the element *Dal-* must be the same
in each. It can hardly be O.E. *dæl* = valley, from the point
of view of either form or, in the case of the second, of meaning. Only once is a form *dal* found in O.E. (Leiden Gloss.)
and *Daldenu* (= valley-valley) is an impossible name. The
forms do not allow of a personal name as the first element,
and one can only suggest that it is O.E. *dāl* (*v.* Dalley *supra*)
found in *dāl-mǣd* = meadow-land held in common. This
word is found in North. M.E. in independent use. N.E.D.
quotes *duas mikel dales* (Newm. Cart.), while in Southern
English we have *dole-land*, *-meadow* and *-moor*. Hence
Dalton and Dawdon are " farm and valley held in common
ownership." For *le v.* Chester-le-Street *supra*. The *dene*
is now called, pleonastically, *Dawdon Dene*. Phonology,
§ 39; App. A, § 1.

Darlington [dɑ·ntən]. Type I: 1104-8 S.D. *Dearningtun*;
c. 1300 D.S.T. *Derningtona*; 1342 Pat. *Dernyngton*; 1583 Wills
Darnton; 1588 Eccl. *Darneton*, N.C.D. *Darington.* Type II:
1197 B.B. *Derlinton*; B.B. *Derlingtona*; 1228 F.P.D.
Derlintone; 1400 D.S.T. *Derlington*, 1507 *Darlyngton.*

Ekblom in dealing with Durnford, Wilts., D.B.
D(i)arneford, quotes *deornan mor* (B.C.S. 1282), *diornan
wiel* (ib. 200) and suggests that all alike contain O.E. *dearne,
*deorne, unmutated forms of O.E. *dierne*, *dyrne*, " secret,
hidden," formed on the analogy of the adv. *dearnunga*,
deornunga. Possibly in Type I we have this adj. applied to
an *ing* (Introd. p. xxvii.). It could hardly be so applied in
the O.E. sense but, at least in later M.E., the word was used
in the sense " dark, sombre, wild, drear " and the meaning
may be " farm by the dark or wild *ing*." Alternatively
De(a)rne might be taken as a nickname from the same adj.
in the sense " underhand, sly, crafty," [1] and *ing* as a
patronymic suffix. Type II is probably derived from Type
I under A.N. influence. Zachrisson (pp. 120 ff) shows how
names with *r——n* gives forms with *l——n* and others

---

[1] Some such name is perhaps found in *dyrnes treow*, B.C.S. 240.

with *n——n* give *l——n*. A new form arising under this
influence would be strengthened by the existence of a genuine
English patronymic *Darling* found in Dalington, Notts.,
earlier *Derlintun* (Mutschmann, p. 40). In Type I the
phonological development seems to have been alternatively
Darnington > *Darnton*, with loss of unstressed syllable
and *Darnington* > *Dar(r)ington*, with assimilation of *n*
and *r*. Cf. Darrington, Yorks., D.B. *Darnintone* (Moorman,
p. 57).[1] It should be said, however, that the ultimate history
of Darlington is possibly the same as that of Darrington.
Moorman gives other forms, *Dernington*, *Darthyngton*,
Dardinton, and suggests that the first element was once
Deornothing, from the personal name *Dēornōth*. In the
absence of any *th* forms for Darlington it is impossible to
speak with certainty on this point.

Darncrook (Gateshead). 1297 Pap. *Dernecroch*.
" Secret, hidden or remote crook of land," *v.* Darlington
supra and cf. Darnall, Yorks., earlier *Dernhale* (Goodall,
p. 116).

Darras (Ponteland). *c.* 1250 T.N. *Araynis* ; 1346
F.A. *Calverdon Darreyne, Calverdon de Arreyns* ; 1360
Pat. *Calverton Darrays* ; 1428 F.A. *Callerton Darres*.
The part of Callerton (*v. supra*) belonging to the family
of Darrayns, from Airaines, Somme Dept.

Davyshiel (Elsdon). 1344 Pat. *Davisel, Daveschole*.
" Davy's shiel," *Davy* being a pet form of David.

Dawdon Hall *v.* Dalton-le-Dale *supra*.

Deanham (Hartburn). 1255 Ass. *Denhum* ; 1268
Denum ; 1276 Ch. *id.* ; 1346 F.A. *Denom* ; 1377 Ipm.
Denam, 1436 *Denom*.

Deanmoor (Alnwick). *c.* 1280 Perc. *Denemora*.
O.E. *denu-hām* and *-mōr*=homestead and swamp in the
valley.

Deerness, R. *c.* 1200 Arch. 2. 25. 62 *Diuerness*.
A Celtic river-name.

Denton (Gainford). 1200 B.M. *Denton*. (Newburn) *c.*
1180 Anc. D. *Dentuna*. (Stannington) 1359 Pat. *Denton*.

[1] There is a spelling *Derington* as early as 1217 (F.P.D.), but this is
probably an error for *Dernigton* = *Dernington*.

Denwick (Alnwick). 1278 Ass. *Denewick*; 1538 Must. *Dennek*.

"Valley-farm and building." Cf. *den-tun* B.C.S. 1322.

Derwent, R. 1259 F.P.D. *Derewente*; 1620 N. vi. 195 *Darwyn*; 1764 Map *Darwen*.

Maclure (p. 186, n. 2) notes that Derwent is the name of several rivers, one of which gave its name to the Roman Station *Derventio*, supposed to be Stamford Bridge, Yorks. *Derv-* occurs also in *Derventum*, now Drevant, in France. It is probable that we have the same river-name in Darenth, Kent, B.C.S. 370 *Diorente* and Dart, Dev., A.S.C. *Dærenta*. Phonology, §§ 8, 56.

Detchant (Belford) [detʃən]. 1166 R.B.E. *Dichende*; 1170 Pipe *Diggenda*; 1249 Ipm. *Dichend*, 1314 *Dychent*; 1336 S.R. *Dichand*; 1346 F.A. *Dychand, Dychant*; 1560 Raine *Ditchand*; 1570 N.C.W. *Ditchin*; 1628 Arch. 1. 3. 95 *Ditchant*; 1715 Arch. 3. 13. 5 *Detchon*.

O.E. *dīc-ende*=ditch (or dyke) end. Cf. *to ðære dicende* (B.C.S. 477) and *dikeshendes* (F.P.D. p. 37). For the spelling *Diggenda* cf. Dissington *infra*. Phonology, §§ 10, 57, 56, 60.

Devils Water (Tynedale). Type I: 1233 N. iv. 45 *Divelis*; 1269 Perc. *Deueles*; 1289 Ipm. *Dyvils*; *p.* 1464 Hexh. Pr. cix. *Ewe Devyls*. Type II: 1577 Holinshed *Dowill*; 1610 Speed *Douols*; 1612 Drayton *Dowell*; 1650 Map *Dowols*.

The explanation of this name is given by Maclure (p. 149 n. 1). " *Glas* is a common river designation among the Celtic people in Great Britain and Ireland and even in Brittany . . . **dubno-glas*=deep stream, Dubglas and Daulas in English-speaking districts have assumed such forms as *Doflisc* (B.C.S. 667), Dawlish, Deviles, Dewlis, Dewlish, Devil's Water." We may add, for further comparison, Dowlish, So., Dowlas, Monm., earlier *Dyueles*, Dulas, Heref. (Bannister, p. 63), Dalch R. Dev., Divelish R. Dors., and we may note that Dewlish, Dors., stands on a stream now called "Devil's Brook." Type II has no early or local confirmation except in an isolated form of Dilston *infra*, and is probably due to antiquarian ingenuity.

Dewley (Newburn). 1251 Ipm. *Deuelawe*; 1296 S.R.

Dewillawe; 1346 F.A. *Deulawe*, 1428 *Deweley*; 1479 B.B.H. *Deulaw*; 1663 Rental *Dewly*; 1739 Newb. *Dula*. **Dews Green** (Whitfield). 12th c. H. 2. 3. 18 *Dewegreane*, 1634 ib. *Dewsgreen*.

" Dew-hill and -green," i.e. where the dew falls heavily. The latter name has developed a pseudo-genitival *s*. App. A, § 2.

Dilston (Corbridge). 1166 R.B.E. *Dovelestone*, 1171 *Develstone*; 1171 Pipe *Deuelestune*, 1176 *Diueleston*; 1273 R.H. *Develiston*; *c.* 1250 T.N. *Diveliston*; 1291 Ch. *Diviles-ton*; 1298 Arch. 3. 2. 21 *Dileston*; 1346 F.A. *Devyleston*; 1650 Arch. 2. 1. 54 *Dilston* alias *Devilston*.

" Devils (water) farm," *v.* Devils Water *supra*. For the sound development, cf. Sc., Nthb. [di·l] for *devil*.

Dingbell Hill (Whitfield). 1386 H. 2. 3. 103 *Vingvell hill* (*sic*); 1613 Whitf. *Dingbell Hill*.

This form is evidently corrupt, and the true name is doubtless " Thingwell Hill," the first element being the same as in Thingwall, Chesh. (D.B. *Thyngwall*), Dingwall, Ross., Tingwall, Shetland, and Tynwald Hill, I. of Man. All these alike go back to O.N. *þing-vellir*=fields of assembly, as in the famous *Thingvellir* in Iceland. It is not probable that a Scandinavian *thing* was ever held in Whitfield. Rather, the hill was so called because it reminded some Scandinavian settler, possibly *Úlfr* of Ouston (*v. infra*), of the hill on some far-distant plain of assembly in his own home-land.

Dinley (Birtley, Nthb.). 1279 Iter. *Dunley*; 1479 B.B.H. *id*.

" Hill-clearing." Phonology, §§ 21, 13.

Dinnington (Ponteland) [dintən]. 1255 Ass. *Duning-ton*; *c.* 1250 T.N. *Donigton*; 1346 F.A. *Donyngton*; 1580 Bord. *Dunengeton*; 1650 Map *Dunnyngton*; 1663 Rental *Dinnington*.

O.E. *Dunning(a)-tūn*=farm of Dunna or of his sons. *o* in the M.E. forms is purely orthographic. Phonology, §§ 13, 59.

Dinsdale-on-Tees. 1086 D.B. *Di(g)neshale*[1]; 1197 Pipe

---

[1] This form refers to the neighbouring Over Dinsdale, Yorks.

Ditleshale; 1267 Giff. *Ditneshale*; 1278 Ipm. *Detinsalle*; *c.*
1300 Kirkby's Inquest *Dedensale, Duttensale*[1]; 1306 R.P.D.
Dytnesale, Dittensale, 1312 *Dytmessale,* 1335 *Ditmishall,*
1338 *Dytinsale,* 1340.31 *Dyconsale, Dytesale, Dicensale*;
1340 R.P.D. *Dydinsale,* 1342 *Dytneshale*; 1479.35
Didensell; 1507 D.S.T. *Detynsall*; *c.* 1570 Eccl. *Dinsdaill*;
1560 V.N. *Dynsell*; 1746 Map *Dunsley.*

The second element refers to the haugh (*healh,* Part II)
on which Dinsdale stands. The first is probably a personal
name, found also in Deightonby, Yorks., D.B. *Dic(h)tenbi*
(Goodall, p. 118). What this is it is impossible to say. No
O.E. names in *Dyht-,* which would suit the phonological
requirements, are on record. There is a noun *dihtnere=*
steward, which might possibly have been used as a personal
name, but it is not very probable, as the word seems to be
learned rather than popular. The unfamiliarity of the name
or its phonological difficulty has led to a great diversity of
forms. The natural development would be *Dihtneshale* >
Dittensale > *Diddensale* (with voicing of *t* before *n*) > *Dinsdale* (with metathesis) > *Dynsell.* Phonology, § 53; App.
A, § 7.

Dipton (Hexhamshire). 1228 Gray *Depedene*; 1479
B.B.H. *Dipden.* (Collierley) 1339.45 *Depeden.*

" Deep-dene." Cf. B.C.S. 520 *to deopan dene.* Debden,
Suff., and Dibden, Hants., are the same names. [*dip*] is Nthb.
for *deep.* Phonology, § 50; App. A, § 1.

Dissington (Newburn). *c.* 1160 Ric. Hexh. *Digentun*;
c. 1190 Godr. *Dichintuna, Discintune*; 1205 Pipe *Dis-
cinton*; 1257 Ipm. *Discington*; 1270 Pat. *Distington* (*sic*)
de *Loval.*

O.E. *Dīcing(a)-tūn*=farm of **Dīca* or his sons. Cf. Dit-
chingham, Norf., D.B. *Dicingaham.* The normal Mod.
Eng. form would be *Ditchington,* but M.E. *ch* gave considerable difficulty to A.N. scribes and speakers, and they wrote
it as *g, c, ss, ch, sc,* resulting sometimes in an actual change
of pronunciation. Zachrisson (p. 21) gives examples in
Messing, Ess., containing O.E. *mæccea*=" match " or
companion, and Whissonsett, Norf., from O.E. *wīcing=*

---

[1] See note on previous page.

viking. *De Loval* from the Delavals who once held the manor. Phonology, § 22.

Ditchburn (Ellingham). 1252 Pipe *Dicheburn*; 1346 Ipm. *Disshburn*.

O.E. *dīc-burna* = ditch-stream or *Dīca(n)burna* = Dica's stream. For *sh v.* Dissington *supra*.

Doddington (*nr.* Wooler) [dɔriŋtən]. 1255 Ass. *Dodington, Dudington*; 1281 Perc. *Dodinton*; 1314 Ipm. *Duddington*; 1346, 1428 F.A. *Dodyngton*; 1764 Ilderton *Dorrington*; 1799 Egling. *id.*

"Farm of Dodda or Dudda or of his sons." For medial *d* >*r* cf. Derrington, Staffs., D.B. *Dodintone*, Derrythorpe, Lincs., F.A. *Dodingthorpe*. *t* has become *r* in Tarrington, Heref., D.B. *Tatintone*, F.A. *Tatynton*. The reverse change from *r* to *d* is found in *paddock*, earlier *parrock* (N.E.D.). Cf. Paddock Wood, Kent, F.A. *Parrok* and dialectal *poddish* for *porridge*. All examples of *d* (or *t*) >*r* show a dental earlier in the word, and the process may, in part at least, be a dissimilatory one.

Doepath Field (Corbridge). *c.* 1290 Perc. *Dapeth*; 1345 N. x. 63 *Dalepeth*; 1594 N. x. 270 *Dawpathe*.

Popular etymology seems to have been at work here. O.E. *dā-pæð*=doe-peth (*v. pæð*, Part II) should give North. M.E. *dapeth*, Nthb. *daypeth*, St. Eng. *doepath*. The 1345 form shows an attempt to associate the first element with *dale*, and the resultant *dalp*, quite regularly becomes *dawp* in the last form. Phonology, §§ 14, 39, 1.

Don, R. (Jarrow). 1104-8 S.D. *Don(us)*.
A Celtic river-name.

Donkleywood (Simonburn) [duŋkli]. 1279 Iter. *Duncliffe*; 1325 Ipm. *Doncliwod*, 1329 *Duncklywode*; 1663 Rental *Donkleywood*; 1833 Map *Dunclay*.

O.E. *dūn-clif* = hill-cliff. *o* is purely orthographic. Phonology, §§ 21, 56; App. A, § 7.

Dotland (Hexhamshire). *c.* 1160 Ric. Hex. *Dotoland*; 1226 B.B.H. *Doteland*, 1287 *Dotteland*, 1479 *Dot(e)land*.

"Dot's land." *Dot(us)* is a man's name in D.B., and Björkman (Z.E.N. p. 29) associates it with O.Sw. *Dote* found in *Dotabotha*. He also mentions an O.Dan. *Dota*

E

(cf. O.N. *Dótta*), a woman's name, which would suit here also. Nielsen (p. 18) gives a name *Dot* in O.Danish on the authority of *Nic. Dotus*, a 12th-cent. name, and Dåstrup, earlier *Dotzthrop*.

Downham (Carham). 1251 Ch. *Dunum*; 1255 Ass. *Dunhum*; *c.* 1250 T.N. *Dunum*; 1542 Bord. Surv. *Downeham*. O.E. *dūn-hām*=hill homestead or, possibly (*æt þǣm*) *dūnum*=(at the) hills. App. A, § 6.

Doxford (Ellingham). *c.* 1150 Vescy *Docheseffordam*; 1230 Pat. *Dockesford*; 1255 Ass. *Doxeford*; 1528 F.F. *Doxworth*; 1539 F.P.D. *Doxforth*.
"Docc's ford." Cf. Doxey, Staffs., D.B. *Dochesig* (Duignan, p. 51). Phonology, § 30; App. A, § 4.

Druridge (Woodhorn). Type I: *c.* 1250 T.N. *Dririg*'; 1296 S.R. *Dryrige*; 1346 F.A. *Dririg*, 1428 *Dryrygge*. Type II: 1354 Pat. *Drurigg*; 1663 Rental *Druridge*. Type III: 1443 Ipm. *Drerigh*.
"Dry-ridge," the three types showing respectively the North, South, and Kent developments of O.E. *drȳge*. It is difficult to understand how a Southern form ultimately survived. Phonology, § 27.

Dry Burn (Carrycoats). *a.* 1182 Newm. *Drieburn*. (Framwellgate) 1382 Hatf. *Driburn*.
"Stream that soon dries up." Cf. *on drygean broc*, B.C.S. 945, *þurrá* in Iceland (N. o. B. ii. 22).

Duddo (Norhamshire). 1228 F.P.D. *Dudeho*; 1447 Raine *Dudhowe*; 1539 F.P.D. *Dodow*. **Duddoe** (Stannington). *c.* 1250 T.N. *Dudden*; 1316 Ipm. *id.*, 1418 *Doden*; 1428 F.A. *Dudden*.
"Duda's *hōh* (Part II) and dene." Phonology, § 36; App. A, § 12.

Dukesfield (Slaley) [duksfi·ld]. 1255 Ass. *Dekesfeud*; 1296 S.R. *Dukesfeld*; 1322 Cl. *Dokesfeld*, 1350 *Duxfeld*; 1535 F.F. *id.*
"Ducc's field." Cf. Duxbury, Lancs. (Wyld, p. 115) and Duxford, Cambs. (Skeat, p. 26). The form *Dukes-* does not agree with the local pronunciation, and is due to a legend (cf. Wallis II, p. 108) that the Duke of Somerset, killed after the Battle of Hexham in 1464, was captured here.

Dunsheugh (Denwick). 1310 Ass. *Dunchehou.* **Duns Moor** (Bingfield). 1479 B.B.H. *Donnismore.* "Dunn's *hōh* (Part II) and moor or swamp." For *o* *v.* Cronkley *supra.*

Dunstan (Embleton) [dustən]. 1244 Ipm. *Dunstan.* **Dunstanburgh** (ib.). 1321 Orig. *Dunstanburgh.* **Dunstanwood** (Corbridge). 1268 Ass. *Dunstanwode.* O.E. *dūn-stān*=hill-rock. "Fort and wood by the hill-rock." In Dunstanburgh the aptness of the name is evident, in the wood it probably refers to a rock on the steep banks of the Devil's Water.

Durham. Type I: 1056 A.S.C. *Dunholm*; 1191 Pipe *Dunolm*; 1227-34 Cl. *Dunholm,* 1343 *Dunolm*; 1307, 1312 R.P.D., *c.* 1380 Coin, *c.* 1490 Coin *id.* and Latinised *Dunolmia,* S.D., Ch. *c.* 1300. Type II: *c.* 750 Bede *Dunelma* (Latinised); 1191 Pipe *Donelme*; Hy. II Coin *Dunhe, c.* 1312 *Dunelm*; *c.* 1300 Ch. *Dunelmia* (Lat.), *c.* 1435, *c.* 1470 Coin *id., c.* 1515 *Dunel*; V.E. *Dunelm.* Type III: *c.* 1160 Gaimar *Dunelme*; *c.* 1170 Jord. *Durealme, Dure(a)-ume*; Hy. 3 Coin *Durh*; 1231 Ch. *Durham*; Edw. I Coin *Dureme*; *c.* 1300 Langtoft *Dureme, Dur(h)am*; 1313-8 Cl. *Durham, Dure(s)m(e)*; 1311 R.P.D., 1323 F.P.D. *id.*; *c.* 1250 Mouskes, Chronique rimée *Duriaume, Durialme*; *c.* 1370 Coin *Dureme, Dorelmie, Durrem, c.* 1470 *Deram(e),* *c.* 1500 *Durham, Dirham, c.* 1505 *Dirham, Derham, c.* 1520 *Durram, c.* 1550 *Durram, Durham*; 1637 Camd. *Duresme.*[1]

The earliest form of the name is commonly given as *Dūn-holm,* i.e. hill-island, a name aptly descriptive of its site, but it should be noted that *holm* in English place-names is unknown apart from Scand. influence, so that this form can hardly date back further than the days of the Vikings, and it may represent an etymologising perversion of some earlier Celtic name. Type II is in part due to the influence of the Lat. adj. *Dunelmensis,* in part to natural weakening of the vowel of the secondarily stressed syllable from *o* to *e.* The

---

[1] The writer has here drawn freely on the wealth of forms quoted in Canon Fowler's paper on the Coins of the Bishops of Durham, and in Zachrisson's books on *Anglo-Norman* and *Latin Influence* (pp. 133–5 and p. 8 respectively).

development of a form -*helm* may also have been influenced
by the occasional use of that element in place-names (*v.*
Part II). Its late survival is doubtless due to the influence
of the familiar Latin adjective. Type III is explained by
Zachrisson as due to dissimilation of *n* to *r* before following
m. Such dissimilation is found in other A.N. spellings,
but has not survived in the modern form of any other place-
name. The suffix, as suggested by the same writer, has
been changed under two influences, (1) the common Fr.
vocalisation of *l* to *u* before *m*, (2) regular reduction of -*elm*
to -*am* in an unstressed syllable as in Brickhampton, Glouc.,
earlier *Brithelmeton.*

Zachrisson further suggests that there is just a possibility
that the change from *Dun-* to *Dur-* may have been assisted
by the common use of Celtic *durus, stronghold, in French
place-names in *Dur-*, and that the spelling -*esme* may be
due, in part at least, to the influences of French place-
names in -*esme*< Celtic -*isma*.[1]

Dyance (Piercebridge). 1207 F.P.D. *Diendes*; 1526.44
Dyaunce; 1765 Gainf. *Dyans.*

A difficult name, which may be of Scand. origin. The
word *dy* is common in Dan. place-names (Steenstrup, p. 91)
meaning a "swamp," and is also found in O.N. (Rygh,
Indl. p. 47). The derivative *dynd*, O.Dan. *dyande*, may
have given M.E. *diende.* The plural *diendes*=swamps,
has been respelled under French influence. Phonology, § 5.

Eachwick (Heddon-on-the-Wall). *c.* 1160 Ric. Hex.
Achewic; 1257 Ipm. *Echewic*; 1475 Newm. *Echewyke.*

Possibly O.E. *ēce-wīc*=lasting, permanent dwelling.
The usual meaning of *ēce* is "eternal," but its use in such
a phrase as *on ēce yrfe*, "as a permanent possession," may
have led to some such development of sense as is here
suggested.

Ealingham (Simonburn). 1279 Iter. *Evelingham, Eve-
lingjam*; 1289 Sc. *Evelingham*; 1296 Ipm. *Ellingeham*;
1653 Comps. *Elingham*; 1663 Rental *Ellingham.*

"Homestead of the sons of *Eofel," a dimin. of *Eof.*

---

[1] The whole discussion of this name by Zachrisson (*loc. cit.*) is in
valuable.

The 1296 and 1663 spellings show assimilation of *vl* to *ll*. Cf. Chillingham *supra*. The modern form is due to an alternative development of *evel* to [i·l], cf. Nthb. [di·l] for *devil* and the famous " dram of *eale* " (=evil) in Hamlet. Phonology, § 34.

Earle (Doddington) [jerl]. 1255 Ass. *Yerdel, Yerdhil*; *c.* 1250 T.N. *Yherdhill*; 1288 Ipm. *Yerdill*; 1346 F.A. *id.*, *Zyerdle* (*sic*), 1428 *Yerdyll*; 1542 Bord. Surv. *Yerdle*; 1579 Bord. *Earlle*; 1663 Rental *Eardle*; 1705 Ingram *Yardhill*, 1709 *Yerle*, 1712 *Erle*.

" Hill marked by a yard or enclosure." Cf. Yearhaugh *infra*. For loss of initial [j] cf. Nthb. [iər] for *year*. Phonology, § 36.

Earlshouse (Sniperley). 1396 Acct. *Erilhous*; 1382 Hatf. *Erlehous*.

" Earl-house," probably so called from a man named *Earle*. This name is probably derived from the title as there is no evidence for an O.E. name *Eorl(a)*. Names like *Erlebald, Eorlebyrht* are of continental origin (Forssner, p. 78).

***Earlside** (Elsdon).[1] 1200 R.C. *Yerlesset*; 1332 Cl. *Erleside*; 1368 Ipm. *Erleyside*; 1368 Cl. *Erlesside*; 1378 Ipm. *Erlsyde*; 1663 Rental *Earlside*.

" Earl's seat " (*sǣte*, Part II). Phonology, §§ 8, 9. There is a *Yarlside* in Cumberland, earlier *Jerlesete*, containing the equivalent Scand. *jarlr*, and Ekwall (p. 33) says that it is fairly common as a hill-name.

Earsdon [jɔ·zən] (Hebron). 1233 Pipe *Erdesdon*; 1261 Ipm. *Herdisdun*, 1335 *Erdesdoun*; 1346 F.A. *Erisdon*; 1436 Pat. *Eresdon*; 1663 Rental *Earsdon*. (Tynemouth) 1203 R.C. *Hertesdona*; 1271 Ch. *Erdisdunam*; 1363 Waterf. *Erdesdon*; 1428 F.A. *Eresdon*.

O.E. *Eardes-dūn*=Eard's hill, *Eard* being short for one of the numerous O.E. names in *Eard-*. Cf. Ardsley, Yorks. (Moorman, p. 10). Phonology, §§ 8, 9, 53. *Hertes* shows inorganic *h* and common confusion of *t* and *d* due to A.N. scribes (Zachrisson, p. 43 n.).

Easington (Belford). *c.* 1250 T.N. *Yesyngton*; 1278

---

[1] Identified by Hodgson (2. 1. 135) with Foulshields or Breadless Row, opposite Byrness on Rede Water.

Ass. *id.*; 1296 S.R. *Yhesington*; 1346 F.A. *Yesington,
Yzesyngdon*; 1538 Must. *Yhessyngton*; 1579 Bord.
Easengtoun. (Co. Durham) *c.* 1050 H.S.C. *Esingtun*; 1197
Pipe *Hesinton*; 1249 Ch. *Esington*; 1539 F.P.D. *Esyngtoune.*
Easington, Co. Durham, is clearly "farm of *Esi* (cf.
L.V.D.) or of his sons." Easington, Nthb., offers difficulties.
No O.E. name *Gesi* is known which might have given rise
to the M.E. forms, with late loss of initial [j] as in Earle
supra and many other place-names. On the other hand,
it is difficult to derive it from O.E. *Esi* with the development
of [j] before the initial vowel because, though there are
plenty of names and words in English which show this in
their modern forms, chiefly in dialect, this is hardly ever
represented in M.E. spellings. *else, ear, earth, even, earn*
show no such spellings before 1500. *earls* is found as *zierles*
as early as 1200, but this may be due to the influence of
O.N. *jarlr* (cf. Earlside *supra*). In place-names we get
this development in Yattendon, Berks., Yenhall, Cambs.,
B.C.S. 1305 *eanheale*, Yarnscombe, Dev., D.B. *Hernescoma*,
Yedbury ib. D.B. *Addeberia*, Yealmpton ib. D.B. *Elintona*,
Yaldham and Yalding, Kent, earlier *Ealdham, Aldinges*,
Yardley, Herts., Yelverton, Norf., Yarnton and Yelford,
Ox., Yarlett and Yarnfield, Staffs., Yearsley, Yorks., D.B.
Eureslage, Youlthorpe ib., D.B. *Aiultorp*, and possibly
Yeadon, D.B. *Iadun*, 1175-85 Yorks. Charters, *Eiadona*,
1283 Kirkby's Inq. *Yedon*, Yaverland, I. of Wt., D.B
Everelant, but only in the case of Yeadon, *u.s.* Yattendon,
D.B. *Etingedene*, 1251 Ch. *Yatingden*, 1368 B.M. *Yatyndene*,
Yedbury, *c.* 1300 Ipm. *Yhaddeburi*, Youlthorpe, Kirkby's Inq.
Yolthorp, Yelverton, D.B. *Ailvertona*, 1346 F.A. *Yelverton*
and Yealmpton, 1309 Ch. *Yhalampton*, have early forms in *y*
been found. If the name is *Esi* we have early development
of [j] before the initial vowel, and later loss of it. Nthb.
dialect develops [j] in some words, e.g. [jerþ], [jel], [jekom]
for *earth, ale, acorn*, and in *Yelderton* for Ilderton *infra*,
and drops it in others, e.g. [iər], [i·ld] for *year, yield*, and
in *Evering* for Yeavering *infra*.

Eastgate (Stanhope). 1457.35 *Estyatshele*; 1637 Camd.
Eastyat.

" East-gate " (*geat*, Part II).

Ebchester. 1230 Pipe *Ebbecestr.*

" Fort of Ebbi." Cf. L.V.D. for this name.

Edderacres (Easington). B.B. *Etheredesacres* (B., C. *Etherdacres*) ; 1314 R.P.D. *Edredakers* ; 1382 Hatf. *Edirdacres* ; 1404 Pat. *Edderdacres.*

" Aethelred's fields " (*œcer*, Part II). Phonology, §§ 41, 53.

Eddys Bridge (Shotley). 1446 D.S.T. *Edisbrigg* ; 1464 F.P.D. *Edisbrig* ; *c.* 1570 Eccl. *Edyedsbridge, Eedesbrig.*

The 1570 spelling *Edyed-* may be mere dittography, but, if any stress is to be laid on it, the first element is the woman's name *Edith*, cf. D.B. *Edd(i)ed, Eddid.* Otherwise the name of the owner or builder of the bridge may have been *Aeddi* or *Ed(d)*, cf. *Eddesford*, B.C.S. 601.

Eden, Castle. *c.* 1050 H.S.C. *Geodene, Iodene* ; 1153-95 F.P.D. *Edene, Iodene* ; 1312 R.P.D. *Eden.* **Eden Burn.** 1270 Ch. *Edeneburn.*

Eden is found as a river-name in Cumb., Kent, and S. Scotland, and the Castle may take its name from the river. On the other hand, there is a Gaelic *Eudan* or *Aodann* = forehead, hill-brow, giving later *Edin* or *Eden* (Matheson, p. 56) which might have given rise to the place-name, and the river-name be derived from it.

Edge Knoll (Witton-le-Wear). 1303 R.P.D. *Edenesknoll* ; *c.* 1300 Lewes Knights *Edisknoll* ; 1382 Hatf. *Ednesknolle* ; 1400.33 *Eddisknoll.*

" Edwin's knoll." Phonology, §§ 49, 53, 31.

Edgewell House (Mickley). 1381 Cl. *Egewelle.*

O.E. *ecg-wielle* = edge-spring, i.e. on the side of a hill, or *Ecgan-wielle* = Edge's spring. Cf. *ecgan croft*, K.C.D. 621.

Edington (Mitford). 1195 Pipe *Idington* ; 1255 Ass. *id.* ; 1322 Ipm. *Ydintoune* ; 1346 F.A. *Edington* ; 1377 Ipm. *Idyngton* ; 1428 F.A. *Edyngton.*

" Farm of Ida or of his sons." Ida is not a common O.E. name, but was borne by the first king of Bernicia. Phonology, §§ 10, 22.

Edlingham (edlindžəm]. *c.* 1050 H.S.C. *Eadwulfincham* ; 1104-8 S.D. *Eadulfingham* ; 1174 D.S.T. *Eduluingeham* ;

1198 Ch. *Edulfingeham*; 1200 R.C. *Edelvingham*; 1233
Pipe *Edelingham*; 1259 Newm. *Edlyngham*; 1346 F.A.
Edlyngeham.
   O.E. *Ēadwulfingahām* = homestead of the sons of
Eadwulf. Phonology, §§ 21, 49, 34.
   Edmondhills (Ancroft). 1318 Acct. *Emotehill*; 1539
F.P.D. *Emodhille*; 1584 Bord. *Emontills*; *n.d.* Raine
Edmondhills, E(y)motehill, Heymotehill.
   No certainty is possible. The first element may be
O.E. *ēamōt* = rivers' meet or *ǣmette*, "ant." Cf. Emmet-
haugh *infra*. Phonology, § 55.
   Edmondsley [Chester-le-Street]. *c.* 1190 Godr. *Edeman-
nesleye*; 1242 D. Ass. *Edmannesleye*; 1297 Pap. *Edmanesley*;
B.B. *Edmansley* (B., C. *Edmondesley*; 1304 Cl. *Edmundesley*;
1312 R.P.D. *Edmanesley, Eadmundesley*; 1433 D.S.T.
Edmundesley; 1727 Houghton *Edomsley*.
   "Edmund's clearing," though perhaps the name should
be O.E. *Eadmann (cf. *Eodman*, Searle) with later change
to the more common *Edmund*.
   Edmundbyers. 1228 F.P.D. *Edmundesbires, c.* 1275
Eadmundbiris.
   "Edmund's byres" (*bȳre*, Part II).
   Egglescliffe. 1085 D.S.T. *Egglescliff*; 1162 Pipe *Egges-
cliua*, 1197 *Ecclescliue, Egglescliue*; 1252 Ch. *Egglesclive*;
1294 Pat. *Ecclescliue*; 1507 D.S.T. *Eglysclyff*.
   "Church-cliff or hill." The first element is probably
that explained by Moorman (Introd. pp. vii., viii.), viz.,
eccles, from Lat. *ecclesia*, through some Celtic form.
Voicing of *c* to *g* may have been helped by the influence
of the Norse name *Egill* (cf. Eggleston *infra*). *Eagles-
cliffe*, the name given to the station, is an unauthorised
corruption.
   Eggleston (Middleton-in-Teesdale). 1197 Pipe *Egleston*;
1260 Pat. *Eggleston*; 1313 R.P.D. *Egleston*; 1336 Ipm.
Eglestoune; 1432 D.S.T. *Eglyston*.
   "Egill's farm," *Egill* being a common O.N. name.
   Eglingham [eglindžəm]. *c.* 1050 H.S.C. *Ecgwulfincham*;
1104-8 S.D. *Ecgwulfingham*; 1135-54 Ty. *Eguluingham*;
1200 Pipe *Eglingham*; 1271 Ch. *Eguluingeham*; 1313

R.P.D. *Eglingham,* 1343 *Eglingeham;* 1596 Bord. *Eglingjham.*

O.E. *Ecgwulfingahām* = homestead of Ecgwulf's sons. Phonology, §§ 49, 34.

Eighton Banks (Lamesley). 1343.31 *Eghton;* 1793 Lowick *Eaton Banks.* " Eh(h)a's farm." Cf. *ehanfeld,* B.C.S. 1282 and *Eccha,* L.V.D.

Eldon (Auckland). 1104-8 S.D. *Elledun;* 1335 Ch. *Eldona.* **Elford** (Bamburgh). 1255 Ass. *Eleford,* 1268 *Elsford;* 1280 Ch. *id.*; 1663 Rental *Elford.* **Ellingham** [elindžəm]. *c.* 1130 F.P.D. *Ellingeham, c.* 1160 *Elinge-ham;* 1252 Pipe *Elingham;* 1255 Ass. *Elingeham;* 1278 Ipm. *Elling(c)ham, Ellincham;* 1346 F.A. *Elyngham;* 1507 D.S.T. *Elyngeham.* **Ellington** (Woodhorn). 1167 R.B.E. *Helingtone;* 1233 Pipe *Elington;* 1268 Ass. *Ellington.*

Probably *"Ella* or *Aella's* hill and ford, the homestead of E.'s sons, E.'s farm." The absence of any forms in *ll* may raise a doubt in the case of Elford, which may have the same history as Yelford, Ox., D.B. *Aieleforde,* Ch. *Eilesford,* 1316 F.A. *Eleforde.* This may contain O.E. *Aebel* or the name *Eli* found in D.B. The forms in *s* show the strong genitive.

Elilaw (Alnham). *c.* 1290 Perc. *Ylylawe;* 1721 Alw. *Ellilaw,* 1746 *Ililaw.* **Ellishaw** (Elsdon) [(e)liʃə]. 1278 Ass. *Illescagh, Illeschawe;* 1291 Tax. *id.*; 1341 Bury *Illeschay;* 1411 H. 3. 243 *Illeshawe,* 1534 *Ellyshawe.*

" Illa's hill and wood " *(sceaga,* Part II). This name is found also in Eleigh, Suff. (Skeat, p. 78). Phonology, § 10. For the local pronunciation with rare shifting of stress from the stem syllable, cf. Heslop (p. 45). " The haugh behind Elishaw catches the floating rubbish that the Rede carries down," hence the local sayings, " He'll be left on the haughs anunder' Lishaw if he dissn't hurry on."

Elrington (Haydon). 1229 Gray *Elrinton;* 1255 Ass. *Elyrington;* 1298 B.B.H. *Elrington;* 1371 Cl. *Elleryngton;* 1663 Rental *Elrington.*

" Aelfhere's farm," cf. Allerdean *supra* or, possibly,

"farm by the elder-covered *ings*" (Introd. p. xxvii.).
Phonology, §§ 1, 50.

Elsdon. 1236 Cl. *Hellesden*; 1244 Ipm. *Ellesden*;
1278 Ass. *Illesden*; 1312 R.P.D. *Ellesden*; 1324 Ipm.
Ellesden, Helvesden; 1432 Pat. *Eluesden*; 1507 D.S.T.
Ellysden; 1663 Rental *Elsden*.
O.E. *Aelfes-denu* = Aelf's valley. Phonology, §§ 1, 53;
App. A, § 1.

Elstob (Stainton-le-Street). 1242 D. Ass. *Ellestobbe*;
1364 R.P.D. *Ellestob*; 1360 Pat. *Ellestubbe*; 1430 F.P.D.
Elstobe.
Surtees (3. 46) tells us that Elizabeth Elstob, "the author
of the famous English-Saxon grammar," had traced back
her descent to Adam de *Elnestobbe*. If this is correct the
name is either (1) O.E. *ellen-stybb* = alder stump, of common
occurrence in old boundary-lists, or (2) "Aelfwine's or Aelf-
noth's stump," cf. Elstow, Beds., earlier *Elnestowe*, Elstead,
Suss., earlier *Elnested* (Roberts, p. 63); otherwise we should
take it to be "Aella's or Ella's stump." Phonology, § 53.

Elswick (Newcastle-on-Tyne) [elsik, elzik]. *a.* 1189
N. viii. 49 *Elstwyc*; 1198 N. viii. 67 *Alsistwic*; *c.* 1205
Coram *Elsissewich*; 1203 R.C. *Alsiswic*; 1271 Ch. *Alliswik*;
1311 R.P.D. *Elsewyk*; 1333 Ipm. *Elstwyk*, 1378 *id.*; 1428
F.A. *id.*; 1628 Freeh. *Elswick*.
"Aelfsige's dwelling." Cf. Alswick in Layston, Herts.,
D.B. *Alsieswiche*, 1303 F.A. *Alswick* and Aliceholt, Hants.,
earlier *Alsiesholt*.
The *t* found in several early forms is probably the same
t which elsewhere developed after *s*, as in such vulgarisms
as *elst* and *elstwhere* (Jespersen, 7. 64; Horn, § 189, n. 1).
Phonology, §§ 1, 53, 49.

Elton. *c.* 1050 H.S.C. *Eltun*; *c.* 1180 B.M. *Eligtun*;
1313 R.P.D. *Elletun*.
Possibly, "farm of *Ella* or *Aella*" (*v.* Elford *supra*), but
the two earliest forms are difficult.

Eltringham (Ovingham) [eltrindžəm]. *c.* 1200 Arch.
2. 1. 64 *Heldringeham*; 1255 Ass. *Heltringham*; 1268
Ipm. *Eltringham*; *c.* 1250 T.N. *Eltrinch'm*; 1296 S.R.
Heltrincham.

"Homestead of the sons of Heltor." Cf. D.B. *Haltor*, (*H*)*eltor* which Björkman (N.P. p. 62) takes to be O.W.Sc. *Hallþórr* or *Halldórr*. Phonology, § 34.

Elvet (Durham). A.S.C. *Aelfetee*; *c*. 1125 F.P.D. *Aeluet(e)*, *Eluete*, *Elfeete*; 1203 R.C. *El(e)uet*; 1228 F.P.D. *Eluet*.

aelfet in the A.S.C. form is Anglian for W.S. *ielfetu*, "swan" (Bülbring, § 180), and *ee* is dat. sg. of *ēa*, river. Hence, "swan-river," a name applied apparently to that part of the Wear on which Elvet now stands. Cf. *ylfethamm* B.C.S. 1307 and *Alptá*, a river-name in Iceland (N. o. B. ii. 20). Phonology, § 1.

Elwick (Belford) [elik]. *c*. 1150 F.P.D. *Ellewich*; 1203 R.C. *Ellewic*; 1296 S.R. *Elwyk*; 1637 Camd. *Ellick*. (Hart) *a*. 1141 B.M. *Ailewic*; *n.d.* S. 3. 90 *Aelwic*; 1214 Pipe *Ellewic*.

"Dwelling of *Aella* or *Ella* and of *Aegel* respectively." The last name is a late form for O.E. *Aeþel*, itself a short form of one of the numerous O.E. names in *Aeþel-*. Cf. Elford *supra*.

Emblehope (Thorneyburn) [emləp]. 1325 Ipm. *Emelhope*; 1330 Cl. *Hemelhop*, 1370 *Hemilhop*; 1686 Elsdon *Emlopp*. **Embleton**. *c*. 1200 R.B.E. *Emlesdune*; 1244 Ipm. *Emildon*; 1255 Ass. *Emeldon*; 1346 F.A. *id.*; 1507 D.S.T. *Emelden*, *Embledon*; 1538 Must. *Emylton*.

The first element is probably O.E. *emel* = caterpillar, hence "caterpillar -hope and -hill" or it may be that word used as a nickname, cf. *emelhyll*, B.C.S. 887 and Emsworth, Hants., earlier *Emelesworth*. Phonology, §§ 37, 55, 36.

Embleton (Sedgefield). *c*. 1190 Godr. *Elmedene*; *c*. 1200 B.M. *Helmedena*; 1340 R.P.D. *Elmeden*; 1351 B.M. *id.*; 1370 S. 3. 54 *Emildon*; 1386 W. and I. *Elmeden*; 1637 Camd. *id*; 1642 Sedgefield *Emleton*.

"Elm-valley" possibly. Cf. Surtees (3. 53) who says that the name is derived from "its deep hollow dene where some remains of an old elm-wood are still seen amongst the hazel copses." The only difficulty in this explanation is the persistence of forms in *Elme-* rather than *Elm,-* which might point to a personal name. The existence of such

a name is clear from Elmington, Northts., D.B. *Elmintone,*
Elmham, Norf., D.B. *Elmenham.* Similarly we have O.Sw.
Almunge, a patronymic from *alm*=elm (Hellquist, p. 7).
Quite late, metathesis of *l* and *m* took place, cf. Embleton,
Cu., earlier *Elmeton,* Amblecote, Staffs., earlier *Elmelecote.*
The reverse change is found in Elmdon, Suss., earlier *Emeldon*
and Elmbridge, Surr., earlier *Emelebrugge.* App. A, § 1.

Embley (Slaley) [emli]. 1359 Pat. *Elmeley*; 1765
N. vi. 347 *Emley.* (Whitfield) 1135 H. 2. 318 *Elmlee.*
" Elm-clearing." Cf. *elmleage,* B.C.S. 235. Phonology,
§ 55.

Emmethaugh [eimitha·f] (North Tyndale). 1169 Pipe
Emmoteshala, 1175 *Hamodeshalch*; 1610 Speed *Emouthaugh*;
1663 Rental *Emmitt-haugh.*

Possibly the first element is O.E. *ēa-mōt*=river-meeting,
for the haugh is at the meeting of Whickhope Burn and
North Tyne, cf. Emmott, Lancs., and Emmotland, Yorks.
(Goodall, p. 132), but the genitival *-es* looks as though we
had to do with a personal name. It might be O.E. *æmette,*
ant, M.E. *emete, emote,* used as a nickname. Cf. *Emmett* as a
surname. Hence " Emmett's haugh." [1] Phonology, § 55.

Eppleton (Houghton-le-Spring). *c.* 1180 F.P.D.
Aepplingdene; 1180 Finch. *Epplindena,* 1153-95 *Hepplig-dene*; 1311 R.P.D. *Epplingden.*

" Valley of Aeppel or his sons." **Aeppel* is a dimin.
of O.E. *Eppa* or *Aeppa.* Phonology, § 1; App. A, § 1.

Erring Burn (St John Lee). 1479 B.B.H. *Eryane,*
Erean; 1547 Hexh. Surv. *Eyren.* **Errington** (ib.). *c.* 1160
Ric. Hexh. *Herintun*; 1280 Wickw. *Eringtone*; *c.* 1250
T.N. *Errington*; 1296 S.R. *Eringtona*; 1479 B.B.H.
Eryngtone.

Errington is " farm on the Erring Burn," with earlier
development of a pseudo-patronymic form, which has in
its turn affected the river-name.

Escombe. *c.* 990 B.C.S. 1256 *Ediscum*; 1104-8 S.D.
id.; B.B. *Escumba*; 1315 R.P.D. *Escum.*

The first element is possibly a personal name, gen. sg.

---

[1] The modern pronunciation does not make the solution of the etymology any easier.

of *Aedd(i)*, cf. Eddy's Bridge *supra*, but it might also be
O.E. *edisc*, Mod. Eng. Dial *eddish*, "park or enclosed park
for cattle, then aftergrowth of grass." The second might
be O.E. *cumb*, "valley," but this is not otherwise found
in Nthb. or Co. Durham, nor does it suit the topographical
conditions. The whole solution is very uncertain.

Esh (Lanchester). *a*. 1196 Finch. *Esse*; *c*. 1200 B.M.
Es; 1312 R.P.D. *Esshe*.

"Ash-tree," probably a prominent landmark. Cf. Ash,
Derbys. and Kent (twice). Phonology, § 1.

Eshells (Hexhamshire). *c*. 1160 Gray *Eskeinggeseles*;
c. 1225 B.B.H. *Eskilescales, Eskingseles,* 1226 *Eskinschel.*

The second element is the common Nthb. *sheles* with
substitution, in one form, of the Scand. loan-word *scales*
(O.W.Sc. *skáli*) for the native English one.

The first element may be either (1) O.W.Sc. *Ásketill,*
M.E. *Askill, Askell, Eskill,* in its alternative form *Asketin(us)*
(Björkman, N.P. p. 17), developing to *Askin* or *Eskin.*[1]
Hence "Asketin's shiel'," or (2) O.W.Sc. *eski,* "ash-tree"
+*eng*=*ing* or grassland (cf. Elrington *supra*), hence "grass-
land with ash-trees in it." *Esking* would often become
Eskin in M.E., and *Eskil* must be explained as due to the
common mistake of anticipating the *l* which is to come later
in the word. Phonology, §§ 1, 59, 53.

Eshott (Felton) [eʃət]. 1186 Pipe *Esseta*; *c*. 1200
Brkb. *Esschet*; 1255 Ass. *Essetet*; 1268 Ass. *Escheyuette*
(*sic*), *Eschette*; 1307 Ch. *Esshet*; 1428 F.A. *Eshette*; 1638
Freeh. *Eshott.*

O.E. *æsc-scēat* = ash-shot, the corner of land marked
by an ash-tree (*scēat*, Part II). Cf. Ashford, Kent, earlier
Ess(ch)et(t)esford and *āc-sceates geat* (K.C.D. 597) with its
modern equivalent *Oakshott.* Phonology, § 1.

Eshott Heugh. 1278 Ass. *Hou.* Cf. Heugh *infra.*

Eslington (Whittingham). 1169 Pipe *Estlinton,* 1176
Eselinton; *c*. 1210 R.B.E. *Eselingtone, Esselintone*; 1231
Pat. *Eslinton*; 1254 Ipm. *Esselington, Es(t)lington*; 1260
Pat. *Estlington.*

---

[1] This may be the source of O.Dan. *Eskin*, which Nielsen (p. 22) con
nects with O.H.G. *Ascvin*, O.E. *Aescwine.*

Cf. Islingham, Kent, B.C.S. 194 *Aeslingaham*, Essendon, Herts., earlier *Eslingadene*. Skeat (p. 20) takes this to be from *Aesclinga*, a dimin. in *ling* from the personal name *Aesc* and to mean " Servants of Aesc," but there is no warrant for the formation of such compounds in O.E., and we must take it rather to be a patronymic from * *Aescel*, dimin. of *Aesc*. The forms in *s, ss* represent a common A.N. spelling of [ʃ], which ultimately determined the pronunciation. Were it not for the forms in *ss* we might take *Es(e)ling* to be a derivative of * *Esel*, dimin. of *Esi, Esa*. Taylor (p. 106) notes *Eslingaford* (D.B.) Islingham, *Eslinghem* (Artois), *Esslingen, Eislingen, Aislingen* (Würtemburg), which may contain the same patronymic.

Esperley (Cockfield). 1230 Cl. *Esperdeslegh*. **Esper Shields** (Bywell St Peter). 1225 Coram *Esperdosele*, 1230 *Estberdesheles* ; 1268 Ipm. *Esperscheles* ; *c.* 1590 Map *Aspersheales* ; 1663 Rental *Esper Shells* ; 1833 Map *Aspershield*.

" Field and sheils of *Aespheard*." This name is not recorded, but cf. *Aesc-heard*. As aspenwood is very soft the name was probably ironical in its original application. Phonology, §§ 1, 53.

Esp Green (Lanchester). 1313 R.P.D. *Espes*. **Espley Hall** (Mitford). 1252 Pipe *Aspele* ; 1257 Ipm. *Espeley*. **Espleywood** (Simonburn). 1279 Iter. *Espeleywode*.

" Aspen-trees " and " Aspen-tree clearing." Cf. Aspley, Beds., Staff. and Warw., Espley in Hodnet, Salop. Phonology, § 1.

Etal (Ford) [i·təl]. 1232 Cl. *Ethale*, 1268 *Ethale* ; *c.* 1250 T.N. *Hethal* ; 1346 F.A. *Etal* ; 1371 Sc. *Ethale* ; 1428 F.A. *Etall* ; 1542 Bord. Surv. *Etayle* ; 1655 Norham *Eatle*.

" Haugh of Eata." *Eata* is a common O. Nthb. name.

Etherley (Auckland). 1437-45 *Ederley*.

" Clearing of *Ēadhere, Ēadred* (cf. Adderstone *supra*) or *Aeþelred* (cf. Edderacres *supra*). Phonology, § 29.

Euden Beck (Bedburn). 1311 R.P.D. *Udeneburn, Yweden* ; 1382 Hatf. *Eudenleys* ; 1441 Finch. *Euedenburn*.

" Yew-valley stream." Cf. *on iwdene*, B.C.S. 927,

and Yeadon, Yorks. (nr. Pateley Bridge), Fountains Chart.
Iwdene. Introd. § 4.

Evenwood (Auckland). 1104-8 S.D. *Efenwuda.*
" Even or level wood."

Ewart (Doddington). 1218 Pipe *Ewurthe*; 1255 Ass.
Ewrth'; 1288 Ipm. *id.,* 1439 *Ewarth*; 1579 Bord. *Eward*;
1589 Wills *Ewertt.*

O.E. *ēa-weorþ*=river-enclosure, the place being encircled
on three sides by the rivers Glen and Till. Phonology, § 43.

Ewart's Hill (Fallodon). 1202 N. ii. 115 *Heworth.*
The same as Heworth *infra.* Phonology, §§ 35, 43. The
modern name has been given a pseudo-possessive form.

Ewesley Burn (Netherwitton) [uˑzli]. 1286 Coram.
Oseley; 1292 Ass. *Oseleyburne*; *n.d.* Newm. *Oselei*, 1547
Vselee; 1701 Hartburn *Yously.*
" Burn by Osa's clearing." *Ōsa* being a short form for
one of the numerous O.E. names in *Ōs-.* Phonology, § 18.

Fairhaugh (Kidland). *a.* 1245 Newm. *Fairhaluh.*

Fairley (Bywell St Peter) [fɛˑrəl]. 1268 Ipm. *Fayrhill*;
1278 Ass. *Fariley*; 1322 N. vi. 197 *Fairhill*; 1385 Pat.
Fayrhils; 1609 N. vi. 198 *Farle,* 1805 ib. *Fairle-hill.*
" Fair-haugh and -hill." In the latter the suffix *-ley*
has replaced *hill* or perhaps rather was added to " Faril,"
the short colloquial form still preserved locally.

Fairnley (Hartburn) [faˑnli]. 1271 Ch. *Farniley*; 1268
Ass. *Farnnilawe*; 1284 Swinb. *Farnylaw*; 1296 Newm.
Farniley; 1436 Ipm. *Farnelawe*; 1671 Arch. 2. 1. 129
Fairnelaw.
O.E. *fearnig*=ferny+*ley* or *law.* Cf. B.C.S. 120 *on
þa fearnige leage.* Phonology, § 8.

Fallodon (Embleton). *c.* 1180 F.P.D. *Faleuedun*; 1233
Pipe *Falewedon*; 1255 Ass. *Fauledon*; 1314 Ipm. *Faleghdon*
alias *Fauledon*; 1323 Ipm. *Faludoun*; 1346 F.A. *Falwedon*;
1663 Arch. 2. 17. 277 *Fallowdoune.* **Fallowfield** (St John
Lee). 1296 S.R. *Faloufeld*; 1350 Pat. *Falughfeld*; 1538
Must. *Fellawfeld*; 1663 Rental *Fallowfield.* **Fallowlees**
(Rothbury). 1388 Ipm. *Falalee,* 1436 *Falowleys*; 1663
Rental *Fallowlees.*
The first element in all these names is probably O.E.

*fealh (oblique case form *fealg-), " ploughed land," later
"ploughed land left uncropped for a whole year or more."
The nom. would give such forms as falugh, falegh (supra),
faugh (Nthb.) and fauch (Scots), pronounced [faf], while
the oblique case forms would give falwe-, falou-, etc. There
is also an O.E. adj. fealo, with alternative form *fealh
(N.E.D. s.v. fauch a²) meaning "pale brownish or reddish
yellow," which early became confused with the word first
discussed, which was primarily a noun, and this adj. may,
at least in part, lie behind the Nthb. names in Fallo(w)-
and be applied to the colour of the soil. The second
elements are obvious.

Fallowlees Burn (Rothbury). a. 1265 Percy Fawley-
burne, Newm. Fauleyburn.
" Burn by the faw-ley," i.e. the clearing of varied
colour. Cf. Fawdon infra. Later the name was changed
under the influence of the neighbouring Fallowlees Farm.[1]

Falstone (N. Tyndale). 1255 Ass. Faleston; 1371 Sc.
Faustan; 1610 Speed Fauston; 1663 Rental Fawstons.
" Fallow-stone," i.e. of dull-coloured yellow or yellowish.

Farglow (Thirlwall). 1279 Iter. Ferglew.
Probably the picturesque perversion of some old Celtic
name. There is no authority for the use of names of this
descriptive type at an early date, though the M.E. form
would admit of such an interpretation.

Farnacres (Whickham). 1278 Ass. Fornacres; 1311
R.P.D. Farnacres, 1312 Fornacres; 1348 F.P.D. Fernacres;
B.B. Farnacres; 1507 D.S.T. Farnacres.
" Fern-fields." [faˑn] is Nthb. and Durh. for fern. The
o of the first and third forms might point to O.W.Sc. Forni
(Björkman, Z.E.N. p. 34), but it would be difficult to ex-
plain the later forms from this except by confusion with
the similar and more common Fern-, Farn-.

Farne Island [fɛˑrən], [fəˑn]. c. 750 Bede Farne; 1257
Pat. Farnealond.
Maclure (p. 170 n. 1) says that the name is probably the
Celtic ferann (ancient stem verann, according to Dr Stokes),

---

[1] Fawley, Berks., goes back to earlier Faleley, Faleleg, id. Hants. to
Falele, but such forms would hardly have given Fawley as early as 1265.

Mod. Irish *fearran*=land, sometimes losing the initial *f* as the Welsh loses the equivalent *gu* and becoming *Arran*. The dissyllabic form is shown by the early forms under Lindisfarne *infra*, and by the local pronunciation. *ealond* looks like an archaic survival of O.E. *ealand*, an alternative form for the more usual compound *ieg-land*. The former would, in early M.E., give *ea-*, *æ-*, or *e-lond*, the latter *eȝ-* or *ey-lond* and possibly *elond*. The words were ultimately completely blended and confused.

*Farnycleugh (Redesdale). *c.* 1250 T.N. *Farinclou* ; 1398 Ipm. *Farneclogh* ; 1586 Raine *Farnycleugh.*
" Ferny clough " (*clōh*, Part II). Phonology, § 8.

Farnham (Alwinton). *c.* 1250 T.N. *Thirnu'* ; 1307 Ipm. *Thirnum* ; 1313 Perc. *Thirhum* ; 1324 Ipm. *Thirnom* ; 1343 Perc. *Thernhamme* ; 1346 F.A. *Thirn(a)ham* ; 1421 Ipm. *Thernhome* ; 1542 Bord. Surv. *Tharnam* ; 1628 Freeh. *Farneham*, 1638 *id.* ; 1649 Comps. *Thernham.*
O.E. *þyrne-hām*=homestead by the thorn bushes or, as suggested by spellings in *-hamme*, *-home*, *-om*, *þyrne-hamm*, i.e. ham or bend of a river marked by a thorn bush. Farnham stands on a sharp curve of the Coquet. The change from initial *th* to *f* is quite common in English place-names : Fingest, Bucks., Furzeleigh, Dev., Farmington, Glouc., Finglesham, Kent, Finedon, North., Fishley, Staffs., all once had initial *th*, and isolated spellings with *f* are found in the case of Thowthorp and Threshfield, Yorks. (D.B. *Fornetorp, Freschefeld*). This change appears sporadically in English dialect, and is of course a common feature of child-speech.

Farnley (Auckland). 1313 R.P.D. *le Farmley* ; 1399 Ipm. *Farmley.*
Probably " farm-clearing," *farm* being descriptive of land held at a fixed rent, rather than farm in the modern sense. The present form is corrupt, cf. Fairnley *supra*.

Farrington alias **Farnton** (Silksworth). 1432.33 *Pharyngton*, 1437 *Feryngdon* ; 1479 B.B.H. *Farendon* ; 1479.35 *Farnton.*
" Farm or hill of Fær or his sons." Cf. Fringford, Oxon (p. 110), earlier *Feringeford*, Faringdon, Berks.,

F

Farringdon, Dors., Hants, Berks., and Farrington, Lancs. and Som. Alexander (p. 110) takes all these to contain a patronymic from *Fær*, a short form of such a name as *Færþegn*, *Færeman*. App. A, § 1 ; Phonology, § 59.

Farrow Shields (Haltwhistle). 1279 Iter. *Ferewithscheles*, *Frewythescheles* ; 1636 Comm. *Farrowsheile*. Possibly " *sheles* of Freyviðr." Lind gives two examples of O.N. *Freyviðr* (O.Sw. *Frøvidh*). This would give M.E. *Frewith* and *Ferwith*. Phonology, § 54.

Fatherless House (Boldon). 1351.31 *Fadreleshous*. Similarly there is a *Faderlesfeld* in Boldon in Hatf. Survey, but why so called it is impossible to say.

Fawdon (Gosforth). 1309 Ipm. *Faughdon* ; 1346 F.A. *Faudon*. (Ingram) 1207 Sc. *Faudon*.

O.E. *fāh-dūn* (North. M.E. *faughdon*)=variegated hill. Cf. Fawside in Scotland, earlier *Fausydde* (Johnston, p. 126).

Fawnlees (Wolsingham). 1359.45 *Fawleys*, 1366.32 *Faulees* ; 1382 Hatf. *Fowleys*.

Faw is probably the same as in Fawdon *supra* and descriptive of the colour of the clearing. *Fow* is a S. form. *Fawn-* is a corruption.

Fawns (Kirk Whelpington). 1302 Ipm. *Faunes*, 1421 *Fawnes*.

Possibly the same as Scots *fawns*, which Jamieson says is used of white spots of moorish ground in Ettrick Forest. The word can hardly be English.

Fawside (Lanchester). 1384.31 *Fauside*, 1349 *Faweside v.* Fawdon *supra*.

Featherstone (Haltwhistle). *c.* 1215 B.B.H. *Fetherstanhalcht*, *Fetherstanehalcht* ; 1222 Sc. *Ferstonehalc* ; 1255 Ass. *Fetherstonelawe*, *Fetherstan* ; 1278 Ass. *Ferstanhallu'* ; 1296 S.R. *Feyrstanhalth* ; 1346 F.A. *Fetherstanehalgh*, 1428 *Fethirstanehaugh*.

Cf. Featherstone, Staffs., 994 *Feotherstan*, D.B. *Ferdestan*, 1271 *Fethereston* (Duignan, p. 60) and Yorks. D.B. *Fredestane*, *Ferestane*, 1122 *Fechrestana*, 1166 *Fetherstan*. There are also Featherston and Featherstall in Lancs. For the former, Wyld (p. 125) gives forms *Fayrstan* (1277) and *ffetherstan* (14th c.), saying that they are clearly unconnected, but the forms of the Nthb. name show that this is not necessarily the case. No forms have been found for Featherstall, though Sephton (p. 172) believes it to be the *Fayrstan* just noted. Duignan takes the first element to be the name *Feader*, found as the name of one of Harthacnut's huscarls (O.Sw. *Fadhir*, O.Dan. *Fathir*, O.N. *Faðir*) and in D.B. as *Fader*. Moorman accepts this for the Yorks. name, and Wyld inclines to it for the Lancs. one. There are, however, two difficulties, (1) the entire absence of any M.E. form in *a* such as one would expect if the name were *Fathir*, even admitting that forms in *e* might, in part at least, be due to M.E. forms such as *feder* for the common *fader* = father ; (2) the impossible coincidence that this personal name, never found elsewhere in English placenames, should three times be found in association with O.E. *stān* = stone or rock.

No compound of *feather* and *stone* is known, though such might conceivably exist, meaning either " moved as easily as a feather " or " marked with feather-shaped forms."

There is an O.E. name *Frið(u)stan* which might become *Fredestan* and *Ferdestan*, but a further metathesis to *Fedrestan* seems unlikely. If it were possible, we might suppose that all these names consisted once of this personal name followed by some suffix but that this was lost later, when the meaning of the first element was forgotten and a name ending in -*stone* seemed satisfactory enough as a place-name. This process has certainly taken place in Featherstone(haugh), Nthb. Phonology, § 44.

*Feathery Haugh (N. Tyndale). 1200 R.C. *Federhaly* ; 1546 N. vii. 470 *Federyhaugh*.

Perhaps so called from the appearance of the trees there. Cf. *Fethreschawe* in Carraw (B.B.H. 1429).

Felkington (Duddo). 1237 Pat. *Felkindon*, 1238 *Felkendon*; *c.* 1250 T.N. *Felkindon*; 1441 Ipm. *Felkyngton*.

" Farm of Feoleca or Filica." Cf. *filican slæd*, B.C.S. 1093. App. A, § 1.

Felling (Jarrow). 1325 F.P.D. *Felling*.

Locally known as " The Felling." Cf. *Fellingen* (N.G.

iii. 272) in Norway, i.e. the felling, or clearing where wood has been felled. Cf. N.E.D. *s.v.*

Mak ȝe in þe plain na duelling.

Til ȝe bi comen to ȝone felling (*Cursor Mundi*).

Felton. 1166 Pipe *Feltona*; 1215 Chron. de Mailros *Feltunia*.

Felton, Heref., West Felton and Felton Butler, Salop, all show the same early forms, viz., D.B. *Feltone* and Felton, Som., is *Feltone* (F.A. 1284). All alike probably go back to *feld-tūn*, i.e. field-farm, " field " being used in its primitive sense (*v.* Part II). This is the sense it must have in *feldbeorg*, B.C.S. 594 and *felddene* ib. 398. Phonology, § 53.

Felton Hill (Carrycoats). 1244 Ipm. *Fyleton*; 1296 S.R. *Filton*; 1303 Sc. *Filton*; 1542 Bord. Surv. *Fylton*.

No O.E. name *Fila* is known. It is just possible that the first element is O.E. *fileþe*, " hay " (Middendorf, *s.v.*). Phonology, § 10.

Fencewood (Mitford). 1253 Ch. *Fencewood*; 1322 Cl. Wood of *le Fense*.

" Enclosed wood."

Fenham (Holy Island). 1125 F.P.D. *Fennum, Fænnum*, 1203 *Fennum*; 1335 Ch. *id.* (Newcastle-on-Tyne) 1375 Cl. *Fenham*.

O.E. (*æt þæm*) *fennum*=(at the) marshes or *fenn-hām*= fen-homestead. App. A, § 6.

Fenrother (Hebron). 1189 Pipe *Finrode*; 1232 Pat. *Finrothre*; 1255 Ass. *Finrother*; *c.* 1250 T.N. *Finrother*; 1257 Ipm. *Fynrother*; 1296 S.R., 1340 Ch., 1428 F.A. *id.*

The second element is perhaps a variant, with unmutated vowel, of the O.E. (*ge*)*ryðer*, "clearing," found in Ryther, Yorks. (Moorman, p. 161). Cf. M.E. *rode*, " to clear from weeds " (N.E.D.), and *rid*, " to clear ground " (O.N. *ryðja*). The first may be the name *Finn*, probably of Scand. origin (Björkman, N.P. p. 40), or it may be the word *fin* discussed under Findon *infra*. No certainty is possible. Phonology, § 10.

Fenton (Wooler). 1291 Tax. *Fenton*. **Fenwick** (Kyloe). 1312 R.P.D. *Fennewik*; 1579 Bord. *Fenneck*. (Stamfordham) 1346 F.A. *Fennewyk*.

" Fen-farm and dwelling." Cf. *fentun,* B.C.S. 1112. Phonology, § 49.

Ferryfield (Stanhope). 1382 Hatf. *Feryfeld.*
" Field by the ferry across the Wear."

Ferryhill (Merrington). 10th c. B.C.S. 1256 *(æt) Feregenne;* L.V.D. *Feregenne*[1]; *c.* 1125 F.P.D. *Ferie;* 1316 Pat. *Ferye on the Hill;* 1646 Map *Ferye on ye mount.*

This would seem to be the somewhat rare O.E. *firgen, fergen,* " wooded hill or mountain." Later a descriptive phrase was added, now shortened to *-hill.*

Fielden Bridge (Auckland). 1303 R.P.D. *Feldyngford;* 1382 Hatf. *Fyldynggate.*

Possibly " Fielding's ford," whatever the origin of that name may be (Weekley, p. 65).

Filbert Haugh (Alnwick). *c.* 1280 Perc. *Hilburhalgh.*

" Hildeburh's haugh." The sound-development is remarkable, but two other similar examples have been noted, viz., Hawkenbury in Headcorn, Kent, B.C.S. 343 *Focgingabyra,* and Falsgrave in Scarborough, Yorks., D.B. *Walescrif,* Ripon Cart. *c.* 1200 *Hwallisgrava.* In the second of these the initial sound was *hw* and development to *f* is not unlikely. Possibly the modern form is merely a corruption.

Finchale [fiŋkəl]. *c.* 1100 Finch. *Finchale, c.* 1190 *Finkale; c.* 1220 D.S.T. *Finkehale;* 1344 R.P.D. *Fynkhal;* 1464 F.P.D. *Fynchall;* 1764 Esh *Fenkle.*

" Finch-haugh " (because frequented by finches) seems the obvious etymology, with North. *fink* for *finch.* Finchale has however been, somewhat doubtfully, identified with the place mentioned in A.S.C. (*s.v.* 788) as *Pincanheale* (D.), *Wincanheale* (E.), where the two forms are due to the common confusion of *þ* and *w* in O.E. script. It is just possible that the identification might be supported on phonological grounds if *þ* is the correct initial consonant. The *finch* is in dialect sometimes known as the *pink,* both names being probably of echoic origin. If the place were originally *Pincanheale,* i.e. Pinca's haugh (cf. Pinkhill, Ox., and *pincan-ham,* B.C.S. 665) it is possible that popular usage

---

[1] This, and not *Foregenne,* is the correct reading according to Björkman's correction of Stevenson's transcript (*Englische Studien,* 1918, p. 245).

associating the place-name with the bird-name might sometimes replace *Pink-* by *Fink-*, a form which ultimately prevailed.

The 1764 form *Fenkle* with *i* lowered to *e* is identical with that of the Nthb. and N. Yorks. *fenkle*, "bend, corner, elbow." Heslop (*s.v.*) suggests that Finchale was named from the "fenkle" in the Wear at this point. The early forms show that if there is any connexion the history must be the other way round, viz., that a sharp bend came to be called a *finkle* or *fenkle* from its resemblance to the well-known bend at 'Finkle' Priory.

Findon Hill (Kimblesworth). 1315 R.P.D. *Fyndon*. Cf. O.E. *finleage*, B.C.S. 627, and *finbeorh*, ib. 992. Middendorf (p. 51) takes these to contain Mod. Eng. dial. *fin*, i.e. fin-weed or rest-harrow, but there is no evidence for this word in O.E. Skeat, in dealing with Finborough, Suff., D.B. *Fineberga*, takes the first element to be O.E. *fīn*, "heap," and explains the name as "heap-barrow," i.e. one artificially constructed. This might also be the interpretation of Findon, Nthb., and Suss., earlier *Findune*, *Fyndon*, though Roberts (p. 67) prefers "hill of Finn." The absence of genitival *e* or *es* from D.B. onwards makes this last very doubtful.[1]

*****Fiselby** (Hartington). 1319 Pat. *Fiselby*; 1378 Ipm. *Fisilby*, 1390 *Fisildene*, 1396 *Fesilby*, 1418 *id.*; 1580 F.F. *Feselby*.

The second element can hardly be the Scand. *-by*, otherwise unknown in Nthb. It is just possible it is O.E. *byge*, "bend, curve," cf. *æscwaldes byge*, B.C.S. 624, though we should then expect M.E. *bye* rather than *by*. The first element may be an English equivalent of Ger. **Fisel* (from *Fiso*), which Förstemann *s.n.* assumes for Veilsdorf, earlier *Fiselestorp*.

---

[1] It is to be regretted that we cannot accept the picturesque explanation of the name given by Prior Fossour. Writing immediately after the Battle of Neville's Cross, which ended here, he says that it was prophetically so called, for "posse dicatur verisimiliter Fyndonne (i.e. presumably Fr. *fin donné)* quasi finem dans vel finem dandus," for the battle, so the Prior thought, would put an "end" to the wars of English and Scots (D.S.T., p. ccccxxxiv.).

Fishburn (Sedgefield). *c.* 1190 Godr. *Fisseburne.*
"Fish-stream" or, possibly, "Fish's." For the former
cf. Fishbourne, Suss., and Fishlake, Yorks. (Goodall, p. 139).
For the latter cf. Fishwick, Lancs., Fishley, Norf. O.E.
fisces-burna, B.C.S. 624, 802 is ambiguous.

Fitches (Witton-le-Wear). 1382 Hatf. *Fychewacke (sic)*;
1392.35 *Fyccheworth.*
A difficult name. Possibly, M.E. *fiche-worth* = vetch-
enclosure. The modern form would then be a shortening
due to the analogy of names like Bells *supra,* where the first
element of a name is used by itself in the possessive
case.

Flass (Lanchester). 1313 R.P.D., 1342.31 *the Flaskes*;
1382 Hatf. *Le Flassh*; 1597 Lanch. *Fflasse.*
"The pools or marshy places." Canon Greenwell
(Hatf. Surv.) says that it takes its name from its low situa-
tion near Deerness Brook. For forms *v.* N.E.D. *flass* is
still used in Nthb. (Heslop *s.v.*) and cf. Flass St. in Durham.

Flatworth (Tynemouth). 1271 Ch. *Flaforda*; 1292
Ty. *Flatford*; 1428 F.A. *Flateford*; 1638 Freeh. *Flatworth.*
"Flat-ford," referring to the shallows on the Dortwick
sands (N. viii. 334). App. A, § 4.

Fleetham (Bamburgh). *c.* 1180 F.P.D. *Fletham*; 1663
Rental *Fleetham.*
O.E. *flēot-hām* = homestead by the fleet or estuary.

Flemingfield (Easington). 1382 Hatf. *Flemyngfeld.*
So called because granted to John le Fleming (Boyle).
For Flemings and Flemish names in England *v.* Forssner,
pp. xxxviii.-xlii.

Flotterton (Rothbury). 1256 Brkb. *Flotewayton*; 1272
Newm. *Flotwaiton*; 1288 Ipm. *Flottewayton*; *c.* 1250
T.N. *Flotwayton*; 1304 Ch. *Flotteweyton*; 1331 Inq. a.q.d.
Flote Watton; 1346 F.A. *Fletwayton, Flotwayton*; *n.d.*
Newm. *Flotwarton*; 1538 Must. *Flotterton.*
O.E. *flote(n)-weg-tūn* = flooded-road-farm, *floten* being
pp. of O.E. *flēotan* (N.E.D. *s.v. flotten*). Flotterton may have
been so called because liable to inundation when Coquet
was in flood. The form has perhaps been influenced by
the neighbouring *Warton.* Cf. Hartington *infra.*

Follingsby (Jarrow). *c.* 1140 F.P.D. *Folete(s)bi, c.*
1180 *Foleteby, Folesceby, c.* 1220 *Folasceby* ; 1335 Ch.
Folethebi ; 1343 J. and W. *Folesceby* ; 1400.45 *Folanceby* ;
1446 D.S.T. *Folauncebey* ; 1539 F.P.D. *Folansbye, Folaunceby* ;
1580 Halm. *Follensbye.*

Cf. Fulletby, Lincs., D.B. *Folesbi, Fullobi,* Lincs. Surv.
Fuledebi, Fuletebi. The first element is a name of the same
type as O.N. *Haf-, Sumar-, Vetr-liði*=sea-, summer- and
winter-traveller. No name *Full-liði* is recorded, but there
may have been such a ŋame from the adj. *full liða,* " well-
provided with troops," " fully able " (Vigfusson and
Fritzner). Cf. Selaby *infra. Foletes* and *Folesce* are
anglicised genitives of this name. For *n, v.* Phonology,
§ 55. Later a pseudo-patronymic form was developed.

Font, R. 1252 Ch. *Funt* ; 1261 Coram. *Font.*

O.E. *font, funta*=fountain, well. Cf. Fovant and Urch-
font, Wilts., Havant, Suss., Mottisfont, Hants, Bedfont,
Midd., and *ceadelesfunta,* B.C.S. 883.

Ford (Nthb.). 1225 Pat. *Forda* ; 1507 D.S.T. *Furde.*
(Bp. Wearmouth) 1361.45 *Forth* ; 1643 Bp. Wearm. *The
foord.* (Lanchester) 1382 Hatf. *Le Forth.*

Self-explanatory. Phonology, § 30.

*Forston. *c.* 1250 T.N. *Forestan* ; 1610, 1645, 1650
Maps *Forston.*

If this identification is correct, *Forestan* was near
Walltown, and the first element may be *forest,* referring
to the Forest of Lowes (*v. infra*). The second might be
either *stān* = stone and the whole name refer to a boundary
stone of the privileged area, or *ton* and the name mean "*forest-
farm.*" App. A, § 7.

Fortherley (Bywell St Peter). 1255 Ass. *Falderleg',
Fauderleg* ; 1346 F.A. *Falderley* ; 1538 Must. *Fawdle* ;
1663 Rental *Fauderlees.*

"Sheep-folder's clearing," *falder* (cf. *Faulder* as a name)
is North. for *folder* (Bardsley). Phonology, § 30.

Foulbridge House (Tanfield). 1403 Acct. *Foulebrigg.*

Self-explanatory. The long vowel shows the late origin
of the name. Cf. names in *Ful-, infra.* Phonology, § 27.

Fourstones (Warden). 1271 Ch. *Forstanes* ; 1278 Ass.

Fourstanes; *c.* 1250 T.N. *Fourstayns*; 1346 F.A. *Foure-stanes*; *c.* 1536 B.B.H. *Fourstones.*
Named, according to Tomlinson (p. 150), from four stones which marked its boundaries. Cf., in a Saxon list of boundaries in B.C.S. 1238, "from the stone to the second stone, and so to the third stone and so to the fourth stone." Phonology, § 14.

Fowberry (Chatton). 1288 Ipm. *Follebiri*; *c.* 1250 T.N. *Folebir*; 1346 F.A. *Folb(u)ry*; 1349 Ipm. *Follebery*; 1428 F.A. *Folbury*; 1538 Must. *Foulbery*; 1542 Bord. Surv. *Fowberye*. (Bamburgh) 1250 Pipe *Fulebrigg*; 1333 N. i. 89 *Fulbrigg.*
The first is O.E. *folan byrig*=foals' *burh* (Part II), i.e. where foals are bred. Similarly Foulbridge, Lancs., earlier *Folric(h) Folrig(ge)*. Wyld (p. 128) rightly rejects all connexion with *foul* and *bridge*. May it not be *foal-ridge*, i.e. hill where the foals are turned out with the mares? The second seems to be "foul-bridge," with later corruption of suffix. App. A, § 12.

Foxton (Alwinton). 1324 Ipm. *Foxden*; 1538 Must. *Foxton*; 1663 Rental *Fowston*. (Sedgefield) *c.* 1170 Reg. Dun. *Foxedene.*
"Fox-valley." App. A, § 1. For *Fowston* cf. Fewston, Yorks., earlier *Fosceton*, still called [faustən]. (Moorman, p. 72.)

Framlington, Long (Felton) [framptən]. 1166 R.B.E. *Franglingtone (sic)*, 1170 *Framelinton*; 1346 F.A. *Framlyngton, Framplington.*
"Farm of Framel or his sons." Cf. Framlingham, Suff., D.B. *Frameling(a)ham* and such names as O.N. *Framarr*, Visigothic *Framirus, Framuldus*, O.H.G. *Framarius* given by Naumann (p. 34). Searle's *Framric* and *Frambeald* are probably continental. There is a rare O.E. *Fram, Froma, Frome*, noted by Redin, pp. 13, 48, 122 (cf. O.W.Sc. *Frami*), which may be a shortened form of such names, or have arisen independently from O.E. *from*=active. From this could be formed dimin. **Framel* (cf. Visigothic *Framila*) and patronymic **Frameling* (cf. Förstemann's *Vramelinsperge* in Lower Franconia). Phonology, § 55, 53, 59.

Frankland (Durham). 1441 Finch. *Frankleyn* ; 1455.34 *Frankleyn Park.*

Perhaps the park was so called from its tenure, and when " park " was dropped the suffix was altered.

Friar's Goose (Gateshead). 1382 Hatf. *le Frergos.*

Perhaps so called because " friar's goose " . (Lat. *eryngium campestre*) flourished here. Cf. Broom and Bushblades *supra* and Bedwyn, Wilts., from *bedwine* or *bedwind* (Ekblom, p. 23).

Friarside (Whickham). 1312 R.P.D. *Frerejohanside* ; 1369.35 *Frerejonside* ; 1382 Hatf. *Freresyde* ; 1768 Map *Fryerside.*

" Friar (John's) hill."

Frosterley (Stanhope). 1239 Cl. *Forsterlegh* ; 1296 Halm. *Frosterley.*

"Forester's clearing." *For(e)ster > Froster.* Cf. Fortherley *supra.* Phonology, § 54.

Fugar House (Whickham). 1297 Pap. *the land of Furgers* ; 1351.35 *Feugerhouses* ; 1382 Hatf. *Fugerhous* ; 1440 Cl. *Foycherhous.*

Granted in 1269 to Wm. de Feugers (S. 2. 245) who belonged to a Breton family, from Fougères (Ille-et-Vilaine dept.), earlier *Feugeriis, Fugires* (Cal. Doc. relating to France).

Fulford (Witton Gilbert). B.B. *Fulford* ; 1382 Hatf. *Fulforth.* **Fulthorpe** (Grindon). 1311 R.P.D. *Fulthorp*, 1313 *Foulthorp.* **Fulwell** (Monkwearmouth). *c.* 1200 F.P.D. *Fulewell.* *(Stamfordham) 1296 S.R. *Fulwell.*

" Foul or dirty ford, thorpe or village, and spring." Cf. *fulanford*, B.C.S. 208, *fulan broces*, ib. 742. Phonology, §§ 21, 30.

Gainford-on-Tees. *c.* 1050 H.S.C. *Geg(e)nford, Geagenforda* ; 1207 F.P.D. *Gainesford* ; *c.* 1200 B.M. *Geynef(f)ord* ; 1307 Ch. *Gaynefford* ; 1311 R.P.D. *Gayne(s)ford*, 1313 1314, 1315, 1344 *id.* ; 1316 R.P.D. *Gayneforth* ; 1400 D.S.T. *Gaynforth*, 1507 *Gaynfurth* ; 1739 Coniscl. *Gainsford* ; Gainf. *Gainforth (passim).*

Possibly *gegn-ford*, " direct or straight ford," with later pseudo-genitival *s*, but such a use of *gegn*, while common in

O.N., is rare in O.E. More probably we have a personal name as in Gainsborough, Lincs., A.S.C. *Gæignesburch*, *Gegnesburh*, Ganstead, Yorks., D.B. *Gagenestad*, and *Geynesthorne*, B.C.S. 1313. This name is probably Scandinavian, cf. O.W.Sc. *Gagni* in Gangstad, earlier *Gaghnastadir* (Lind *s.n.*), *Gagnstorp* and *Gagnesjön* (Falkmann, p. 218), and the name *Gegnir* once common in Iceland (Lind). Phonology, § 30.

Gallow Hill (Corbridge). *c.* 1290 Perc. *Galueside.* "Gallows side or hill," a fairly common name.

Gamelspath (Coquet Head). 1380 Ipm. *Kenylpethfeld*, 1411 *Kemylespathe*; 1456 Raine *Kemblepeth*; 1473 Ipm. *Gammyllespeth*; 1542 Bord. Surv. *Kemlespeth*; *c.* 1580 Map *Kemblespeth*, 1724 *Gemblespeth.*
The name of the old Roman road to Ad Fines camp (N. x. 461). The first element may be the M.E. name *Gamel* (Mod. Eng. *Gamble*) from O.W.Sc. *gamall*, "old." Cf. Björkman N.P. p. 45, Z.E.N. p. 35. For initial *k* cf. *K(A)M(A)L* for *GAMAL* in a Runic inscription in Furness (Collingwood in *Saga-Book of the Viking Club*, vol. iii. p. 139).
Gamel certainly did not build the path, and is probably not the name of its sometime owner. Why then so called? Ancient roads and earthworks are often thought by primitive people to be of demonic origin (cf. Devil's Dyke and Causeway), and the name *Gamel* may, by some Scandinavian settler, have been applied colloquially to the Devil in the same way that we speak of "the old one." If so, *Gamelspath* would mean "Devil's road."

Garden House (Bellingham). 1279 Iter. *Gardino.*
O.North. Fr. *gardin* = garden.

Garmondsway (Bp. Middleham). 1104-8 S.D. *via Garmundi*; 1230 Pipe *Garmundeswaye*; B.B. *Germundesweya* (B., C. *Garmondeswaye*).
"Garmund's road." *Gārmund* is a rare O.E. name, and here it may be an anglicising of the more common O.Dan. *Germund*. Cf. Björkman, Z.E.N. p. 36. The road is the ancient road along which King Cnut went barefooted to the shrine of St Cuthbert (*Hist. Dunelm. Eccl. c.* 8).

Garretlee (Longhorsley). 1296 S.R. *Gerardesley*; 1443 Ipm. *Garartlee*; 1637 Camd. *Garretlee*. **Garret Shiels** (Elsdon). 1290 Abbr. *Gerardscheles*; 1378 Ipm. *Garareschell*; 1590 Bord. *Garrett Sheiles*.

"Gerard's clearing and shiels." Searle gives *Gerhard* (D.B.) and *Gerardus* (a 7th-cent. Bp. of London). These all go back probably to O.G. *Gerard* (Forssner, p. 65). For *Garrett* cf. Crosby Garrett, Westm., earlier Crosby Gerard and Garret Hostel (= Gerard's Hostel), Cambridge. Phonology, §§ 8, 57.

Gateshead-on-Tyne. *c.* 750 Bede *ad caput caprae*; *c.* 1000 O.E. Bede *æt Ræge heafde*; 1104-8 S.D. *ad caput caprae*; *c.* 1190 B.B. *Gatesheued*; 1228 F.P.D. *id.*; B.B. *id.*; 1378 J. and W. *Gaytesheued*; 1507 D.S.T. *Gateshevid*; 1610 Allen *Gateside*; 1637 Camd. *Gatesende*.

Probably a name in which an original Celtic name has been transformed by folk-etymology. Bede's "at the she-goat's head" looks like an attempt to give some intelligible interpretation of a Celtic name. Gateshead has by some been identified with *Gabrosenti* in the *Notitia*, the *Gabrocentes* of the Ravenna Geographer. If this is correct, we can see how Bede's form might have been suggested by the initial *Gabro-*, the British cognate of Lat. *capro-*. Whether this identification is true or not, popular opinion laid hold of the interpretation of the name found in Bede, and its English form **gāte-hēafod* survives in M.E. *Gatesheued* with the more usual gen. in *es*. If folk-etymology has been at work, we need not trouble to give it an intelligible meaning as applied to the site of Gateshead. If we have no connexion with an earlier Celtic name to explain, this place-name may be an example of the type discussed by Bradley (*Essays and Studies* u.s. vol. i. p. 31) in which places are named after animals' heads. Bradley suggests that these names point to a custom of setting up the head of an animal, or a representation of it, on a pole, to mark the meeting-place of the hundred.

The form in O.E. Bede is a translation of Bede's Latin made by someone with no knowledge of the English name which was already developing. He translated Bede back into

O.E. and used O.E. *ræge*, " wild she-goat," instead of *gāte*. The *ad* (or *æt*) is a relic of the idiom whereby a place was not called " X," but " at X." Cf. A.S.C. *s.a.* 552 "the place which is called æt Searobyrg (i.e. *at Salisbury*)," and the form quoted in Note 1 on Alnmouth *supra*. App. A, §§ 7, 12. Later corruptions are due to association with North. *gate* = road.

Gatherick (Lowick). 1281 Pat *Gateriswyk*; 1287 Ass. *id.*; 1538 Must. *Gaderyk*; 1539 F.P.D. *Gaderwike*; 1560 Raine *Gathericke*.

" Dwelling by the *gaiter* or wild dogwood-tree." Its M.E. forms are *gaitrys, gattris, gaytre*, and in the 16th c. *gadrise*. Phonology, § 29.

Gaunless, R. *c.* 1170 F.P.D. *Gauhenles*; *c.* 1230 *id.*; 1242 D.Ass. *Gawenles*; 1291 R.P.D. *Gaunles*, 1312 *Gaounles*.

A pre-English river-name.

Gellesfield Hole (Whickham). 1444.34 *Gellesfeld*.

" Field of Gell." *Gell* is O.W.Sc. *Gellir*, originally a nickname meaning " loud-voiced " (Lind. *s.n.*). Cf. *Gell-tofta* in Skane (Falkman, p. 127).

Gibside (Whickham). 1339 Boyle *Gippeset*; 1375.45 *Gibset*, 1396 *Gibsete*.

If *Gippe*- is the original form cf. Gibsmere, Notts., D.B. *Gipesmare*, Gipton, Yorks., earlier *Gipetuna*, Gipping, Suff., all of which contain some personal name *Gippe* otherwise unknown. More probably the original form was *Gibb(e)*, the common pet-form of Gilbert. Hence " Gib's seat." *sǣte*, Part II. App. A, § 8.

Gilden Burn (Amble). *c.* 1200 N. v. 262 *Gildenes dene*.

" Gildwine's valley." Phonology, § 49.

Girsonsfield (Otterburn). 1331 Ipm. *Grenesonesfeld*; 1586 Bord. *Girsonsfeilde*, 1590 *Gressounfeild*; 1663 Rental *Grissonsfeld*.

" Greenson's field." Cf. *Greeneson Hesills* (Ipm. 1378) in the same district. Phonology, §§ 54, 53.

Glantlees (Felton). 1200 R.C. *Glanteleia*; *c.* 1250 B.M. *Glanteley* alias *Glenteley*; 1255 Ass. *Glanteley*; *c.* 1250 T.N. *Glenteley*; 1346 F.A. *Glantly*, 1428 *Glantlees*.

Glanton (Whittingham). 1210 R.B.E. *Glentedone*; 1219 Pipe *Glantendon*; 1278 Ass. *Glantedone, Glentendon*; 1311 Pat. *Glantesdon*; 1320 Ipm. *Glantoune*; 1346 F.A. *Glanton*; 1399 Ipm. *Glaunton*.

Cf. Glentham, Lincs., D.B. *Glandham, Glentham*, Lincs. Surv. *Gle(i)ntheim*, and Glentworth, ib. D.B. *Glenteuurde*, Lincs. Surv. *Glenteworda*. This *Glent-* or *Glant-* must be allied to Teut. **glint, *glant*, found in Sw. Dial. *glänta, glenta*, " to slip, slide, flash, gleam," in O.H.G. *glanz*, " bright, clear," and perhaps in O.N. *glettr, gletta*, " banter, railing " (N.E.D. *glent* vb.). In Danish a hawk is sometimes called *glente*, so also in Swedish it is known as *glänta*, probably from its swift gliding motion (Falk. and Torp, *s.v. glente*). No M.E. adj. or noun of this form is found, though the vb. *glent*, " to move quickly," is quite common. Probably in these place-names we have some personal name,[1] ultimately a nickname, derived from *glente* or *glänta*, a hawk. " Hawk " itself is a common Scand. name. Hence " Hawk's clearing or hill." Phonology, § 56. App. A, § 1.

Glen, R. *c.* 750 Bede *Gleni*; 1255 Ass. *Glene*.

A Celtic river-name: cf. O.Ir. *glenn*, " valley," and Glen, R. Lincs.

Glendale. 1179 Pipe *Grendal*, 1182 *Grendala*; 1558 V.N. *Glendell*.

" Glen valley." For *l—l > r—l v.* Zachrisson, p. 121.

Glendue (Hartleyburn). 1239 B.B.H. *Glendew*.

" Black glen." Cf. Glendoo, I. of Man and Ireland (Joyce ii. p. 483), Glendui (Milne, p. 178), Glen Dubh (Watson, Index). The glen is one of the narrowest and darkest in S. Tyndale and until recent years was thickly overgrown with trees.

Gloster Hill (Warkworth). *a.* 1178 Newm. *Gloucestre*; 1637 Camd. *Gloucester-hill*; 1691 Warkw. *Glowster-hill*.

A Romano-Celtic name, perhaps the same as the more famous Gloucester, A.S.C. *Gleaweceaster*, M.E. *Glowcester*.

Gofton (Simonburn). 1279 Iter. *Goffedene*; 1329 Pat. *Goseden* (*sic*), 1358 *Gofden*; 1663 Rental *Gofton*.

" Gof's valley." Cf. *gofesdene*, K.C.D. 641 and the

---

[1] They are so near that they are probably named from the same man.

name *Goffe* (R.H.) which Forssner (p. 119 n.) takes to be of continental origin. It is identical with the Frisian *Goff(e)* which Winkler (p. 131) takes to be short for *Goffert* < *Godferd*. Phonology, § 50 ; App. A, § 1.

Golden Pot (Redesdale). *c*. 1230 H. 2. 116 n. *Golding-pot*.

" Pot of Golda or his sons." Cf. Dixon, *Upper Coquet-dale* (p. 8). " Standing about a mile apart on the moors . . . are two freestone blocks . . . the Outer Golden Pot and the Middle Golden Pot. They were probably boundary or guide-stones, and earned their name because hollowed out at the top."

Goosecroft (Wolsingham). 1382 Hatf. *Gosecroft*.

O.E. *gōs-croft*, " goose croft " or *gōsa-croft*, " croft of the geese," *v. croft*, Part II.

Gorfen Letch (Fenrother). 1270 Perc. *Gorsfen*; *n.d.* Newm. *Gorfen*.

O.E. *gorst-fenn*=marsh land overgrown with furze or, possibly, *Gores-fen, Gor-* being short for some such name as *Gormund* or *Gornōþ*. Phonology, § 53.

Gosforth (Newcastle-on-Tyne). 1166 R.B.E. *Goseford*; 1278 Ass. *Goseforth*; 1378 Ipm. *id.*; 1448 Pat. *Gosseford*; 1663 Rental *Gosford*; 1699 Woodhorn *Gosworth*.

O.E. *gōs(a)-ford*=ford of the geese or goose-ford, i.e. where they are often seen. Cf. *doccena ford*, B.C.S. 888= ducks' ford, Enford, Wilts., B.C.S. 905 *enedford*, i.e. duck-ford, Gosforth, Cumb. and Gosford, Warw., Oxon., Som., Suff. The O.E. name *Gōsa* inferred from *gosanwel*, B.C.S. 754, is very doubtful. Forssner does not think it is English at all (p. 124), and in any case it is impossible to believe that seven fords should happen to be owned by a man with this very rare name.

Gosforth stands on the Goose Burn. Nearer its mouth this stream is called the *Ouseburn* (*v. infra*), and the whole stream must once have borne this name. The present name of the stream must be due to a process of back-formation. Phonology, §§ 21, 30. App. A, § 4.

Goswick (Holy Island). 1228 F.P.D. *Gosewic(h)(e)*; 1237 Cl. *Gosewic*; 1323 B.M. *Gossewyk*.

O.E. *gōs(a)-wīc* = goose-dwelling or "dwelling of the geese." Cf. *gatawic*, B.C.S. 834, *oxenawic*, ib. 904, *sceapwic*, ib. 620. Phonology, §§ 21, 49.

Greatham (nr. Hartlepool). 1228 F.P.D. *Gretham*; 1693 Bp. Wearm. *Greetham*; 1702 Sedgf. *id.*

O.E. *grēot-hām*=gravel-homestead. Cf. Girton, Cambs., F.A. *Grettone*, Gretton, Northts., D.B. *Gretone* and Griesheim, Hesse (Sturmfels, p. 30). Surtees describes it as "cheerfully situated on a rise of dry gravelly soil."

Greencroft (Lanchester). B.B. *Grencroft.* **Greenhaugh** (Tarset). 1325 Ipm. *le Grenehalgh.* **Greenhead** (Haltwhistle). 1289 Sc. *le Greneheued.* **Greenlee** (ib.). 1285 Swinb. *Greenleye.* **Greenley Lough** (ib.). 1285 Swinb. *Wigglesmere.* **Greenridge** (Hexham). 1304 Cl. *Grenerig.* **Greenwell**[1] (Wolsingham). 1304 Cl. *Grenwell.*

Self-explanatory. The old name of Greenley Lough is "Mere of *Wiggel*," that name being a dimin. of O.E. *Wigga*. Cf. Winkler (p. 439) who gives *Wig(ge)* and *Wiggele*. Greenhead is the high ground at the watershed between Irthing and Tipalt (Heslop, p. 365).

Greymare Hill (Shotley). 1307 N. vi. 90 *Graymere*; 1768 Shotley *Graymarehouse.* **Greystones** (Haughton-le-Skerne). 1313 R.P.D. *Graystanes.* **Greyside** (Neubrough). 1479 B.B.H. *le Graysyd.*

"Grey *mere* or boundary mark (O.E. *mǣre*), stones and hill." Cf. Mereburn *infra* and *to þǣm grǣgan stane*, B.C.S. 985. Phonology, § 14.

Grindon (Bp. Wearmouth). *c.* 1190 Godr. *Grendune*; 1507 D.S.T. *Grynden.* (Norhamshire), B.B. *Grendona*; 1539 F.P.D. *Gryndone.* (Warden) 1279 Iter. *Grendon*; 1403 Ipm. *Grindon.*

"Green hill." Cf. *on grenan dun*, B.C.S. 565. Phonology, §§ 1, 7.

Grindstone Law (Bingfield). 1479 B.B.H. *Gryndstan-law.*

"Grindstone-hill," i.e. where they are quarried.

Grottington (St John Lee). *c.* 1160 Ric. Hex. *Grotten-*

---

[1] The personal name *Grinwell* in Lanchester Registers *passim* indicates the old pronunciation. Phonology, § 7.

dun; 1298 B.B.H. *Grotinton*, 1479 *Grotyngton*; 1663 Rental *Groteington*; 1676 St John Lee *Groatington*.

" Hill of Grott(a) or his sons." Cf. *grottes graf*, B.C.S. 216 and Gretton, Salop, D.B. *Grotintune*. Teut. strong grade **greut* and weak grade **grut* (cf. O.E. *grēot*, " gravel " and *grot*, " groats ") give two series of names, (1) Goth. *Greutingi*, E. Frankish *Griuzing*; (2) O.N. *Grytingr*, O.S. *Gruting*, O.H.G. *Grutilo* (Naumann, p. 41, and Schönfeld, pp. 113-4). Cf. also Winkler (p. 137) who gives *Grote*, with patronymic *Grotinga*, and place-name *Grottyngha*. App. A, § 1.

Gubeon (Morpeth) [guˑbiən]. *c.* 1200 Newm. Wm. de *Gobyon*; 1663 Rental *Gudgeon*; 1668 H. 2. 2. 39 *Gubeon* alias *Gudgeon*; 1676 Mitford *Gudgeon*.

Named from a member of the Gobyon, Gubiun (or Gubbins) family. In T.N. Hugh Gubiun held the neighbouring Hepscott and he was sheriff of the county in 1296. This family has left its name in a large number of manors. Morant mentions four in Essex. Clutterbuck, *History of Herts* (vol. ii. p. 216) gives a *Gubions* or *Gobions*, and there is a Yardley Gobion, Northts.

Gunnerton (Chollerton). 1169 Pipe *Gunwarton*; 1269 Ipm. *Gonewerton*; *c.* 1250 T.N. *Gunwarton*; 1296 S.R. · *Gunewarton*; 1428 F.A. *Gunwarton*; 1479 B.B.H. *Gunwardton, Gonwarton*.

" Farm of *Gunnvarðr* (*m.*) or *Gunnvǫr* (*f.*) " *v.* Björkman, N.P. pp. 54-9. The former is very rare and may be a hybrid of English origin, the latter is found in L.V.D. as *Gunnwara*. Phonology, § 49.

Guyzance (Shilbottle). 1240 Newm. *Gsynes* (*sic*); 1252 Pipe *Gynes*; 1266 Ipm. *Gysinis*; 1296 S.R. *Gysings*; 1314 Ipm. *Gysins*; 1346 F.A. *Guisnes, Gysnes*, 1428 *Gysyns*; 1586 Raine *Guisons*; 1663 Arch. 3. 1. 261 *Guison*.

Cf. Guines, nr. Calais, which has early forms *Gisnes, Gysnes, Gynes*. The place must have been named from a land-holder deriving his name from Guines. Cf. Guisnes Court, Ess., earlier *Tholishunt Gynes*, and Puncherton *infra*.

G

Hackford (Hexham). N. iv. 11 *Hackeford* ; 1479 Eng. Misc. (Surtees Soc., vol. 85, p. 37) *Hakefurth*.

Cf. Hackford, Norf., D.B. *Hacforda, Hakeforda,* Hackforth, Yorks., earlier *Hakford, Hacford, Hackeford,* and *Hacfordland* (Pat. 1389) nr. Wooler.[1] These probably contain O.W.Sc. *Háki,* found also in Hackness, Yorks. (Lindkvist, p. lxiii. and Björkman, Z.E.N. p. 43). The only difficulty lies in the fact that so many fords happen to be owned by a man bearing a not very common name. *hack* might be a dialect form of *hatch* and the name be descriptive of a ford at which there is a *hack* to stop animals from being carried down stream. The Nthb. form is, however, *heck* rather than *hack.* Phonology, §§ 11, 30.

Hadston (Warkworth) [hadsən]. 1189 Pipe *Hadeston* ; 1255 Ass. *Haddeston* ; 1676 Warkw. *Hadsen.*

" Farm of *Hadd,*" a pet form of names in *Heaþu-.* Phonology, § 53.

Haggerston (Ancroft). 1228 F.P.D. *Hagardestone* ; 1268 Ass. *id.* ; *c.* 1250 T.N. *Hardgareston* ; 1278 Ass. *Haggarston.*

" Farm of Heardgar." This name is not found in O.E., but has its equivalents in other Teutonic languages. Phonology, §§ 53, 54.

Hagg Wood (Ellingham). 1342 N. ii. 240 *le Hagg.*

hagg=" a cutting or felling, a portion of a wood marked off for cutting " (Jamieson) *hagwood*=" a copse wood fitted for having a regular cutting of trees in it " (ib.), " a fenced place, a wood into which cattle are not admitted " (Heslop). The last usage is probably not from the same word, all the others may be referred to O.N. *hǫgg,* " cutting, opening for cutting trees " (Rygh. *Indl.* p. 58). Cf. O.N. *hǫgg-skógr*= wood of felled trees and *v.* Björkman in *Englische Studien,* vol. 44, p. 252.

Haining (Elsdon). 1304 Pat. *Hayning.* (Herrington) 1309 Halm. *le Hayninge.*

haining = " the preserving of grass for cattle, pro-

---

[1] There is another Hackford on the Devil's Water, said to be so called (N. iv. 66) from the " hackwood " or birchberry. No old forms have been found.

tected grass, any fenced field or enclosure, or separate place for cattle " (E.D.D.), and is in common use in Northern England. Cf. also The Haining, near Selkirk. It is the Dan. *hegning* of *Hegningen, Heiningen,* which Steenstrup (*Indledende Studier,* p. 274) explains as used of enclosed as opposed to common land. Cf. Dan. *hegn,* M.E. *hain,* "hedge," "enclosure" (Björkman, *Scand. Loan Words,* p. 242) and Hainton, Lincs., D.B. *Haintone.*

Hall Garth (Coatham Mundeville). 1382 Hatf. *le Halgarth.*
" Hall-enclosure," *v. garth,* Part II.

Hallington (St John Lee). 1247 Gray *Halidene* ; 1255 Ass., 1479 B.B.H. *id.* ; 1547 Hexh. Surv. *Hallidene,* 1608 *Hallendon* ; 1637 Camd. *Haledon* ; 1663 Rental *Hallington.*
O.E. *hālig-denu*=" holy-valley," from its identification with the site of Bede's *Hefenfelth* or Heavenfield (III. 2), the scene of the great victory of St Oswald in 634. Leland (*Itinerary,* vol. v. p. 61) says, " There is a Fame that *Oswald* wan the Batelle at *Halydene* . . . and that *Haliden* is it that Bede calleth *Hevenfeld.*" Phonology, § 22 ; App. A, § 1.

Halton (Corbridge). 1161 Pipe *Haultone,* 1177 *id.* ; 1247 Ch. *Hawelton* ; *c.* 1250 T.N. *Hawilton* ; 1254 Arch. 2. 1. 47 *id.* ; 1273 R.H. *Halton* ; 1273 Pipe *Halweton* ; 1286 Ipm. *Hawelton* ; 1296 Ch. *Haulton* ; 1318 Inq. a.q.d. *Ha(u)lghton,* 1322 *Halton* ; 1377 Ipm. *Haulton* ; 1428 F.A. *Halghton.*
Probably O.E. *healh-tūn,* farm on the *healh* (Part II), which is variously found in later English as Halton, Yorks. (2) and Salop, Haughton (*v. infra*), Hallaton, Leic., Halloughton, Notts. and Warw. The following are among the spellings found for these names :—*Haluton, Haloghton, Halecton, Halghton, Halluton, Hawledon.* The persistent early *w* might, however, point to O.E. *halig(a)tūn*=holy-farm. Cf. Halstock, Dev., 1285 F.A. *Halghestok,* 1379 B.M. *Halwestoke,* 1386 *Halghenstoke,* Halliford, Midd. 962 B.C.S. 1085 (*to*) *halganforde,* F.A. *Halgheford,* Hallatrow, Som., D.B. *Helgetreu,* F.A. *Halwe-, Halu-, Hale-, Halgh-tre.*

Haltwhistle [ho·təsəl]. 1240 Sc. *Hautwisel* ; 1279 Iter.

Hautwysel; 1278 Ass. *Hawtetwysill*; 1291 Tax. *Haut-wisill*; 1307 Ch. *Hautwisel*; 1311 R.P.D. *id.*, 1313 *Haut-wysell*, 1338 *Hautwesele*, 1340 *Hautetwysel*; 1372 Swinb. *Hautwysel*; 1479 B.B. *Haltewesyll*; 1507 D.S.T. *Haut-wesyll*; 1516 Raine *Hautewesill*; 1542 Bord. Surv. *Haute-wysle*; 1595 Bord. *Hawtwissell*; 1610 Speed *Haltwesell*; 1655 Corbr. *Hoatewhisle.*

A hybrid compound of O.Fr. *haut*, "high," and M.E. *twisel*, O.E. *twisla*, "fork of a river or road" (Part II), descriptive of the position of Haltwhistle on steeply rising ground between Haltwhistle Burn and S. Tyne. For the prefix cf. Alkborough, Lincs., earlier *Hauteberg, Alta Berga*, and Ault Hucknall, Derbys., earlier *Hault*=High Hucknall (Walker, p. 145). The *l* in the later forms is a learned respelling like that in *fault* and has similarly affected the pronunciation, for [hɔ·lt] now commonly replaces [ho·t].

Ham Burn (Hexhamshire). 1225 Gray *Hamburne*; 1287 B.B.H. *Hameburne.*

Probably O.E. *hām-burna*=stream by the homestead. Phonology, § 21.

Hamsteels (Lanchester). 1242 D.Ass. *Hamstele*; 1297 Pap. *Hamesteles*; 1382 Hatf. *Hamstels*; 1479 B.B.H. *Hamstell*; Esh (*passim*) *Hamstels.*

O.E. *hām-steall* = home buildings or sheds. Cf. *on deopan hamsteall*, B.C.S. 216. The lengthening of vowel in mod. *-steels* is probably due to the influence of the common dialectal *steel* (*v.* Steel *infra*). Phonology, § 21.

Hamsterley (Auckland). *c.* 1190 Godr. *Hamsteleie*; 1307 R.P.D. *Hamsterley.* (Lanchester) 1382 Hatf. *Hamsterley.*

This name is difficult. Winkler gives a Frisian personal name *Hamstra* (p. 143), cf. also *ha: r*, "corn-weevil," borrowed in early Mod. Eng. from Germ.

Hanging Leaves (Cockle Park). 1262 Ipm. *Hengan-delley*, 1264 *Hengandeles.*

"Hanging or sloping fields," with North. M.E. pres. part. form.

Hanging Wells (Stanhope). 1458.35 *Hyngyngwell.* **Hang-well Law** (Ellingham). 1266 N. ii. 277 *le Hengandewelle.*

"Hanging well or spring," descriptive of one spouting

from an overhanging rock. Cf. *Hengandewelleside* (N. i. 285) and *Hangandewell* in Wolviston (F.P.D. p. 371).

Harbottle (Holystone). 1220 Sc. *Hirbotle*; 1244 Ipm. *Hyrbotle*, 1283 *Hirbotel*, 1324 *Hirbotil*; 1430 F.P.D. *Herbotill*; 1430 Pat. *Herbotell*; 1479 B.B.H. *Harbotell*, *Hirbotle*; 1539 F.P.D. *Harbotell*.

O.E. *here-botl*=army-building or, as Holland's Camden puts it (p. 812), "In the English Saxons tongue *herbottle* . . . is the station of the army." For *Hir-* cf. Harlow *infra* and *v*. Morsbach, § 107. Phonology, § 8.

Harbour House (Durham). 1311 R.P.D. *Harbarwes*; 1343.31 *Harebarouhous*; 1382 Hatf. *Harebarowes*; 1432.45 *Harbarhous*.

M.E. *harbarwes*, pl. of *hereberȝe*=shelter, harbour, lodging. This became [harbərəs] and then the suffix was altered as in Crookhouse *supra*. Phonology, § 8; App. A, § 6.

Hardwick (Heselden). 1324 F.P.D. *Herdewyk juxta mare*; 1364.32 *Herdewyk on Sea*. (Sedgefield) *c.* 1150 F.P.D. *Herdwich*; 1403.33 *Herdewyk nigh Segefeld*. (Stockton) 1413.33 *Herdewyk nigh Norton*.

A very common English place-name, first found as *heorde-wic*, K.C.D. 653. Skeat takes this to be "dwelling of the herd" from *heord* (gen. sg. *heorde*) "flock." N.E.D. takes the first element to be O.E. *hierde*, "shepherd, herdsman," but the form in K.C.D. is against this. Vinogradoff (*Growth of the Manor*, p. 224) says that it refers sometimes to a pastoral settlement, but usually signifies the grange and stable in a small manorial settlement as opposed to *berwick* (*v. supra*), "the farm."

Harehope (Eglingham). *c.* 1150 Perc. *Harop*; 1252 Pipe *id.*; *c.* 1250 T.N. *Har(r)op*, *Harhop*; 1289 Ipm. *Hayropp*, 1308 *Harhop*; 1628 Arch. I. 3. 94 *Hareupp*. (Wolsingham) 1382 Hatf. *Harehopeleys*. **Harelaw** (Glendale). 1296 S.R. *Heyreslaw*. (Kirkharle) 1358 Pat. *Harelaw*. (Pelton) 1382 Hatf. *Harelawe*. (Stanhope) ib. *Harlaugh*. (Wolsingham) ib. *Harelaw*.

The first element is probably the word *hār* discussed under Harsondale *infra*, meaning "boundary." Harehope in Eglingham and Harelaw in Kirkharle and Pelton are on

the boundary of their respective parishes. It might, of
course, be O.E. *hara*=hare in some cases. The form of
Harelaw in Glendale points to a different history, and the
first element may be the personal name *Hegær* found in
L.V.D.

Harlow Hill (Ovingham). 1244 Ipm. *Hyrlawe*; 1278
Ass. *Hirlawe*; 1329 Ipm. *id.*; 1346 F.A. *Herlawe*, 1428
Herlow; 1538 Must. *Harlawe.*

O.E. *here-hlāw*=army-hill. Cf. Harbottle *supra*. Phonology, § 8; App. A, § 12.

Harnham (Bolam). 1271 Ipm. *Hernham*; 1285 Pat.
Herneham; *c*. 1250 T.N. *Harnaham*; 1346 F.A. *Harnam.*

O.E. *hyrne-hām*, "homestead in the corner of land."
O.E. *hyrne* is a derivative of *horn*. Wallis (ii. 538) says, "It
stands on an eminence . . . a range of perpendicular rocks
on one side and a morass on the other. The entrance is
by a narrow declivity to the North." Phonology, § 8.

Harpath Sike (Cheviot). 1304 Pat. *Epprespeth* (*sic*),
1307 *Erriespeth.*

This may be O.E. *heriges-pæð*=path of the army, an
alternative to the more common *here-pæþ*, discussed in
Crawford Charters, ed. Napier and Stevenson, pp. 46-7,
v. Herpath in Heslop. It is possible, however, that the
forms given above are corrupt and should be referred to
Yarnspath *infra*. Phonology, §§ 8, 1.

Harperley (North Bedburn). B.B. *Harperleia*; 1382
Hatf. *Harplye.*

Cf. Harpurhey, Lancs. (Sephton, p. 78, no early forms)
and *Harpermor* in Bp. Middleham (Hatf. Surv.). Probably
from the common word *harper* used as a personal name.
" Harper's clearing."

Harraton (Chester-le-Street). *c*. 1190 Godr. *Hervertune*;
1297 Pap. *Herverton*; 1447.34 *id.*; 1562 Wills *Harraton.*

O.E. *Herefrið-* or *Herefær-tūn. Herefær* is not found,
but cf. *Uilfares dun* (Sweet. O.E.T. p. 472) and the numerous
O.N. names in *fari* (Lind. *s.n.*). For *rf > rv* cf. Harvington,
Worc., earlier *Herefordtun*. Phonology, §§ 8, 51.

Harrowbank House (Stanhope). 1382 Hatf. *Harew-
bank.*

The first element may be O.E. *hearg*=heathen grove, temple, as in Harrow, Middx., B.C.S. 304 *æt hearge* or M.E. *harewe*=harrow. No certainty can be attained.

Harsondale (Haydon). 1255 Ass. *Harestanesden*; 1368 Ipm. *Harsenden*; 1663 Rental *Harsondale*.

Mathieson (*Place-names of Elginshire*, p. 187) explains *Harestanes* as a boundary wall with notches like a hare's lip, and Lindkvist (p. 56) suggests for *Haresteinegate*, Yorks., connection with M.E. *hare*, "hare." Both suggestions are incorrect. *Harestane* is O.E. *hār-stān*, "grey" or "boundary stone" often found in O.E. charters. In the S. and Midl. it becomes *Hoarstone*. Cf. Duignan, *Worcestershire Place-names*, p. 70, and N.E.D. *s.v.* The same boundary stone is referred to in *Harstanley* in Staward (Coram 1362). Phonology, §§ 14, 53; App. A, § 11.

Hart. 1292 Ch. *Hart*; 1312 R.P.D. *Harte*.

Either (1) O.E. *heorot*, "stag," or (2) *heorte*, "heart." If (1), the second element may have been lost, but cf. *Heorot* as the name of the hall in *Beowulf*, supposed to be so called from the antlers on the gables. For (2) cf. the use of O.N. *hjarta* (Rygh. *Indledning*, p. 55 and N.G. xvi. 91, 158). Names like *Herten* are supposed to have been given from some fancied resemblance of the site to a heart. It may be noted that names such as *Hjartøen* and *Hjartholmen* are often reduced in Norway to simple *Hjert*.

Hartburn (Nthb.). 1203 R.C. *Herteburne*; 1284 De Banco *Hertburgh*; 1507 D.S.T. *Hertburn*; 1663 Rental *Harbourne*; 1798 Bothal *Harburn*. (Stockton) *c.* 1190 Godr. *Herteburna*. **Hartford,** (Horton) [harfəd]. 1203 R.C. *Hertford super Blitham*; 1663 Rental *Harford*. **Harthope Burn** (Cheviot). 1305 Ipm. *Herthop*. **Hartley** (Earsdon). 1166 Pipe *Hertelawa*; 1573 N. ix. 96 *Hartley*. **Hartley Burn** (S. Tyndale). 1479 B.B.H. *Hartely-burne*.

Obvious compounds of O.E. *heorot*=hart, stag. Cf. *heorot burna*, B.C.S. 247, Hertford, Herts., A.S.C. *Heorotford*, Harford, Glouc., and *heoratleg*, B.C.S. 260. Phonology, § 53; App. A, § 10.

Hartington (Hartburn). 1170 Pipe *Hertweiton*; 1255 Ass. *Hertwayton*; 1318 Inq. a.q.d. *Hertewarton*; 1346 F.A.

Hertwatton ; 1436 Ipm. *Hartwayton* ; 1542 Bord. Surv.
Harterton ; 1663 Rental *Hartington* ; 1680 Elsdon
Harterton.

O.E. *heorotwegtūn*=stagpath-farm. Cf. *horsweg, swin-
weg* (B.C.S. 299, 801). The development is peculiar, but
cf. Flotterton *supra.*

Hartlepool [hɑˑtlipuˑl]. *c.* 750 Bede *Heruteu, id est insula
cerui* ; *c.* 1196 Finch. *Herterpol* ; 1200 R.C. *Hertelpole*, Pipe
Hertepol ; 1306 Ch. *Hertelpol* ; 1307 R.P.D. *Hertpoll'*,
1312 *Hartrepoll*, 1313 *Hertrepoll*; 1316 *Hertelpol* ; 1430
F.P.D. *Hertilpole* ; 1479 B.B.H. *Hertyllpull* ; 1539 F.P.D.
Hartylpole.

The earliest form seems clear enough and is applicable
to the site of Hartlepool on a peninsula (*v. ea*, Part II),
though grammatically we must interpret the name as
" stag-island " rather than " island of the stag." The
difficulty is to connect this with the forms that arise in the
12th and 13th cents. Here the suffix is clearly *pool*, but
what is the relation of the *Herter-, Hertel-* to the old name ?
The confusion of *r* and *l* can be explained as due to Anglo-
Norman scribes (Zachrisson, p. 142), and either *r* or *l* may
be the original consonant. If *r*, the history might be that
Heruteu > M.E. *Hert-e*, and that Hartlepool was originally
Hert-e-pol, i.e. pool by the stag-peninsula, and that then an
inorganic *r* developed (cf. Hartington *supra* and forms in
Zachrisson, p. 145). Original *l* is less probable but might
have developed in anticipation of the *l* of the final syllable.
In any case the ultimate prevalence of *Hartle-* may have
been helped by the existence of an O.E. name *Heortla*,
found in Hartlebury, Worc. (B.C.S. *Heortla(n)byrig*). Were
it not for Bede's form we should naturally explain Hartlepool
as containing this name.

Harton (Jarrow). 1104-8 S.D. *Heortedun* ; *c.* 1125
F.P.D. *id.*, 1203 *Hertendune* ; 1296 Halm. *Herton* ; 1335
Ch. *Herteden* ; 1446 D.S.T. *Harton.*

Cf. Hartington, Derbys., with earlier first element *Herten-,
Hertin(g)-, Harting-*, and *heortingtun*, B.C.S. 553. We have
apparently here a personal name *Heorta* derived from the
animal name. Hartlebury, Worc., earlier *Heortlanbyrig*

(Duignan, p. 77), shows a diminutive derived from this name. Phonology, § 51 ; App. A, § 1.

Hartside (Ingram). 1255 Ass. *Hertesheved* ; 1663 Rental *Hartside*.

" Hart's head," i.e. stag's headland or, possibly, in the sense noted under Gateshead *supra*. App. A, § 7.

Harvey Hill (Wolsingham). 1382 Hatf. *Horbe.* Unexplained.

Harwood House (Hartburn). *c.* 1155 B.M. *Harewud* ; 1268 Ass. *Hartwode* ; 1278 Ipm. *Harewode* ; 1356 Pat. *Harewod* ; 1421 Ipm. *Harewood*.

Harwood Shiel (Hexhamshire). *a.* 1214 Dugdale vi. 2. 886 *Harewode*.

" Boundary wood," *v.* Harelaw *supra*. Harwood House is on the boundary of Hartburn and Redesdale parishes (H. 2. 1. 288), Harwood Shiel on that of Shotley High Quarter and Hexhamshire High Quarter. The 1268 form is probably due to the influence of the neighbouring Hartburn and Hartington.

Haswell (Easington). 1131 F.P.D. *Hessewella* ; *c.* 1190 Finch. *Hesewell*, 1180 *Essewella*, 1200 *Hess(e)well* ; 1253 Ch. *Hessewell* ; 1313 R.P.D. *id.* ; 1539 F.P.D. *Heswell*.

The first element is probably a personal name. No O.E. one of this form is known, but cf. Heintze (*s.v. Hasse*), who gives old forms *Hasso, Hesso*, later *Hasse, Hesse*, referring probably to men of Hessian origin. Perhaps the name was borne by some continental settler in England.

Haughstrother (Haltwhistle). 1312 Ipm. *le Haukstrothre*.

" Marsh on or by the corner of ground," *v. healh* and *strother*, Part II.

Haughton (Simonburn). 1177 Swinb. *Haluton*, 1267 *Haluchton* ; 1279 Iter. *Haluton, Halchtona* ; 1284 Swinb. *Halghton* ; 1318 Ipm. *Haulktoune* ; 1610 Speed *Haughton*. **Haughton-le-Skerne.** *c.* 1050 H.S.C. *Halhtun* ; B.B. *Halctona, Halghtona* (B., C. *Halughton*) ; 1507 D.S.T. *Haughton*.

v. Halton *supra. le Skerne* because on the river of that name, cf. Chester-le-Street *supra*.

Hauxley (Warkworth).[1] [ha·ksli]. 1203 R.C. *Haukeslawe*; 1271 Ch. *Hauekeslowe*; 1428 F.A. *Hawkeslawe*; 1638 Freeh. *Hauxley*; 1697 Warkw. *Haxlee, passim.* " Hawk's hill or (perhaps) his barrow." Cf. *hafeceshlæw*, B.C.S. 687. App. A, § 2.

Hawden (Newbrough). 1330 Cl. *Hauden.* O.E. *haga-denu=* haw-valley, i.e. where haws abound.

Hawick (Kirkharle). 1284 Ipm. *Hawik*; 1296 S.R. *Hawyk*; *c.* 1250 T.N. *Hawic*; 1346 F.A. *Hauwyk.* O.E. *haga-wīc* = dwelling with a *haw* or hedge or, possibly, where haws abound. Cf. *wiðigwic*, B.C.S. 700, *ðornwic*, ib. 707.

Hawkhill (Alnwick) [hɔ·kəl]. 1177 Pipe *Hauechil*; 1288 Ipm. *Hauckill*; 1346 F.A. *Haukhull, Haukell, Hawkill*; 1538 Must. *Hawkell.* **Hawkhope** (Falstone) [hɔ·kəp]. 1325 Ipm. *Haucop*; 1603 Rental *Hauckup.* **Hawkuplee** (Whitfield). 1374 Ipm. *Haucopley*; 1610 Speed *Hawcople.* **Hawkwell Hall** (Stamfordham). 1249 Ipm. *Haukewell*, 1268 Ass. *id.*; 1346 F.A. *id., Hauk(is)well*, 1428 *Haukeswell*; 1479 B.B.H. *Haukewell*; 1663 Rental *Hawkwell.* Obvious compounds of *hawk*, used either of the bird or of a man so named, cf. *hafochyll*, B.C.S. 936, *heafocwyll*, ib. 246.[2] Phonology, §§ 36, 49.

Hawthorn (Easington). *c.* 1190 Godr. *Hagelhthorn* (*sic*), *Haithethorn*; 1155 F.P.D. *Hagethorn, c.* 1220 *Hauthorn*, 1539 *Hawthorne.* O.E. *haga-þorn* = hawthorn. Cf. Broom *supra.*

Haydon[3] (Warden). 1255 Ass. *Heiden*; 1346 F.A. *Haydon*; 1479 B.B.H. *Hayden.* " Hay-valley." Cf. *heigdun*, B.C.S. 282, *hegcumb* 627, *heglea* 1307.

Hazeldean[4] (St John Lee). 1298 B.B.H. *Knitel-hesell*, 1328 *Knytel-hesil.*

---

[1] This has been identified with *Hafodscelfe* in H.S.C. Either the identification is wrong, or the form should be *hafocesscelf* = Hawk's shelving ledge, with later change of suffix.

[2] So similarly Hauxwell, Yorks., D.B. *Hauocheswelle*, and not " Jacob's well," as some would have it, referring to the activities of James the deacon.

[3] There is also an unidentified *Hayden* in Ellington, 1265 and 1270 Ipm.

[4] This identification is made in N. iv. 96.

" Cnytel's hazel-bush." *Cnytel* (dim. of *Cnut*) is found
once in O.E. There is an O.E. *cnyttels* (glossing Lat. *nervus*),
dialectal *knittle*, " a string to tie a sack with," but there is no
evidence that it was ever applied to the hazel. Phonology,
§ 2 ; App. A, § 1.

Hazelrigg (Chatton) [hezlrig]. 1288 Ipm. *Heselrig* ;
1296 S.R. *Hessilrig* ; 1428 F.A. *Hesilryge* ; 1663 Rental
Heslerig.

O.E. *hæsel-hrycg* = hazel-ridge. Cf. *hæsel-hyll*, B.C.S.
674. Phonology, §§ 2, 27.

Hazon (Shilbottle). 1169 Pipe *Heisende* ; *c.* 1250 T.N.
Heysanda ; 1266 Ipm. *Haysand*, 1334 *Hysaund* ; 1428
F.A. *Haysand* ; 1538 Must. *Hasande* ; 1628 Arch. 1. 3. 94
Hayson ; 1638 Freeh. *Hason* ; 1663 Rental *Hazon*.

O.E. *heges-ende*=hedge's end, referring to some boundary.
Cf. Detchant *supra*. Phonology, § 56.

Headlam (Gainford). *c.* 1190 Godr. *Hedlum* ; 1207
F.P.D. *id.* ; 1316 Cl. *Hedlem*, 1317 *Hedelom* ; 1335 Ipm.
Hedlem ; 1341 R.P.D. *Hedelham*, 1344 *Hedlame* ; 1382
Pat. *Hedelham*.

" Homestead of * *Heddel*," a dimin. of O.E. *Hæddi*.

Headshope (Elsdon). *n.d.* Newm. *Heuedshope* ; 1618
Redesd. *Headshope*.

" Head's hope." For *Head* as a name cf. Weekley, p.125.

Headworth (Jarrow). 1104-8 S.D. *Heathewurthe* ; *c.*
1125 F.P.D. *He(a)thewrthe* ; 1335 Ch. *Hethewrthe* ; 1430
F.P.D. *Hedworth*.

Possibly O.E. *hǣþ-weorþ* = heath-enclosure or " Haethe's
enclosure," cf. *Hæthe*, L.V.D., but we should not expect
early spellings in *ea*. The name must remain doubtful.
Phonology, § 42.

Healey (Bywell). 1268 Ipm. *Heley* ; 1570 N. vi. 170
Temple Helay. (Netherwitton) 12th c. Newm. *Helay*.
(Rothbury) 1100-35 Brkb. *Heley, Over Heley* ; 1309 Ipm.
Grenehelay.

O.E. *hēa(n)-lēage* (dat.) = high clearing. [hi·] is Nthb.
for *high*. Loss of *n* gives *Healey* in contrast to *Henley*
and *Hanley* found elsewhere. Healey in Bywell belonged
to the Knights Templars.

Healeyfield (Lanchester). B.B. *Heleie*; 1382 Hatf.
Heley Aleyn; 1464 F.P.D. *Helayfeld.*
Healey, *u.s.* *Aleyn* must have been the owner, perhaps
the marshal who owned Allenshiel or Allensford *supra.*

Heatherley Clough (Wolsingham). 1432.33 *Hethereclogh.*
" Hæðhere's clough " (*v.* Heatherslaw *infra*) rather than
" heather-clough," for the M.E. form of that word is *hather*
or *hadder.*

Heatherslaw (Ford). 1175 Pipe *Hedereslawa*; 1254
Ipm. *Hedereslau*; 1255 Ass. *Herdeslawe*, 1278 *Herders-
lawe*; 1314 Inq. a.q.d. *Haddreslawe*; 1346 F.A. *Hed(d)res-
lawe*, 1428 *Hederslawe*; 1579 Bord. *Heytherslaw.*
" Hæðhere's hill." The name is not found in O.E.,
but is a possible compound of *Hæð-* (*v.* Searle). It probably
forms the first element in Hatherley, Glouc., earlier *Haider-
leia, Hedrelega*, though Baddeley (p. 80) gives a different
explanation. Phonology, § 41; App. A, § 2.

Heatherwick (Elsdon) [haðərwik]. *c.* 1250 T.N. *Hather-
wick*, 1331 Ipm. *Hatherwick*; 1618 Redesd. *id.*; 1673 Elsdon
Heatherweek, Hadderweek (*passim*); 1751 Edl. *Hatherwick.*
" Heather-dwelling." Cf. Heatherley Clough *supra.*
Phonology, § 41.

Heathpool [heθpul]. 1249 Ipm. *Hethpol*; 1290 Ch.
id.; 1542 Bord. Surv. *Hethepol.*
Probably " pool under Hetha," the name of a hill above it.
The map form is corrupt.

Heaton (Newcastle-on-Tyne). *c.* 1200 Vescy *Hactonam*;
1296 S.R. *Heton juxta Castrum.* (Norham) B.B. *Hetona.*
" High farm." Cf. Healey *supra.*

Hebburn (Jarrow) [hebərən]. *c.* 1104-8 S.D. *Heabyrm*;
c. 1125 F.P.D. *Heabyrine, Heberine*; 1334 Ch. *Heberne*;
1539 F.P.D. *Hebbarine, Hebarn*; 1696 N.C.D. *Heberon.*
Clearly not of English origin.

Hebron (nr. Morpeth). 1251 Ch. *Heburn*; 1264 Ipm.
id., Heborin; 1346 F.A. *Heburnne*; 1663 Rental *Hebbourn.*
O.E. *hēah-burna* = high-burn. Phonology, §§ 21, 51

Heckley (Embleton). *c.* 1250 T.N. *Hecclive*; 1283 Perc.
Hecclif; 1307 Ch. *Heckelive*; 1346 Ass. *Hecclif, Hecley*;
1353 Perc. *Hetcliffe* (*sic*); 1663 Rental *Heckley.*

Possibly O.E. *hēah-clif* = high cliff. Cf. Scots. *he(y)ch* = high. Phonology, § 56 ; App. A, § 7.

Heddon, E. and W. (Heddon-on-the-Wall). 1177 Pipe *Hidewine,* 1187 *Hiddewin* ; 1255 Ass. *Hydewyn* ; *c.* 1250 T.N. *Hydewin* ; 1298 Arch. 3. 2. 3 *Hidwyn* ; 1346 F.A. *Hidwin, Hiddewyn,* 1428 *Hydwyn* ; 1538 Must. *Hedwyne* ; 1580 Bord. *Hedwen* ; 1638 Freeh. *Heddon.* This name is probably pre-English, and certainly different from Heddon-on-the-Wall. *win(n)* or *wyn(n)* is fairly frequent in O.E. place-names ; cf. *winburne* (A.S.C.), *wynnabæc,* B.C.S. 233, *wynford,* 721, *wynnawudu,* 931, *wynne mæduan,* 683, *wynne dun, wynnefeld* (K.C.D. 710), but its meaning is very uncertain and it is only found as a first element (*v.* Middendorf, p. 155 and Bosworth-Toller *s.v. wyn*). The name is perhaps Celtic, with the suffix *-wen* commonly found in Welsh names. Its sound development has been influenced by Heddon-on-the-Wall. Phonology, § 49.

Heddon-on-the-Wall. 1175 Pipe *Hedun* ; 1262 Ipm. *Hedon, Heddun* ; *c.* 1250 T.N. *Hedon super murum* ; 1291 Tax. *Heddon.* **Heddon, Black** (Stamfordham). *c.* 1250 T.N. *Nigram Heddon.* **Hedley Hill** (Lanchester). *c.* 1190 B.B. *Hethleia,* B.B. *Helley* (B., C. *Hedley*). **Hedley** (Lamesley). 1382 Hatf. *Hedley.* **Hedley-on-the-Hill.** 1255 Ass. *Hedley* ; 1307 Newm. *Heddeley* ; 1275 Ass. *Karlhedley,* 1292 *id.*

"Heath-hill and clearing." Cf. *hæðdun* B.C.S. 801, and *hæðlege,* 455 = Headley, Worc. Phonology, §§ 21, 51, 42. *On the Wall,* because on the line of the Roman Wall ; *Black,* probably from the soil ; *Carl,* perhaps because once in the possession of a man named *Karle* (Björkman, N.P. p. 77), or of some *carls* (cf. Carlton *supra*).

Hedgeley (Eglingham) [hidžli]. *c.* 1150 Perc. *Hiddesleie* ; 1247 Sc. *Hiddesley* ; 1255 Ass. *Hydesleg,* 1278 *Hygeley* ; 1289 Ipm. *Hydesley* ; 1296 S.R. *Hegeley* ; *c.* 1250 T.N. *Hiddesley* ; 1306 Sc. *Hygele* ; 1334 Perc. *Higgeley* ; 1498 H. 3. 2. 127 *Hegeley.*

"Hiddi's clearing." Cf. *Hiddi,* L.V.D. Phonology, §§ 7, 31.

Hedley, Black (Shotley). 1262 Ipm. *Blakedeley* alias *Blakhedley*; 1296 Orig. *Blakedesleye*; 1312 Ipm. *Blackhedreley*; 1307 Abbr. *Blakdesle*; 1313 Cl. *Blakehedreleie*; 1318 Inq. a.q.d. *Blachedley.*

" Black Hæðhere's clearing." Cf. Heatherslaw *supra.* " Black " from the colour of the soil. The name was probably modified under the influence of the neighbouring Hedley-on-the-Hill.

Hefferlaw (Embleton). 1283 Tate II. 379 *Heforside*; 1346 Ipm. *Heffordlawe*; 1353 Perc. *Heforthlawe*; 1649 Comps. *Heffordlawe.*

" High-ford hill." For the sound development cf. *heifer* < O.E. *hēahfore.* Phonology, §§ 21, 51, 30.

Heighington [haintən, haiiŋtən]. 1228 F.P.D. *He(h)ington*; B.B. *Heghyngtona*; 1362 D.S.T. *Heynton*; 1599 Lanch. *Highington.*[1]

" Farm of Heaha or his sons." *Hēaha* is a shortened form of an O.E. name in *Hēah-* (Redin, p. 50). Phonology, §§ 36, 59.

Heighley Hall (Gainford). 1404 S. 4. 37 *Heighle.*

" High-clearing." The form is perhaps of later origin than *Healey.*

Helm (Felton). 1255 Ch. *Helm*; 1390 Ipm. *Helme*; 1663 Rental *Helm-on-ye-Hill.* **Helme Park** (Wolsingham). *c.* 1050 H.S.C. *Healme*; 1104-8 S.D. *Helme*; 1299 Acct. *id.*; 1382 Hatf. *le Helme park.*

Helm in Felton stands on a well-marked rounded hill, and is probably so called from its resemblance to a helmet (O.E. *helm*) or from its being on the top of a hill. Cf. *Hjelmen* Hill (N.G. xv. 99), *Hjelmen*, " a little, high island " (N.G. xi. 48), and the island of *Hjelm* on the E. coast of Jutland. The same word is found in The Elms, Heref., earlier *Heaume*, *The Helm* (Bannister, p. 68).

Hendon (Bp. Wearmouth). 1382 Hatf. *Hynden.*

O.E. *higna-denu* = valley of the monks or *hind-denu* = hind-valley. Phonology, § 10 ; App. A, § 1.

Henknowl (Auckland). B.B. *Henknolle*; 1313 R.P.D. *Henneknolle.*

---

[1] Personal name.

"Hens' knoll." Cf. Hinding Flat *infra, henna leah*,
B.C.S. 677, Henmarsh, Glouc.

Henshaw (Haltwhistle). 12th c. B.B.H. *Hedeneshalch*;
1262 Ch. *Hethingishalt*; 1279 Iter. *Heinzhalu*; 1298
B.B.H. *Hetheneshalgh*; 1326 Ipm. *Henneshalgh*; 1371
Pat. *Hentishalghe*; 1479 B.B.H. *Hennishalgh*; 1597 Bord.
Henshaw.
Cf. Hensall, Yorks., which Moorman (p. 96) takes to be
O.E. *hǣðenes healh* = heathen's corner of land, so named
from a heathen Danish settler, singled out by his Christian
neighbours. More probably the first element is O.W.Sc.
Heðinn (Björkman, Z.E.N. p. 45). Cf. Heynstrup, Den-
mark, earlier *Hethensthorp* (Nielsen, p. 46). Phonology,
§ 44; App. A, § 6.

Hepburn (Chillingham). *c.* 1050 H.S.C. *montem Hybbern-
dune*;[1] *c.* 1250 T.N., 1319 Ipm. *Hibburn*; 1346 F.A.
Hilburn; 1352 Cl. *Hibbourn*; 1377 Ipm. *Hibbirn*; 1428
F.A. *Hibburn*; 1542 H. 3. 2. 209 *Hebburne*; 1628 Arch.
I. 3. 94 *Hebborne*.
The form in H.S.C. suggests that the later ones are
corruptions of an original Celtic one, otherwise we might
suggest O.E. *hyllburna* = hill-stream, or, rejecting the 1346
form, *hidaburna* B.C.S. 825, a river-name found as Head-
bourne, Hants. Its origin is unknown. Bates (p. 50)
attributes the present form to the Ordnance Survey.

Hepden Burn (Kidland). 1233 Newm. *Heppeden*.

Hepple (Rothbury). 1199 Pipe *Hepedal* (*sic*); 1229 Pat.
Hyephal; *c.* 1250 T.N. *Hephal, Heppal*; 1252 Ch. *Hephale*;
1280 Ipm. *id.*; 1346 F.A. *Happale, Heppale, Hephale*,
1428 *Heppell*.
O.E. *hēope-denu* and -*hēale* = dog-rose valley and haugh.
Cf. Hipbridge, Lincs., B.C.S. 1270 *heopebricge*. There is
a name *Heppo* in D.B. but it is probably of continental
origin. Forssner (p. 147) takes it to be O.H.G. *Herpert*
or *Herprant*. Phonology, § 36.

Hepscott (Morpeth). 1257 Ch. *Heppescotes*; 1288 Ipm.
Hebbescotes; *c.* 1250 T.N. *Hebscot*; 1310 Ch. *Heppscot*;
1313 R.P.D. *Heppescotes*; 1428 F.A. *Hepscotes*.

[1] Referring to Hepburn Bell.

" Hebbe's cotes " (*cote*, Part ii). Bardsley gives a name *Hebba* which might be a pet form of O.E. *Hēahbeorht*. Phonology, § 51.

The Hermitage (St John Lee). 1496 N. iv. 144 *Armytage*, 1568 *Tharmitag* ; 1663 Rental *The Hermitage*. The reputed haunt of St John of Beverley (N. iv. 143). Cf. Armitage, Staffs., earlier *Hermitage*. Phonology, § 8.

*Hernehouse (Redesdale). 1398 Ipm. *Hirnhous* ; 1618 Redesd. *Hernehouse*. " House in the corner of land." Cf. Harnham *supra*.

Heron's Close (Fenrother). 1255 Ch. *Heyrun*, 1340 *Heyroun* ; 1653 Comps. *Heron's Close* ; 1663 Rental *Hearon's Close*. Ground once held by Wm. Heron of Hadston (H. 2. 2. 131). *heron* < M.E. *heiroun, heyroun* < O.Fr. *hairon* = heron.

Herrington (Houghton-le-Spring). 1197 Pipe *Erinton* ; 1260 F.P.D. *Heringtona* ; 17th c. *passim*, Bp. Wearm. *Harrington*. Possibly " Hering's farm." Cf. *Hering*, a personal name once found in O.E., and *heringesleah*, B.C.S. 543, *hæringæs gæt*, K.C.D. 739. Phonology, § 22.

Hesleden, Monk (Easington). *c.* 1050 H.S.C. *Heseldene* ; *c.* 1125 F.P.D. *id.*, *Haseldene, Hæseldene* ; 1344 R.P.D. *Monkheselden* ; 1541 Allen *Hasylden Monachorum*. **Hesley-hurst** (Rothbury). 1268 Ass. *Heselyhyrst*. **Hesleyside** (Bellingham). 1279 Iter. *Heselyside*. " Hazel valley, hazely wood and hill " (*hyrst*, Part ii). *Monk* because it belonged to the monks of Durham. Phonology, § 2.

Hetchester (Throckerington). *n.d.* Newm. *Heichester, Haichester* ; 1272 Newmn. *Haycesters*. The *chester* with a " hay " or hedge (O.E. *hege*) or, possibly, where " hay " is made. Cf. Haydon *supra*. Roman remains have been found here. The modern form seems to be corrupt.

Hetherington (Wark-on-Tyne). *n.d.* Swinb. *Hetherin-tun* ; 1291 Ipm. *Hetherrinton* ; 1610 Speed *Hatherinton* ; 1663 Rental *Heatherington*. **Hetherslaw** (Stamfordham). 1479 B.B.H. *Hethreslaw, Hedderslaw*.

"Farm of Hæðhere or his sons," "hill of man of the same name," *v.* Heatherslaw *supra* and cf. Harrington, Northts., earlier *Hetherington.*

Hett (Merrington). *c.* 1168 F.P.D. *Het*; 1369 Halm. *Hett in Spen*; 1539 F.P.D. *Hette.*

Possibly *hett* is here a dialectal form of *hat,* and the place was so called from some fancied resemblance of the ground to a hat. Cf. Steenstrup, *Indledende Studier,* pp. 275-6, where we have Dan. *Hætten,* referring to a smaller wood jutting out of a larger one, and *Munkehætte* = monk's hat, applied to a little wood. The difficulty of form is greater than that of meaning. *het* is not the Nthb. or Durh. form of *hat,* though we find such a sound-development in *peth* and *efter.* Possibly the name is Scand. rather than English. Cf. O.N. *hette,* dat sg. of *høttr.* "hat," and O.N. *hetta* = hood.

Hetton-le-Hole and **le-Hill** (Houghton-le-Spring). 1180 Finch. *Heppedun, c.* 1200 *id. Heppeden*; *c.* 1230 F.P.D. *Hepedon*; 1315 R.P.D. *Hetton,* 1344 *Hepdon*; 1535 Finch. *Hepton-in-Valle*; 1539 F.P.D. *Heptone super montem*; 1637 Camd. *Hetton-in-the-Hole.*

O.E. *hēope-dūn* = dog-rose hill (cf. Hepple *supra*), descriptive of the hill at the foot of which stands Hetton-le-Hole. *le* here has no early justification, and must have been introduced on the analogy of other names with a second qualifying element. Phonology, § 51; App. A, § 1.

Hetton (Chatton). 1162 Pipe, 1288 Ipm. *Hetton*; 1289 Cl. *Hethton*; *c.* 1250 T.N. *Hetton*; 1296 S.R. *Heddon*; 1346 F.A. *Heldon* (*sic*), *Hetton,* 1428 *Heddon.*

O.E. *hæð-tūn* = heath-farm. Cf. Hetton, Yorks. (Moorman, p. 97). Phonology, § 51; App. A, § 1.

Heugh (hjuf] (Esh). 1411.33 *le Hough.* (Quarrington) 1382 Hatf. *le Hough.* (Stamfordham) 1276 Ipm. *Hough*; 1298 Cl. *le Hogh*; 1346 F.A. *le Hugh*; 1628 Arch. I. 3. 94 *Heugh,* Freeh. *Hugh.*

v. hōh, Part II. For the sound *v.* E.D.G. pp. 138-9.

Heworth (Aycliffe). 1091 Cart. Will. Reg. *Hewarde*; B.B. *Heworth*; 1435.33 *Heworth by Acle.* (Jarrow) *c.* 1125 F.P.D. *Hewrth.*

O.E. *hēah-weorþ* = high enclosure. Cf. Surtees 2.83

H

on the view of the vale of Tyne from Heworth in Jarrow.
Phonology, §

Hexham-on-Tyne. *c.* 750 Bede *Hagustaldensis ecclesia* ;
c. 1000 O.E. Bede *Agostaldes ea, Heagostealdes ea* ; *c.* 1200
A.S.C. *Hagustaldes-ea, -ee, -ham, Hagstd ee, Hagusteald* ; *c.*
1154 Hist. Reg. *Hestaldesige* ; 1187 Pipe *Hextoldesham* ;
c. 1160 Ric. Hex. *Hesteldesham, Hestoldes-, Hestaldesham* ;
1228 F.P.D. *Extildham* ; 1232 Ch. *Hextildesham,* 1239
Hexteldesham ; 1267 Giff. *Exhildesham* ; 1273 R.H. *Exildes-
ham* ; 1283 Ch. *Hextildesham* ; 1312 R.P.D. *Hextildeham* ;
1351 Hexh. Pr. *Hexham,* 1535 *Hextildesham.*

Richard of Hexham (Bk. I. ch. i.) says that the place
was called *Hestoldesham, quasi prædium Hestild* from a small
stream of that name.[1] This may be a piece of etymologising on Richard's part and *Hestild* be really a back-formation
from the town-name, but before rejecting it we should
remember that (1) many town names do take their rise from
rivers, (2) no other example of so early a back-formation
is known.

The forms in A.S.C. show the suffix *ea,* dat. *ee* or *ie,*
" river " (cf. Elvet *supra*). *-ige* shows confusion with the
allied O.E. *īeg,* island. The first element is apparently
gen. sg. of O.E. *hago-steald,* " bachelor, young warrior,"
which is found in the variant form *hægsteald* in *hægstel-
descumb,* B.C.S. 476, *hegestuldessetl,* 887. Cf. Germ. *Hagas-
taldeshusen, Hagstedt* (earlier *Hagastaldstedi*) in Förstemann
s.n. Against this is the improbability of such a name as
" Bachelor's river." Far more probably, as in *Eoforwic*
(Lat. *Eburacum*) and *Searoburh* (Lat. *Sorbiodunum*) we have,
by a process of folk-etymology, the anglicising of some earlier
Celtic river-name. By this process the stream and later the
town came to be called *Hagostealdes ea.* When by ordinary
phonological process the first element in the river-name
became *Hextild,* all trace of its meaning was lost and the
stream became simply *Hextild.* The town, on the other
hand, was early changed to *Hagustaldesham,* a name
yielding better sense and provided with a more common
suffix.

[1] This is probably the stream now known as Cowgarth Burn.

In the later development *Hest-* and *Hext-* go back to
O.E. *hægsteald* rather than *hago-steald*. The latter would
have given *ha(w)st-*, the former *he(y)st-* rather than *Hext-*.
The last may perhaps be explained by the influence of the
common M.E. *hexte* = highest.

Higham Dykes (Ponteland). 13th c. Newm. *Heyham,
Heiham* ; 1289 Ipm. *Hecham* ; 1663 Rental *Higham Dykes*.

Highlaws (Hartburn). *c*. 1250 T.N. *Heylaw*. (Mitford).
1292 Q.W. *Heghelawe* ; 1489 Ipm. *Heghlawe* ; 1637 Camd.
Highley ; 1663 Rental *Highlies*.

" High-homestead and -hill." App. A, § 2.

Hinding Burn and Flat (Alnwick). 1275 Tate *Hen-
neden-burne, -flat*.

" Hens' valley." Cf. *henna dene*, B.C.S. 1080. Phonology, § 10.

Hindley (Bywell St Peter). 1255 Ass. *Hyndelegh*. (Henshaw). 1328 Ipm. *Hyndley*.

O.E. *hind-lēah*=hind-clearing, so called from the animal.
Cf. Hindley, Yorks. and Lancs.

Hirst (Woodhorn). 1268 Ipm. *Hyrst*.

" Wood." (*hyrst*, Part II).

Hisehope Burn (Muggleswick). 1153-95 F.P.D. *Histes-
hope* ; 1260 F.P.D. *Hystleyhopeburne*.

Perhaps these contain a name *Hest* or *Hist* from O.W.Sc.
hestr, " horse," used as a nickname. Cf. *Bjarni hestr*
(Fritzner, *s.v.*) and *Hest(s)fjǫrðr* and *Hestvík* in Iceland
(Jónsson in *Namn og Bygd*, 1916, pp. 76, 80). " Hest's
hope and clearing." Phonology, § 7.

Hitchcroft (Shilbottle). 1445 Pat. *Hitchecroft*.

" Hicca's croft." Cf. *hiccan thorn*, B.C.S. 1143.

Hobberlaw, earlier **Birtwell** (Alnwick). 1296 S.R. *Berte-
welle* ; 1454 Pat. *Bartewell* ; 1569 Tate ii. 262 *Byrtwell*
or *Uberlow*.

O.E. *beorhte wielle*=bright or clear spring. Cf. Brightwell
Baldwin, Oxon. The later name cannot be explained.

Holburn (Lowick). *c*. 1250 T.N. *Hoburn'* ; 1278 Ass.
Houburne ; 1361 Cl. *Hulbourne* ; 1539 F.P.D. *Holbo(u)rne* ;
1663 Rental *Howbourn*.

Holdforth (Auckland). 1382 Hatf. *Hol(le)forth*.

" Hollow stream and ford." Cf. *on holan baec, ford,*
B.C.S. 945. Phonology, §§ 39, 30.

Hole Row (Shotley). 1318 Inq. a.q.d. *Holes*; 1396 Ipm.
le Holerawe; 1663 Rental *Holrow.*

" (Row in the) hole(s) or hollow(s)." Phonology, § 16.

Holford (Shotton-in-Glendale). 1342 Cl. *Holford*; 1379
Holforth. v. Holdforth *supra.*

Hollingside (Whickham). 1382 Hatf. *Holynsyde.*
Holme Hill (Muggleswick). 1446 D.S.T. *le Holme.* **Holm-**
side (Lanchester). 1214 Pipe *Holneside*; 1297 Pap. *Holm-*
syde; B.B., 1339, R.P.D., 1358 Pat. *Holneset*; 1382 Hatf.
Holmeset; 1423.45 *Holmset.*

All alike are probably from O.E. *holegn*=holly-tree,
(cf. Hulne *infra*), dialectal *holm*. App. A, § 8.

Holstone House (Stockton). 1343 Hatf. *Holstanmore.*

Probably O.E. *(æt) hola(n)stane*=(at the) hollow stone
or rock, possibly some old boundary-stone.

Holy Island. *c.* 1125 F.P.D. *Haliæland*; 1255 Ass.
Halieland; 1273 R.H. *Halilaund.*

" Holy " from its association with early Christian
missionaries. For *eland, v.* Farne Island *supra*. Phonology, §§ 14, 5.

Holystone. 1240 Newm. *Halistane*; 1314 R.P.D.
Halistan; 1426 Sc. *Halystan*; 1539 Arch. 3. 4. 114 *Haly-*
stone; 1604 ib. 118 *Hollistones, Haliston*; 1658 ib. 121
Hallistan; 1724 ib. 122 *Holystone, Hallyston*; 1833 Map
Halystan.

O.E. *hālig-stān*=holy-stone. Leland tells us (v. 62)
" some hold opinion that at *Halistene* or in the River
Coquet thereabout over 3000 were christenyd in one day."
The legend may or may not be true, but the meaning of the
name is clear. *Halli-* and *Holli-* show shortening of the
vowel of North and South M.E. *haly* and *holy* respectively. Cf. *holiday* and *Halliday*, Holywell *infra* and
Holywell-st [holiwel], London. Phonology, §§ 14, 22.

Holywell (Earsdon) [haliwel]. 1218 Pipe *Halewell*;
c. 1250 T.N. *Haliwell*; 1346 F.A. *Halywell*; 1429
Ipm. *Halliwell*. (Wolsingham) [holiwəl]. 1361.45 *Haly-*
well.

"Holy spring." Cf. *halgan wyll*, B.C.S. 299. For local pronunciation cf. the name of the sulphur spring near the Steel in Hexhamshire, Holy Well on the ordnance map, but Halliwell locally. Phonology, §§ 14, 22.

Homer's Lane (Warden). 1479 B.B.H. *Hollemarsse* now *Holmerscrofte*. "Hollow-marsh." Cf. Owmers *infra* and *Holmers* in Eshott (N. vii. 327). Phonology, § 39.

Hooker Gate (Spen). 1587 Ryton *Huckergaite*, 1596 *Hookegate*, 1602 *Huckergayte*, 1611 *Howkeryeat.* "Huckster's road (*gata*, Part II) or gate (*geat*, Part II)." *Hukker* is once found in this sense (N.E.D. *s.v. hucker*). Later corrupted to the more common name *Hooker*.

Hoppen (Bamburgh). 1255 Ass. *Hopum*; 1296 S.R. *Hopune*; 1314 Ipm. *Hepon* alias *Hopene*; 1346 F.A. *Hopoun*; 1638 Freeh. *Hoppyn.* Possibly O.E. (*æt þǣm*) *hopum*=at the hopes (*v.* Part II), but the short vowel is a difficulty and the topography makes it unlikely.

*****Hopperclose** (Harbottle). 1331 Ipm. *Hoperesfeld*; 1618 Redesd. *Hopperclose.* An early example of *Hopper* used as a personal name.

Hoppyland (Hamsterley). 1342 Ipm. *Hopiland*; 1382 Hatf. *Hopyland.* Cf. *Hoppilegh*, Heref., Bannister, p. 97. An unsolved problem.

Horden (Easington). *c.* 1050 H.S.C. *Hore-tune, -dene*; 1260 Pat. *Horden*; 1313 R.P.D. *Hordon*, 1314 *Horden.* O.E. *hor(h)* or *horu-tūn* or *-denu*=filth-farm or valley. App. A, § 1.

Horncliffe (Norhamshire) [haˑkli]. *c.* 1250 T.N. *Hornecliff*; B.B. *Horcliva* (B., C. *Horneclyffe*); 1560 Raine *Horclife, Horkliffe*; 1580 Bord. *Harkley*; 1639 N.C.D. *Harcley.* Either "horn-shaped cliff" or "cliff on a horn of land," (Cf. Woodhorn *infra* and O.H.G. *Hornberc* in Förstemann), or "Horn's cliff." Cf. *Horn child* and O.N. *Horni.* Phonology, § 56; App. A, § 6.

Horsley (Ovingham). 1346 F.A. *Horsleye.* **Horsley,**

Long. 1197 Pipe *Horselega*. **Horsleyhope.** *c.* 1190 F.P.D. *Horsleihope*.

Cf. O.E. *horsa-lēah* (Middendorf, p. 75) for this obvious name.

Horton (Blyth). 1270 Ch. *Horton Shirreve*; 1300 De Banco *Horton Guyschard*. (Doddington) *c.* 1250 T.N. *Horton Turbervill*; 1346 F.A. *Horton Turbilwyle*. (Ponteland) 1346 F.A. *Horton*.

A very common place-name. Cf. *hortun*, B.C.S. 1158 = filth-farm, dirty farm. Horton in Blyth was so called from *Guiscard* de Charron, *Sheriff* of Nthb. (1267-70). Cf. Whisker Shiels *infra*. Horton in Doddington from Wm. Turberville [1] (cf. T.N.).

Houghall (Durham) [hɔfəl]. 1226 F.P.D. *Hocchale*, 1291 *Howhal(e)*, 1342 *Hochale*, 1539 *Houghalle*; 1446 D.S.T. *Hoghall*.

A difficult name. Possibly O.E. *hōh-hēale* (*hōh, healh*, Part ii), i.e. haugh of land at the foot of the heugh, a name descriptive of its actual position.

Houghton (Heddon-on-the-Wall). 1279 Ass. *Hochton*; 1663 Rental *Houghton*. **Longhoughton.** 1281 Perc. *Howton, c.* 1325 *Hoghton*. **Houghton-le-Side** (Gainford). 1200 B.M. *Hoctona*; *n.d.* R.P.D. *Hoghton*. **Houghton-le-Spring.** 1307 R.P.D. *Houghton*.

"Farm on the *hōh* (Part ii) of land." Cf. *hohtun*, B.C.S. 64. *le side*, because on a hill (cf. Chester-le-Street *supra*); *le Spring*, apparently from its owner. In Bp. Kellaw's Register (*s.a.* 1311) we read that Houghton belonged to Albreda, "relicta domini Henrici Spring." Introd., p. xxiii.

Housty [2] (Allendale). 1233 Gray *Hoggesti*, *n.d. Hoxsti*; 1608 Hexh. Surv. *Houstie*. **Houxty Burn** (Wark). 1304 Ass. *Houstyes*.

Possibly O.E. *hogges-stig(u)* = hog's sty, or, if *stigu* is used in the wider sense of any wooden enclosure or hall (cf.

---

[1] Turberville is an O.Norman name from *Torberville, Thouberville, Trublerville* (v. Fabricius, *Danske Minder i Normandiet*, pp. 205, 268), the *ville* of Ðorbjörn, found in D.B. as *Torber(n), Turber(n)*. For *Turbil-* v. Zachrisson, p. 120.

[2] Cf. also *Hokesti* (N. vi. 197), unidentified.

Bosworth Toller, s.v.) it may mean "Hogg(e)'s farm." In the former case the name was probably given in contempt. For the sound development cf. Foxden *supra*.

Houtley (Hexhamshire). 1243 Pat. *Holtolaye*; 1296 S.R. *Holteley*.

"Holte's clearing." Cf. *Holt* (D.B.) and O.N. *Holti* in Lind. Phonology, § 39.

Howburn (Carham). 1346 F.A. *Houb(o)urn*. **Howden Dene** (Corbridge). c. 1290 Perc. *Holden*.

"Hollow burn and valley." Phonology, § 39. Cf. Holburn *supra*.

Howick [houik]. Type I: c. 1100 N. ii. 359 *Hewic*. Type II: 1230 Pat. *Hawic*; 1278 Ass. *Hawick, Hawyk*; 1374 Acct. *Hawyk*. Type III: 1281 Wickw. *Howyk*; 1288 Ipm. *Howick*; 1291 Tax. *Howyk*; 1311 R.P.D. *Houwyk*; 1318 Inq. a.q.d. *Howyke*; 1340 Pat. *id.*; 1359 Cl. *Houwyk*.

Types II and III are explained by Lindkvist (p. 182) as showing alternative forms *hár* and *hór* of O.W.Sc. *hár*=high. *wick* he takes to be O.W.Sc. *vik*=creek, inlet, bay. Type I, if not due to a mistake, shows the influence of the English *hē(a)h*=high.

Howl (Ferryhill). c. 1350 Robt. de Grayst. *Howall*; 1362 D.S.T. *Howell*.

Howsdon Burn (Alwinton). 1290 Ch. *Hollisdon*.

Cf. Hollesley, Suff., and *holingaburna*, K.C.D. 722 from which Skeat (p. 79) infers an O.E. name *Hol*, "Holl's hill." Phonology, § 39.

Howtel (Kirknewton). 1226 Pipe *Holthale*; 1255 *id.*, *Holtele*; 1346 F.A. *Holtall*; 1480 Ipm. *Hotell*; 1542 Bord. Surv. *Howttyll*.

O.E. *holt-hēale* (dat.)=wooded-haugh or "Holt's haugh." Cf. Houtley *supra*. Phonology, §§ 39, 36.

Hudspeth (Elsdon). 1252 Ch. *Hodespeth*; 1297 Ipm. *Hodispeth*, 1324 *Hodespith*; 1628 Freeh. *Hudspeth*.

"Hod's path" (*pæð*, Part II). Cf. Hoddesdon, Herts., D.B. *Hodesdone* and Hodsock, Notts, B.C.S. 1282 *hodesac*. There was also an O.E. *Hudd*, cf. *huddesig*, B.C.S. 801. For such variant forms cf. M.E. *coss* and *cuss* (=kiss), *þrostle* and *þrustel*, and v. Morsbach, § 120 n. 3, Luick, *Hist. Gramm.*,

§ 78 n. 2. These variants explain *Hudesak* for Hodsock, *Hoddeswell* and *Huddeswell* for Hudswell, Yorks., *Hodenknole* for Huddeknoll, Glouc., and may help to explain the variation between *Hodere-* and *Hudere-* in the difficult name *Huddersfield.*

Hulam or **Holam** (Monk Heselden). *c.* 1050 H.S.C. *Holum*; *c.* 1200 F.P.D. *id.*; B.B. *Holome*; 1304 Cl. *Holum*; 1339 R.P.D. *id.*; 1539 F.P.D. *Holome*; 1756 Staindrop *Hullum.*

O.E. *holum* (dat. pl.)=(at the) hollows. Cf. *on holun*, K.C.D. 741, *of ðan holum*, B.C.S. 491. For the phonology, cf. Nthb. and Durh. [(h)uəl] for *hole.*

Hulne (Alnwick) [hul]. 1271 Pat. *Hol*; 1283 Perc. *Holne*; 1288 Ipm. *Holin*; 1295 Perc. *Holne*; 1296 S.R. *Holen*; 1334 Perc. *Holne*; *c.* 1590 Bord. *Hull*; 1790 N.C.D. *Hull.*

O.E. *holegn*=holly, Nthb. [holn]. Phonology, §§ 12, 56.

Humble Burn (N. Tyndale). 1302 Ass. *Suthumbleburne.* Probably so named from Humble Hill, *v.* Humbledon *infra.*

Humbledon Hill (Bp. Wearmouth). Type I: 1382 Hatf. *Hameldon.* Type II: 1303 R.P.D. *Homelmore*; 1408.35 *Homildon.* **Humbleton Hill** (Doddington). Type I: 1169 Pipe *Hameldun*; 1229 Pat. *Hameldon*; 1255 Ass. *id.*; *c.* 1250 T.N. *Hamildon.* Type II: 1296 S.R. *Homeldon*; 1346 F.A. *id.*; 1402 Sc. *Holmedon*, 1405 *Homeldone*, 1428 F.A., 1538 Must. *Homyldon*; 1579 Bord. *Homiltoun*; 1628 Freeh. *Homleton*, 1638 *Hombleton.* Type III: 1403 Pat. *Humbledon*; 1580 Bord. *Humbleton*; 1628 Freeh., 1638 *id.* (Westwick) *n.d.* F.P.D. *Homeldona.*

This and other names containing the same elements are fully discussed in an article by the present writer in *Namn og Bygd*, 1920 volume, and it is there shown that all alike probably contain an O.E. adj. *hamel*, "mutilated," which might be used of a hill of some particular shape or outline.[1] Forms in *o* are probably due to nasalising of *a* to *o* before *m* (Bülbring, § 123, Morsbach, § 88) and to confusion of this word with Scots. and North. dial. *hommyll, homill, hummell, humble*=

---

[1] Humbleton, in Doddington, is a hill with a well-marked cleft in it.

hornless, dodded, a word which is itself related to *hamel.*
Cf. Dodd Hill used in Scotland and elsewhere (Maxwell,
p. 157). Humbleton Hill (Bp. Wearmouth) is a well-rounded,
" dodded hill." Phonology, § 55 ; App. A, § 1.

Humshaugh (Simonburn). 1279 Iter. *Hounshale* ; 1307
Pat. *Hounshalgh* ; 1318 Ipm. *Homeshalk* ; 1358 Pat.
Homysalgh ; 1373 Orig. *Hounshalgh* ; 1386 N. 2. 3. 21
Homsalgh ; 1580 Bord. *Hemshaugh* ; 1663 Rental *Hums-haugh.*
" The haugh of Hun." O.E. *Hūn* >M.E. *Houn.* *n* >*m*,
perhaps by a process of dissimilation. Phonology, §§ 21, 52.

Hunstanworth [huntənwud]. B.B. *Hunstanwortha* ;
1694 Stanh. *Hunsonworth,* 1697 *Hunsenwood,* 1727
Husenwood.
" Hunstan's enclosure." Phonology, § 53. Local
tradition explains its own pronunciation as " hunting wood
(or forest of the monks of Blanchland)." App. A, § 3.

Hunterley Hill (Muggleswick). 1311 R.P.D. *Hunterlaw.*
" Hunter-hill." The modern form is pleonastic. App.
A, § 2.

***Huntland** (Wark and Simonburn). 1177 Swinb.
Hunteland. **Huntlaw** (Whalton). 1279 Iter. *Huntelaw.*
Huntshield Ford (Stanhope). 1458.35 *Huntsheleford.*
" Hunter's or hunt-land, -hill and -shiel." Cf. M.E.
hunte=hunter or hunt, and Huntlands, Heref. (Bannister,
p. 100).

Hunwick (Auckland). 1104-8 S.D. *Hunewic* ; 1446
D.S.T. *Hunwyke.*
" Hun's dwelling." Cf. Humshaugh *supra.*

Hurbuck (Lanchester). 1303 R.P.D. *Hurthebuck,* 1312
Hurtebuckside.
Possibly the same as O.N. *hurðarbak*=back of the door,
space behind it. This seems to have been used in place-
names, but exactly with what sense is not clear. Fritzner
(*s.v.*) mentions three such in Norway. Kålund, in the index
to his *Historisk-topografisk beskrivels af Island,* gives three
in Iceland, and Jónsson, *Bæjanöfni á Islandi,* has others.
Some Scandinavian settler may so have named his farm by
simple transference of the name without its having neces-

sarily any direct application to the English place-name. Phonology, § 53.

Hurworth-on-Tees. *c.* 1190 Godr. *Hurdevorde*; 1252 D.S.T. *Hurthewrth*; 1311 R.P.D. *Hortheworth*, 1312 *Hurtheworth*; 1400 D.S.T. *Hurrworth*. **Hurworth Bryan.** 1438.34 *Hurworth Bryan*, otherwise called *Hurworth-on-the-Moor*.

The first element is perhaps O.E. *hurð, " wickerwork, hurdle." (Cf. O.N. *hurð*, Goth. *haurds*=door, O.H.G. *hurt*=wicker-work), dim. *hyrdel*=hurdle. This is found in German place-names (Förstemann, col. 1514), and the whole name would mean "hurdle enclosure." Cf. *tuneweorð*, B.C.S. 994 and *tunles weorð*, 820, meaning "hedge- and hedgeless enclosure."

Hutton Henry (Monk Heselden). *c.* 1050 H.S.C. *Hotun*; 1307 R.P.D. *Hoton*; B.B. *id.* (C. *Hotton*); 1430 F.P.D. *Huton*; 1446 D.S.T. *Hoton*.

Probably O.W.Sc. *hór*=high and *tún*=farm. Cf. Howick *supra*. The village stands on high ground (S. i. 58). " Henry " from its owner Henry de Eshe (Hatf. Surv.)

Hylton (Monkwearmouth). 1312 R.P.D. *Hilton*; 1335 Ch. *Helton*; 1539 F.P.D. *Hylton*.

" Hill-farm." Phonology, § 10.

Ilderton (nr. Wooler). 1189 Abbr. *Hilderton*; 1228 F.P.D. *Ildertone*; 1255 Ass. (*H*)*ilderton*; 1291 Tax. *Hildirton*; 1311 R.P.D. *Ildirton*; *c.* 1250 T.N. *Hildirton*; 1336 Ch. (*H*)*ildreton*; 1346 F.A. (*H*)*ildreton, Hillerton*, 1428 *Ilderton*; 1538 Must. *Yeld'ton*.

" Hild's farm," *v.* Lindkvist, pp. 10-11. *Hilder*=O.W.Sc. *Hildar*, gen. sg. of *Hild* (*f.*). Cf. Hinderclay, Suff., D.B. *Hilderclea*. Phonology, §§ 35, 9.

Ingleton (Staindrop). 1104-8 S.D. *Ingeltun*.

" Ingeld's farm." O.E. *Ingeld* should give Mod. Eng. *Inyeld*. The *g* must be due to the influence of O.W.Sc. *Ingjaldr*, M.E. *Ingald* or *Ingold* found in Ingoldmells, Lincs., T.N. *Ingoldemol*, Ingoldisthorpe, Norf., T.N. *Ingaldesthorp*.

Ingoe (Stamfordham). 1229 Pat. *Hinghou*; 1244 Ipm.

c. 1250 T.N. *id*; 1304 Ch. *Inggou*; 1324 Ipm. *Inghow*;
1346 F.A. *Yengew, Ingowe*; 1524 Raine *Yngoo.*
  "*Inga's hōh*" (Part II), *Inga* being a short form for one
of the O.E. names in *Ing-.* Phonology, § 36.
  Ingram. 1244 Cl. *Angreham*; 1255 Ass. *Angram*[1];
1283 Ipm. *Hang(e)rham, Angeharm (sic); c.* 1250 T.N., 1291
Tax., 1313 Perc. *Angerham*; 1324 Ipm. *Angra(ha)m*;
1333 Ch. *Angreham*; 1346 F.A. *Angham, Angram,* 1428
Ayngrame; 1507 D.S.T. *Yngram*; 1538 Must. *Ingreme.*
  The first element may have either of the meanings
suggested for Angerton *supra.* The change from *Ang-* to
Ayng-, Ing- is difficult. Cf. Nthb. [θeŋ], [teŋz] for *thong,
tongs* (O.E. *þwange, tange*).
  Irthing, R. 1278 Sc. *Erthingge*; 1402 Pap. *Hirthenam*;
1479 B.B.H. *Yrthin.*
  A Celtic river-name.
  ***Isehaugh** (Mitford). 1370 Pat. *Ineshaulgh*; 1456 Ipm.
Isehaugh.
  "Haugh of *Ine* or *Ini.*" Phonology, § 53.
  Islandshire. 1107 F.P.D. *Ealondscire*; B.B. *Elandshire*;
1539 F.P.D. *Elaundshier.*
  The shire (*scīr*, Part II) grouped around Holy Island,
one of the outlying parts of the patrimony of St Cuthbert
and, until 1844, one of the liberties of the Bishopric of
Durham. *Island* for *Eland* (cf. Ponteland *infra*) under
the influence of St. Eng.
  Island Farm (Bp. Middleham). 1491.36 *Eland. v.*
Islandshire *supra.*
  Ivesley (Knitsley). 1382 Hatf. *Ivesleyburdon*; 1757
Lanch. *Isley.* **Iveston** (Lanchester) [aistən]. 1297 Pap.
Yvestan; B.B. *Ivestan*; 1303 R.P.D. *Ivestane*; 1637
Camd., 1646 Map *Iseton.*
  "The clearing and rock of *Ifa* or *Ivo.*" *Ifa* and its
patronymic *Ifing* are found in O.E., *Ifa* being probably
a shortened form of such a name as *Ifweald.* For *Ivo, v.*
Forssner, p. 168. App. A, § 7.
  Jarrow-on-Tyne. *c.* 750 Bede *In Gyruum*; *c.* 1104-8

---

[1] Angram, Yorks., earlier *Angerum*, is taken by Goodall (p. 59) to be
dat. pl. of the word *anger.*

S.D. *Gyruum, Girwe, Girvum*; *c.* 1125 F.P.D. *Gyruum, Gyrwe, Girue,* 1203 *Girwuum,* 1228 *Jarwe*; 1335 Ch. *Gyrue*; 1345 R.P.D. *Jarou*; 1396 D.S.T. *Jarrow.*
"(Among the) Gyruii," a tribal name found elsewhere in Bede for a people between Mercia and East Anglia. (Cf. Chadwick, *Origin of the English Nation*, p. 8). For [j] > [dʒ] cf. Jesmond *infra* and Jevington, Suss., earlier *Yevinton.* (Roberts, p. 96). Dr Fowler quotes me "Yarrow Monastery" from an engraving dated 1728, showing a late survival of [j].

Jesmond (Newcastle-on-Tyne). 1204 Pipe *Gesemue*; 1242 Pat. *Jesemuth*; 1255 Ass. *Gesemue*; 1254 Pat. *Jesemuth*; 1297 Ipm. *Yesmewe*; 1312 Inq. a.q.d. *Jesemuth*; 1333 Ipm. *id.*; 1346 F.A. *Zesemuth*; 1378 Ipm. *Jesemuthe*; 1414 Inq. a.q.d. *Gesmond*; 1428 Ipm. *Jesmuth* alias *Jesmund*; 1449 Pat. *Jessemond, Jessemuth*; 1514 Arch. 2. 1. 31 *Jesmound,* 1556 *Gesmonde*; 1711 Long Benton *Jazment*; 1772 Ponteland *Jasemond.*
"Mouth of the Ouseburn." The old name for this stream was *Yese* and initial *y* has become *j* [dʒ] as in Jarrow *supra, v.* Zachrisson, pp. 57 ff. Jesmond is a mile from the mouth, but cf. Stourmouth, Kent. For the change of suffix Zachrisson (p. 62) suggests substitution of -*mond* from A.N. *mont, mond,* "hill," possibly following on spellings of *Gesmonth* with *n* for *u* or, alternatively, an introduction of *mond*=mouth of a river, common on the Continent, as in Termonde, Belgium. There is no authority for the local legend of "Jesus' mound."

*Karswelleas (Redesdale). 1360 Pat. *Cresswelle Leghes*; 1618 Redesd. *Karswelleas.*
"Fields by the cress-spring," *v.* Cresswell *supra.*

Kearsley (Stamfordham). 1244 Ipm. *Kerneslawe*[1]; 1273 R.H. *Kerneslau*; 1278 Ass. *Kirneslawe,* 1278 *Kermeslawe*; 1346 F.A., 1361 Cl. *Kereslaw*; 1361 Cl. *id.*; 1454 Ipm. *Careslawe*; 1638 Freeh. *Kearsley.*
"Hill of Kjarni or Crin." O.W.Sc. *Kjarni* (Jónsson, p. 314) is probably found in Carnforth, Lancs., earlier *Chreneford, Kerneford* (Wyld, p. 86). *Crin* (D.B.) or *Crina,* the name of a moneyer of Cnut, would, by metathesis, give

---

[1] In R.P.D. there is a *Kirneschaw,* which may contain the same name.

Kern-, Kirn-. Cf. Mutschmann's explanation of Kersall, Notts., earlier *Kyrnessale* (p. 73). Phonology, § 53 ; App. A, § 2.

Keenleyside (Allendale). 1230 Gray *Kenleya* ; 1343 Pat. *Kynley* ; 1547 Hexh. Surv. *Keneley* ; 1552 Bord. Laws *Keynleye* ; 1608 Hexh. Surv. *Kinleyside* ; 1610 Speed *Kineleyside* ; 1637 Camd. *id.* ; 1663 Rental *Kenley.* " Hill by *Cēna's* clearing." Phonology, §§ 21, 7.

Keepwick (St John Lee). 1279 Iter. *Kepwike* ; 1298 B.B.H. *Kepwyk*, 1479 *Kepewyk* ; 1653 Comps. *Keepicke.* " Kepe's dwelling." This name is not found in O.E., but has its parallel in Frisian. *v.* Winkler (p. 212) who gives also a patronymic *Kepynga.*

Kellah (Featherstone).[1] 1279 Iter. *Kellaw* ; 1479 B.B.H. *Kellaw, Kelloue.*

Kelloe. *c.* 1170 Reg. Dun. *Kelflau* ; 1312 R.P.D. *Kellawe* ; 1400 D.S.T. *Kellow* ; 1679 Houghton *Kelley.* O.E. *cealf-hlāw*=calf-hill. Phonology, § 53. *-low* is S. and Mod. English. App. A, § 12.

Kenners Dene (Tynemouth). 1295 N. viii. 223 *Kenewaldesden.* " Cenwald's valley." Phonology, §§ 22, 49. O.E. *Cēnwald.*

Kenton (Gosforth). 1255 Ass. *Kynton, Quenton* ; *c.* 1250 T.N. *Kinton* ; 1309 Ipm. *Kynton* ; 1346 F.A. *Kyn(g)-ton*, 1428 *Kynton* ; 1432 Pat. *Kyneton* ; 1537 F.F. *Keynton* ; 1550 V.N. *id.* ; 1638 Freeh. *Kyn(e)ton* ; 1651 Comps. *Kineton.*

Cf. Kineton or Kington, Warw., found twice, for which Duignan (p. 76) gives forms *cyngtun* (B.C.S. 1234), *Cintone, Quintone* (D.B.) in the one case, and *Kynton* (14th c.) in the other, Kineton in Temple Guiting, Glouc., 1330 Ch. *Kyngton*, Keinton Mandeville, Som., D.B. *Chintune, c.* 1300 B.M. *Kyngton*, 1428 F.A. *Keinton.* The forms may in part be due to alternative O.E. forms *cyne-tun*=royal-farm and *cyning-tun*=king-farm, in part to *ng* becoming *n* before *t.* Phonology, § 10.

---

[1] Hugo de *Calflawe*, who signed an agreement between the convents of Hexham and Lambley (*Hexham Priory*, ii. 48), may well have come from here.

Kepier (Durham). *c.* 1310 R.P.D. *Kypier, Kypyer, Kypiyer, Kypere.*

The second element may be dial. *yare,* " wear or dam thrown across a river and often used for taking salmon in their upward course " (Greenwell, Glossary to *Hatf. Surv.*). Cf. Yearhaugh *infra.* Greenwell further suggests that the name means " yare which *keeps* or catches the fish." The M.E. forms are against this.

Ketton (Aycliffe). 1091-2 F.P.D. *Cathona, c.* 1125 *Cattun, Chettune,* 1135-54 *Chettune,* 1228 *Kettone*; 1335 Ch. *Ketton*; B.B. *Kettona.*

Cf. Ketton, Rutl., D.B. *Chetene,* Ketford, Glouc. (Baddeley, p. 95). All alike may contain the name *Kett,* which is perhaps identical with M.E. *ket,* " flesh," O.N. *kiǫt,* " flesh," used as a nickname. Forms in *Cat* are perhaps due to the influence of that more common name, *v.* Catton *supra.*

Keverstone (Staindrop). 1306, 1317 Pat. *Kevrestone*; *c.* 1330 D.S.T. *Kewreston.*

Cf. Keresforth, Yorks., earlier *Keuerisforth,* and the name *Cheure* (D.B.) noted by Goodall (p. 189). Moorman (p. 112) thinks that the latter is a Scand. form of O.E. *ceafor,* " beetle," but there is no evidence for the use of any cognate in the Scand. dialects. There does seem to have been a S. English name which may be the same as this with palatalised initial consonant. Cf. Charingworth, Glouc., D.B. *Chevringaurde,* Cheston in Ugborough, Dev. F.A. *Chevereston,* and possibly Cheverton in Brading, I. of Wt. D.B. *Cevredone.* App. A, § 7.

Keyhirst (Ewesley) [ka·(h)əst]. 1292 Ass. *Kahirst*; 1745 Netherw. *Kehirst.*

" Jackdaw-wood." Cf. *cafeld,* B.C.S. 1052 and Cawood, Yorks. B.C.S. 1102 *kawudu.* O.E. *cā* > North. M.E. *ka* > Nthb. *kae.*

Kibblesworth (Lamesley). 1185 F.P.D. *Kybbleswurth.*

" Kibble's enclosure." Cf. Kibblesworth, Warw. B.C.S. 455 *cybles weorðig.* *Cybbel* is a dimin. of *Cybba.* Cf. *cybbanstan* B.C.S. 1002.

Kidland (Holystone). 1271 Ch. *Kideland*; 1292 Q.W.

Kidelaund; 1663 Rental *Keednall*; 1704 Alnham *Kidlin*; *c.* 1760 Map. *Keedland.*

O.E. *Cyda(n)land=* Cyda's land. Phonology, §§ 21, 56.

Kielder (North Tyndale). 1309 Sc. *Keldre*; 1325 Ipm. *Keilder,* 1329 *Keldirheies*; 1330 Fine *Kailder*; 1370 Cl. *Keldreshays, Keldre*; 1542 Bord. Surv. *Keylder*; 1663 Rental *Keilder.*

Pre-English and probably by origin a river-name, cf. Calder, R. Cu., Yorks., Lancs. (2), of which the earlier forms are *Kaldre* and *Keldre.* -*hayes* = hedges (O.E. *hege*), enclosures.

Kilham (Kirknewton). 1176 Pipe *Killum*; 1216 B.M. *Kyllum*; 1227 Ch. *Killum*; 1255 Ass., *c.* 1250 T.N. *id.*; 1323 Ipm. *Kylnom*; 1335 Ch. *Killum*; 1442 Ipm. *id.,* 1480 *Kilholme*; 1542 Bord. Surv. *Kylham.*

O.E. *cylnum* (dat. pl.) = (at the) kilns. Phonology, § 50; App. A, § 6.

Killerby (Heighington). 1091 F.P.D. *Culuerdebi*; 1197 Pipe *id.*; 1207 F.P.D. *Kiluerdebi*; 1312 R.P.D. *Kyllewardby,* 1313 *Kilverby*; B.B. *Killirby* (B., C. *Kylwerby*).

Cf. Kilwardby, Leic., Leic. Surv. *Culverteb'*, Killerby, Yorks. (twice). Björkman (Z.E.N. p. 54, N.P. p. 81) takes these to contain a hybrid personal name compounded of O.W.Sc. *Ketill* (Late O.E. *Cytel*) and English -*weard.* Phonology, §§ 49, 53. " Ketilweard's *by* " (Part II).

Killingworth (Long Benton). 1251 Ch. *Killingworth*; *c.* 1250 T.N. *Killingworth*; 1255 Ass. *Cullingwurth*; 1346 F.A. *Killyngworth.*

"Farm of Cylla or his sons." Cf. Killinghall, Yorks. (Moorman, p. 114) and Kilnwick, Yorks., earlier *Killingwyk.*

Kimblesworth (nr. Witton Gilbert). 1216-72 B.M. *Kymliswrth, Kimleswrthe*; 1312, 1315 R.P.D. *Kym(b)elesworth.*

"Cymel's enclosure." **Cymel* is dimin. of *Cyma.* Phonology, § 55.

Kimmerston (Ford). 1244 Ch. *Kynemereston*; 1254 Ipm. *Kenemeriston*; 1340 Ch. *Kynmerston*; 1346 F.A. *Kinmerston, Kylmerston,* 1428 *Kymerston.*

"Cynemær's farm." The same name is found in

Kempsford, Glouc., and Kilmersdon, Som., Kemerton, Glouc. Phonology, §§ 57, 52.

Kingswood (Whitfield). 1135 H. 2. 3. 8 *Kingeswood*. Self-explanatory.

Kipperlynn (Bywell St Peter). Type I : 1307 N. vi. 190 *Skitterlyn* ; 1620 N. vi. 195 *Skitterinlyn* ; 1663 Rental *Skitterlyn*. Type II : 1719 N. vi. 96 *Lyndeen* alias *Skipperline*. There is a vb. *skite*, "to void excrement," with derivative *skitter*, "to void their excrement," and the term *skittering* is a term of contempt. *Skitterlyn* probably means therefore "trickling stream" (*lyn*. Part ii). Cf. *Skytteren, Skytra*, river-names in Norway (N.G. ii. 285), *Skitterick*, R., Yorks. (Goodall, p. 259), *Skitermyln* in Heworth (D.S.T.), Skitter, Lincs. c. 1150 B.M. *Scitra, Schitere*. *Skipper-* and *Kipper-* are probably due, the first to squeamishness, the second to humour.[1]

Kirkharle. 1177 Pipe *Herle* ; c. 1250 T.N. *Kyrkeherle* ; 1346 F.A. *Kyrkherll*, 1428 *Kirkehirle*.

Harle is probably one of those rare place-names in which the gen. sg. of a personal name is used by itself. Cf. Bell Shiel *supra*. It is from O.E. **Herela*, inferred by Skeat for Harlton, Cambs. (p. 10) and found also in Harlthorpe, Yorks. *Kirk* because marked by a church.

Kirkhaugh (S. Tyndale). 1236-45 Swinb. *Kyrchalu* ; 1279 Iter. *Kirkehalghe* ; 1507 D.S.T. *Kirkhaugh*. **Kirkheaton.** 1296 S.R. *Kyrkeheton*.

"The haugh and the *Heaton* (*v. supra*), marked by a church."

Kirkley (Ponteland). 1175 Pipe *Crikelawe* ; 1255 Ass. *Grekelawe* ; 1257 Ch. *Crickelawe*, 1267 *Crekellawe* ; 1275 Cl. *Kirkelawe* ; 1278 Ass. *Creckelawe* ; 1289 Ipm. *Crekkelawe* ; 1291 Ch. *Creckelawe* ; 1298 B.B.H. *Crekelagh* ; c. 1250 T.N. *Crekelawe* ; 1311 Ipm. *Creklawe*, 1342 *Criklawe* ; 1346 F.A. *Kirklawe*, 1428 *Kirkelawe* ; 1479 B.B.H. *Craklawe* ; 1638 Freeh. *Kirkley*.

Cf. Johnston, *Place-Names of England and Wales*

---

[1] O.E. *sciteres -flod and -stream*, B.C.S. 129, 1200, are perhaps similarly primitive in their suggestion. We may note an Icelandic parallel to this name—*Mígandi á* (N. o. B. ii. 27).

(p. 220), where *s.n. Creech*, he quotes *collem qui dicitur brittanica lingua Cructan apud nos Crycbeorh* (B.C.S. 62). O.E. *cryc-beorg*=North. M. Eng. *crikelawe*. Cf. also Creech Hill, Som., B.C.S. 112 *crichhulle*. In all alike the second English element translated the first Celtic one, so that the name is really "hill-hill." App. A, § 2.

Kirk Merrington (nr. Auckland). *c.* 1125 F.P.D. *Mærintun, Meringtonas*[1]; *c.* 1200 Joh. Hex. *Merringtun.* "Farm of Mæra or his sons." *Mæra* is a shortened form of an O.E. name in *Mær-*. Cf. Meering, Notts., D.B. *Meringe.* Phonology, § 22.

Kirknewton. 1336 Ch. *Niweton in Glendala.* Distinguished by its church or by its position in Glendale.

Kirkwhelpington. 1182 Pipe *Welpinton*; 1267 Ch. *Whelpinton, Welpington.*

D.B. gives a name *Welp* which may go back to O.E. *hwelp* (cf. *hwelpes dell*, B.C.S. 596) or, more probably, to O.W.Sc. *hvelpr*, used as a nickname (cf. the Orkney earl named *Hvelpr eða Hundi Sigurðarson*, i.e. Whelp or Hound, son of *Sigurðr*). "Farm of Whelp."

Knar (Knaresdale). *c.* 1275 Anc. D. *Knar*; 1325 Ipm. *Knarre.*

Knaresdale [naˑzdəl]. *c.* 1240 Swinb. *Cnaresdale*; 1255 Ass. *Gnaresdale*; 1291 Tax. *Knaresdale*; 1798 St Mary le B. *Knarsdale.*

Before discussing this name it should be noted that in addition to the farm-name *Knar*, there is a *Knar* stream, "a rough mountain torrent which intersects the western portion of it (i.e. Knaresdale) from west to east" (H. 2. 3. 78). Further certain other English and Scandinavian names call for notice.

In Yorkshire we have Skelden, earlier *Chenares-, Kenares-, Neresford* (D.B.), *Cnarresford, Knarford* (*c.* 1300), and Knaresborough, earlier *Cnardesburc* (1159), *Chenaresburg* (D.B.), *Knaresburgh.* These probably contain O.W.Sc. *Knǫrr* (gen. *Knarar*), a personal name (Björkman, Z.E.N. p. 55).

In Norway *Knardal* and *Knarredalen* are of fairly

---

[1] Pl. because there were E., W., and Mid. Merrington (F.P.D. 1539).

I

frequent occurrence, and Rygh (*G.P.* pp. 162-3) believes these to contain the same personal name. Similarly in Iceland we have *Knarartunga, Knararnes, Knararhöfn* which clearly contain this name (Jónsson, *Bæjanöfn*, pp. 487, 493, 514). There are also place-names *Knörr* (Jónsson *u.s.* p. 572), *Knarberg, Knarfjeldet* (N.G. i. 199) which probably are derived from O.W.Sc. *knǫrr,* " a large kind of ship," used also apparently of a piece of land or a hill of that shape.

With these points before us alternative solutions may be offered :—(1) that the valley was first called " Knǫrr's dale," then the river was named Knar by a process of back-formation, and finally the farm took its name from the river, or means " Knǫrr's farm," with suppression of the second element (cf. Bell Shiel *supra*) ; (2) the farm was called *Knar* by some Scandinavian settler after a *Knörr* in his own home, the valley was then called *Knaresdale*, with pseudo-genitival *s*, and finally the stream named after the farm or by the process of back-formation suggested above ; (3) the farm and valley were named after the same man, and that the farm name lost its suffix, while the river was named after the farm. On the whole the first seems the most likely solution.

Knitsley (nr. Consett). 1303 R.P.D. *Knyhtheley*, 1312 *Knycheley*, 1313 *Knyghteley*, 1382 Hatf., 1453 F.P.D. *Knycheley* ; 1587 Wills *Knitchley* ; 1621 Esh *Knitsley* ; 1637 Camd. *Knichley* ; 1768 Map *id.*

O.E. *cnihtes-leage* (dat.)= knight's clearing, *cniht* being used either as a personal name (cf. *Cniht,* a moneyer to Cnut) or in its old sense of servant or young warrior (cf. *cnihta land* B.C.S. 917). Phonology, § 40.

Kyloe. Type I: *c.* 1170 D.S.T. *Culei* ; 1335 Ch. *Culeia*. Type II: 1228 F.P.D. *Killey* ; *c.* 1250 T.N. *Kylei* ; 1344 R.P.D. *Kylay* ; 1460 H. 3. 1. 30 *Kilay* ; 1539 F.P.D. *Kylow,* · *Kylay* ; 1550 H. 3. 2. 207 *Kylo* ; 1560 Raine *Kylhowe, Killowe,* 1636 *Kilo* ; 1637 Camd. *Killey* ; 1724 Chatton *Keiloe* ; 1758 Alnham *Keiley* ; 1771 Ilderton *Kylo*.

" Cow or kye clearing." Type I from O.E. *cū* = cow ; Type II from pl. *cȳ*.

Kyo (Lanchester). *c.* 1200 D.S.T. *Kyhou*; *c.* 1240 Finch. *Kyhow*; 1382 Hatf. *Kyowe*; 1673 Ryton *Kia.*

hōh[1] (Part II) on which the "kye" pasture. Phonology, § 36.

Ladley (Wolsingham). 1242 D.Ass. *Laddeley*; 1366.32 *Ladley*; 1422.45 *Ladle.*

"Ladda's clearing." Cf. the signature "Godric *Ladda*" quoted in the N.E.D. *s.v. lad*, from an 11th cent. document.

Lambley (Knaresdale). 1201 R.C. *Lambeley*; 1542 Bord. Surv. *Lamley.*

Lambton (Chester-le-Street). 1297 Pat. *Lampton*; 1314 R.P.D. *Lambeton*, 1334 *Lampton*; 1698 Sherb. *id.*

"Lambs'-clearing and farm." Cf. *lambaham*, B.C.S. 402. Phonology, § 51.

Lamesley (Chester-le-Street). 1297 Pap. *Lamelay*; 1312 R.P.D. *Lamesley, Lomesley*; 1340 R.P.D. *Lamesleye.*

A difficult name. Possibly from O.E. *lama, loma,* "lame," may have been derived from a nickname *Lame, Lome,* hence "Lame's clearing."

Lampart (Haltwhistle). 1291 Ipm. *Lythel lampard,* 1328 *Lampard*; 1329 Fine *id.*; 1372 Swinb. *Parva Lamparde*; 1564-94 Map *Lamprade.*

Clearly not of English origin.

Lanchester. 1197 Pipe *Langecestre*; 1345 R.P.D. *Langechestre.*

"Long chester or fort." Some have identified it with the *Longovic(i)o* of the *Notitia* (M'Clure, p. 114). Phonology, §§ 6, 51.

Landieu (Wolsingham). 1228 F.P.D. *Landa Dei*; 1637 Camd. *Landew.*

A Fr. form. Cf. F.P.D. p. 216, "locus qui vocatur Landa Dei . . . concedimus et confirmamus in perpetuum sacristariae Dunelmensi."

Langhope (Hexhamshire). 1229 Gray *Langhop*; 1663 Rental *Langupp.* **Langley** (Haydon). *c.* 1175 H. 2. 3. 366 *Langalea.* (Lanchester) B.B. *Langleia.* **Langton** (Gainford). 1104-8 S.D. *Langadun*; 1313 R.P.D. *Langeton.*

---

[1] Kyloe Registers (by the kindness of the Rev. W. C. Harris) 1691 *Keylloe,* 1695 *Kiloe,* 1701 *Keillo,* 1710 *Kyloe.*

Lanton (Kirknewton). 1255 Ass. *Langeton* ; 1638 Freeh. *Lanton*.

" Long hope, clearing, farm or hill." Phonology, §§ 6, 36, 51 ; App. A, §

Layton (Sedgefield). *c*. 1190 Godr. *Latune* ; 1284 Finch. *Laton*.[1]

Cf. Layton, Lancs. (Wyld, p. 171), with the same forms. An unsolved problem.

Leadgate (Chopwell). 1590 Ryton *Lidgate*, 1605 *The Lide Yate*, 1612 *Lidge yeat*, 1613 *Lidyate*, 1617 *Leadgait*.

O.E. *hlid-geat*, " swing-gate," found in dialect either as [lidžit] (Lincs.) or *liggate*, *ligget* (Scotl.). The modern form is corrupt.

Leam (Redesdale). 1175 Pipe *Leum* ; *c*. 1250 T.N. *Lem* ; 1297 Ipm. *id.* ; 1327 Orig. *la Lene* ; 1331 Ipm. *Le Leme* ; 1346 F.A. *Leme* ; 1359 Pat. *Leem* ; 1618 Freeh. *Overleame*. **The Leam** (Heworth). *c*. 1200 F.P.D. *le Lem* ; 1365 Halm. *le Leme*.

O.E. *lǣg-hām* = fallow, unploughed homestead, farm laid down to grass. Cf. *lea-rig* (N.E.D.). Phonology, § 36.

Leamside. 1380 Halm. *le Lemside*.

" Hill by *The Leam* (*u.s.*)."

Learchild (Edlingham). 1247 Sc. *Leverilcheld* ; 1252 Pipe *Luerescheld* ; 1255 Ass. *Leverichull* ; *c*. 1250 T.N. *Levericheheld* ; 1428 F.A. *Leverchyld* ; 1586 Raine *Lurchild* ; 1628 Arch. i. 3. 94 *Leerchild*. **Learmouth** (Carham). 1176 Pipe *Leuremue*, 1226 *Livermue* ; 1251 Ch. *Levermue* ; 1255 Ass. *id.* ; 1346 F.A. *Levermuth* ; 1461 Ipm. *id.* ; 1542 Bord. Surv. *Leremouthe*.

" Spring (*celde*, Part II) and mouth or estuary of Leofhere." For *Leuer* and *Luuer*, *v.* Wyld on Liverpool (p. 177) and cf. Lorbottle *infra*. For *Lear*, cf. *Lerpoole* = Liverpool, for *Luuer* cf. Loversall, Yorks. Learmouth may possibly be O.E. *lefer-mūþ* = estuary overgrown with *levers* or *livers*,

---

[1] It is tempting to take this as M.E. *lagh-tun* = low farm, but the absence of forms with *gh* or *h* is difficult to explain. Note, however, Layton, Yorks., D.B. *Lastun*, *Latton*, Kirkby's Inq. *Laton*, which seems to go back to this.

a species of yellow flag. Cf. Livermere, Suff. (Skeat, p. 83).
Phonology, § 45. For -*mue*, *v.* Zachrisson, pp. 82-3.
Leas Hall (Catton). 1255 Ass. *Leyes.* **Lee Hall** (Bellingham). 1415 Ipm. *La leye.* **Lees** (Haydon). 1368
Ipm. *Leghes. v. lēah*, Part II.

Leighton, Green (Hartburn) [gri·nlaitən]. 1252 **Pipe**
Litendon; 1255 Ass. *Lightdon, Lutedon*; 1268 Ass.
Lychecedon; *c.* 1250 T.N. *Lythedun*, 1273 R.H., 1288
Ipm. *Lithedon*; 1305 Ch. *Litendon*, 1307 *Lityndon*; 1324
Ipm. *Lightyndon*; 1346 F.A. *Lichdon*; 1360 Pat. *Grene-
lighton*; 1378 Ipm. *Lighton*; 1411 Inq. aqd. *Lyghton*;
1428 F.A. *id.*; 1663 Rental *Greenligton.*
"Lihtwine's hill." Cf. *Lihtwine* (D.B.) and *lihtenes-
ford*, B.C.S. 1117. Phonology, §§ 49, 59; App. A, § 1.

Lemmington[1] (Edlingham). Type I: 1157 Pipe *Lemetun*,
1185 *Lemechton*; 1200 R.C. *Lemocton*; 1229 Pat. *Lemoke-
ton*; 1255 Ass. *Lemmocton, Lem(m)ecton*; *c.* 1250 T.N.
Lemotton; 1289 Ipm. *Lemoton*, 1308 *Lemothon*; 1309
Ch. *Lemothton*; 1334 Perc. *id.*; 1395 Ipm. *Lematon*; 1428
F.A. *id.*; 1538 Must. *Lamadon*; 1583 N. vii. 167 *Lea-
mockdon.* Type II: 1247 Sc. *Lemontone*; 1278 Ass.
Lemanthon, Lemangton; 1402 Ipm. *Leman(g)ton*; 1589
Bord. *Lemmanton*; 1628 Arch. I. 3. 94 *Leamondon*; 1663
Rental *Leamendon*; 1722 Edl. *Lemonden*, 1724 *Lemingdon,
Lemington.*

Type I is O.E. *hleomoc-tūn* = brook-lime farm, one where
this species of speedwell grows. O.E. *hleomoc* > M.E.
lem(e)ke, lemoke, leomeke. Cf. also *Lemetheley* in Sturton
(N. v. 241), *t* being a common mistake for *c*. Type II is
probably developed from Type I (*Lematon*) by the introduction of *n* in an unstressed syllable (Phonology, § 55). It
was further influenced by association with *leman* = sweet-
heart, and ultimately given a pseudo-patronymic form.
App. A, § 1.

Lesbury. *c.* 1190 Godr. *Lechesbiri*; 1228 F.P.D.

---

[1] The earliest forms found for Lemmington in Newburn are 1649 Ryton
Leamadon; 1692 Newb. *Lementon*; 1696 Ryton *Laminton*; 1725 Newb.
Lemmington. These forms look as if the name was identical with
Lemmington in Edlingham. Possibly the one was named from the other.

Lescebr' ; 1255 Ass. *Lescebyr*, 1278 *Lastebir* ; 1280 Ch.
Lessebury ; 1288 Ipm. *Lessebiry* ; 1291 Tax *Lecebyr* ; 1307
Ch. *Lescebiri, Lascebiri* ; 1313 R.P.D. *Letebyri* ; 1336
S.R. *Lescebiry* ; 1378 Ipm. *Lestebury* ; 1507 D.S.T. *Lesbery.*
O.E. *Lǣces-byrig* (dat.) = Leech's *burh* (Part II). Cf.
Letchworth, Herts., D.B. *Leceworde*, Laysthorpe, Yorks.,
D.B. *Lechestorp*. Skeat (p. 56) rightly assumes an early use
of O.E. *lǣce* = leech, doctor, as a personal name (cf. *Leech,
Leitch*). Cf. also *lǣcesmere, lǣcesford*, B.C.S. 894, 932.
For spellings and pronunciation cf. Dissington *supra*.

Lewisburn (Wellhaugh) [luzbɔ·rn]. 1318 Ipm. *Lusbur'*,
1326 *Lusburn* ; 1327 Orig. *Lusseburn* ; 1357 Sc. *Lus-
burne* ; 1536 Raine *Lushburn* ; 1542 Bord. Surv. *Luse-
burne*.

A difficult name. The first element is apparently *lush*.
It may be cognate with Bav. *lusche*, " swamp," Schlesw.
lusche, " slough " (Middendorf, *s.v. lus*), and is perhaps
found in O.E., cf. B.C.S. 1029 *be ðære lusce (lucs = lusc)*.
Cf. also Luston, Heref., D.B. *Lustone*. Hence " stream
through swampy land." Another possibility is that it is
connected with the vb. *lush*, " to rush, dash," still used in
Cumb., hence " rushing stream." The modern form is
corrupt.[1]

Light Birks (Haydon). Type I, 1296 S.R. *Littelbirkes.*
Type II, 1328 Ipm. *le Lythbirkes, Litghbirkes*, 1368 *Light-
byrkes*.

Type I is " little birches " and probably a mistake.
Type II is " light " birches. Cf. Lighthazels, Yorks.
(Goodall, p. 199) and Lighthorne, Warw. (Duignan, p. 81).

Lilburn (Eglingham). 1177 Pipe, 1203 R.C., 1271
Ch., 1334 Perc. *Lilleburn* ; 1346 F.A. *Lillebourn, Lilborn,
Lylburn* ; 1428 F.A. *Lilburn*.

Lilswood (Hexhamshire). 1233 Gray *Lilleswrth* ; 1233
N. iv. 45 *Lilleswude* ; 1663 Rental *Litsewood* (*sic*).

" Lilla's stream and wood," the second showing the strong
form of the name. Cf. *Lillesham*, B.C.S. 479. Middendorf
(p. 89) takes the first element in *lylleburnan*, B.C.S. 779,

---

[1] It is possible the name may be Celtic, cf. Water of Luce (earlier
Luss) in Galloway (Maxwell, p. 246).

to contain *lylle*, a by-form of *lilie*, "lily," but this is not authenticated. App. A, § 3.

Linacres (Wark-on-Tyne). 1279 Iter. *Linacres*. O.E. *līn-æceras* = flax-acres or -fields. Phonology, § 22.

Linburn Beck (Witton-le-Wear). 1382 Hatf. *Lynburn*. "Burn with the lynn or pool" (*lin*, Part II) *v.* Introd. § 4.

Lindisfarne. *c.* 750 Bede *in insula Lindisfarnensi*, *ad ecclesiam Lindisfaronensis*; *c.* 1000 O.E. Bede *Lindisfearena eae*; *c.* 1120 A.S.C. *Lindesfarena ee*. Simeon of Durham (I. 5) writes as follows: "vocatur autem Lindisfarne a fluviolo scilicet Lindis excurrente in mare, qui duorum pedum habens latitudinis, non nisi cum recesserit mare videri possit." This microscopic stream cannot now be identified. It seems too small to be the R. Low which has to be crossed by pilgrims to the island. Cf. Lindsey, Lincs., earlier *Lindisse*, which Maclure (p. 170 n. 1) connects with Irish *lind*, O.W. *linn*, Bret. *lin* = pool, marsh. *v.* Farne *supra* and *ea*, Part II.

Linnolds (Hexham). 1251 Ipm. *Linelis*; 1269 Perc. *Lynel*; 1334 *le clos de Lynels*; 1649 Arch. 2. 1. 53 *Linnells*; 1714 Corb. *Linolds*. "Linel's (farm)," cf. Kirkharle *supra.* **Linel* is a dimin. of *Līna*, itself a shortened form of *Līnbeald*. Phonology, § 55.

Linsheeles (Holystone). 1292 Q.W. *Lynsheles*; 1314 Pat. *Lyndesele*; *c.* 1250 T.N. *Linesl'*; 1324 Ipm. *Linesheles*; 1346 F.A. *Lynsheles*; 1618 Arch. 1. 2. 327 *Lynshields*. "Shiels by the linn or pool" (*lin*, Part II). The *d* in the 1314 form is a difficulty. The climate makes O.E. *lind* = lime-tree, very unlikely.

Linton (Woodhorn). 1251 Ipm. *Linton*. Farm on the R. Lyne (*v. infra*).

Lintz Ford (Tanfield). 1138-59 Newm. *vadum de Lince*, *Lincestrete*; 1242 D.Ass. *via de Linz*; *c.* 1300 Newm. *Ly(n)chesforde, Lyncheclouh, Lynchestrete*; 1313 Newm. *vadum de Lynce*; 1389 Pat. *Lyns*; 1419.33 *Lynthys*; 1445.34 *Lyntes*.

A compound of O.E. *hlinc* = link, rising ground, ridge, bank, giving *hlinc-ford, -strǣt, -clōh*, or with the gen. sg. *hlinces* as in *hlinces-broc*, B.C.S. 691. For the A.N. spellings *v.* Zachrisson, pp. 18 ff. The *ts* or *z* is difficult. It may represent an A.N. pronunciation of *c* which has replaced O.E. *ch*. The modern form has probably been affected by the tradition of a settlement of German sword-makers at Lintz Green.

Lipwood (Haydon). 1178 Pipe *Lipwude*; 1255 Ass. *Lipwode, Lypwode*; 1346 Ipm. *Lippwode.* "Lippa's wood." Cf. *lippan dic*, B.C.S. 924, and Winkler (p. 236), who gives a name *Lippe* and a place-name *Lippenwoude*. It should be noted, however, that there is an unexplained *hlyp-* often found in O.E. place-names (*e.g. hlypcumb*, K.C.D. 643) which might give rise to this name, *v. Crawford Charters*, pp. 54-5 and Lypiatt, Glouc. (Baddeley, p. 104), and cf. *Liprigs*, Nthb. (H. 2. 3. 383).

Little White (Brancepeth). 1360.35 *Litilwhite*. Unexplained.

Lodge Hill (Bearpark). *n.d.* Acct. *Loge Hill.* Self-explanatory.

Lokenburn and -dene (Alnwick). 1260 Tate ii. 385 *Lokensenburne*, 1405 *Lokenfenburne*. No solution can be offered.

Long Framlington and Longhorsley, *v.* Framlington, Long, and Horsley, Long.

Longhirst (Bothal) [laŋəst]. 1297 Pipe *Langhurst*. **Longlee Moor** (Ellingham). 1442 N. ii. 303 *Langeley.* **Long Newton** (Teesdale). 1335 Ipm. *Langeneuton*. **Longshaws** (Stanton). 1253 H. 3. 2. 140 *Langsævæ*; 1253-90 Perc. *Longesaue*; 1434 R.C. *Lanshaes.*
Self-explanatory. Phonology, §§ 6, 51.

Longwitton, *v.* Witton, Long.

Lorbottle (Whittingham). Type I: 1176 Pipe *Leuerboda*, 1178 *Leuerbotle*; 1253 Ch. *Liuuerboth*; 1368 Pat. *Leyrbotel.* Type II: 1200 R.C. *Luuverbotr'*; 1236 Cl. *Luuerbatte*; 1268 Ass. *Lowerbotre*; 1273 R.H. *Louirbotdil*; 1280 Ipm. *Lurbotil*; 1291 Ch. *Louerbothel*; *c.* 1250 T.N. *Lov(e)rbothill*; 1309 Ch. *Lourbotel*; 1327 Ipm. *Lourbotill*; 1360 Cl. *Lourbotell*; 1428 F.A. *id.*; 1650 Arch. 2. 1. 56 *Lorbottle*; 1663 Rental *Lurbottle.*
"Leofhere's building" (*botl*, Part II). For the types *v.*

Learchild *supra* and cf. Lurley, Dev., F.A. *Luverlegh, Lever-legh.* For *-botre, v.* Zachrisson, pp. 120 ff. Phonology, § 45.

*Lowes, Forest of. 1329 Orig. *foresta de Lowes.*
Cf. Leland (vii. 64) " The Forest of *Loughes* is in Tindale, on the West syde of Northe Tyne, betwixt the *Tynnes* armes." There it is marked on maps till the 18th c. It was so named from the Nthb. *loughs* or lakes—Crag Lough, Littlelow, Greenley, and Broomley Loughs north of the wall, and Grindon south of it (H. 3. 2. 327), *v. luh*, Part II.

Lowick. 1180 Pipe *Lowich* ; 1228 F.P.D. *Lowic* ; 1239 Ipm. *Louwyk* ; 1346 F.A. *Lowyk* ; 1542 Bord. Surv. *Lawyke.*
Lowlynn (Lowick). 1237 Cl. *Leulin* ; *c.* 1250 T.N. *Loulinne* ; 1539 F.P.D. *Lowlyne* ; 1610 Speed *Lowlyn.*
" Dwelling and pool (*lin*, Part II) on the R. Low."[1]

Lucker (Bamburgh). 1169 Pipe *Lucre* ; 1255 Ass. *id.* ; 1288 Ipm. *Locre* ; *c.* 1250 T.N. *Lukre* ; 1290 Abbr. *Loker* ; 1298 Cl. *Lucker* ; 1314 Ipm. *Louker* ; 1346 F.A. *Loker* ; 1379 Ipm. *Lokere* ; 1538 Must. *Lowker* ; 1663 Rental *Lucker.*
Cf. *Luker* (N.G. iii. 195) which the editors connect with O.N. *lúka,* " the hollow of the hand," found also in the compounds *Lukmoen, Luktorpet, Lukevandet.* M.E. *Lucre, Lukre* may represent O.N. *lúkar,* the pl. of this word, and mean " the hollows."

Ludworth (Pittington). 1267 F.P.D. *Ludeworthe* ; 1391 D.S.T. *Luddeworth* ; 1430 F.P.D. *Ludworth.*
" Luda's enclosure."

Lumley (Chester-le-Street). *c.* 1050 H.S.C. *Lummalea* ; *c.* 1190 Godr. *Lummesleie* ; *c.* 1196 Finch. *Lumleia* ; 1223 Pipe *Lumenele* ; 1304 Cl. *Lomelay* ; 1312 R.P.D. *Lumley, Lomley,* 1316 *Lummeleye,* 1345 *Lomley.*
Cf. Lumsden, Co. Berwick, earlier *Lumesdene.* Both names probably contain a Scand. personal name. Cf. *Lum* and *Lumi,* which Nielsen (p. 63) postulates for certain Danish place-names—*Lumsas, Lumsthorp, Lumelef, Loma-lunda.* This was probably by origin a nickname taken from O.W.Sc. *lómr*="" loom "" or "" ember-goose."" In Iceland

---

[1] The obvious etymology with *low* (adj.) is impossible. The M.E. forms would certainly show North Eng. *law.* The 1542 form is perhaps due to an attempt to associate the name with such a form.

we have *Lómatjörn* from the bird and *Lómstaðalækur* from the man's name (Jónsson, *Bæjanöfn*, pp. 507, 433). Cf. *loom sb.*[2] in N.E.D. with M.E. forms *lumb, lumme.*

Lutterington (Auckland). B.B. *Lutringtona.* Cf. Lutterworth, Leic., D.B. *Lutresurde. Luter* is from O.E. *Lēodhere* or *Hlothere.* For the former name cf. D.B. *Loderus* and M.E. *lude* < O.E. *lēod*; for *d > t, v.* Zachrisson, p. 43 n. and cf. D.B. *Letmarus* for *Lēodmær.* For the latter, *v.* Moorman's explanation of Lotherton, Yorks., earlier *Luttringtun* (p. 25). Hence " farm of Leodhere or Hlothere or of his sons."

Lyham (Chatton). 1268 Ass. *Leyham,* 1278 *Leyum, Lium*; 1288 Ipm. *Lyhum*; 1296 S.R. *Leyum*; 1313 Cl. *Lyham*; 1346 F.A. *Lyam, Lyome*; 1380 Ipm. *Lyham*; 1558 V.N. *Lyme.*
Cf. Leam *supra* and Leigham, Dev., F.A. *Leyham* but with a different sound-development. For the alternatives cf. O.E. *hnǣgan* > M.E. *neyen* and *nyen,* to neigh, St. Eng. [nei], Dial. [nai].

Lynch Wood (Brinkburn). 1200 R.C. *Linchwiteburne*; 1248 Brkb. *Linchewood.*
O.E. *hlinc-wudu*=ridge-wood. Cf. Lintz Ford *supra.*

Lyne, R. *c.* 1050 H.S.C. *Lina*; 1297 Newm. *Lyne.*
Cf. Lyne Water in Peebles (*c.* 1190 *Lyn*), Lyn, R., Dev., and Welsh *llyn*=pool or stream (*lin,* Part II).

Lynmouth. 1278 Ass. *Lymu*; 1342 Ipm. *Lynmuth.*
" Lyne-mouth." Phonology, § 21.

Lysdon (Earsdon). 13th c. N. ix. 253 *Lidisdene*; 1533 id. 135 *Lysden*; 1628 id. 202 *Lysdon.*
" Valley of *Lida* or *Hlyda.*" Cf. *lidanege,* B.C.S. 1282, *hlydan pol,* K.C.D. 1309. App. A, § 1.

Lynesack (Auckland). 1307 R.P.D. *Lynesak.*
" Lin's oak." Cf. Linnolds *supra.* Phonology, §§ 14, 23.

Maggleburn (Wingates). 1261 Coram. *Macgild*; 1208 Newm. *Maggild.*
A Celtic river-name.

Mainsbank (Stamfordham). 1479 B.B.H. *leȝ mayns de Stanfordham.*

" *Mains*, demesne lands " (Heslop). It is very common in Scotland.

Mainsforth (Bp. Middleham). 1296 Halm. *Mayn'ford*; B.B. *Maynesford*; 1304 Cl. *id.*; 1391 D.S.T. *Maynesforthe*; 1539 F.P.D. *Mansforthe*; 1701, 1779 Bp. M. *Mensforth.* " Mægen's ford." [1] *Mægen* is found as the first element in some O.E. names. Phonology, § 30.

Manywaygoburn (Haydon). *c.* 1150 H. 2. 3. 383 *Manuggawburn.*
Corrupt beyond recovery.

March Burn (Slaley). *c.* 1275 N. vi. 377 *Marchen-, Merching-burne.* ***Marchingley.** [2] 1262 Ipm. *Merchingley,* 1312 *Merchenley*; 1347 Inq. a.q.d. *id.* " Stream and clearing of Merc or his sons." *Merc* is perhaps short for O.E. *Merc-helm.* Phonology, § 8.

Marden (Tynemouth). 1294 N. viii. 251 *Merden*; 1316 N. viii. 17 *id.*; 1668 N. viii. 241 *Mardon.* **Marley** (Whickham), B.B. *Merleia.*
O.E. *mǣr-denu*=boundary-valley (cf. *on mǣrdenum*, B.C S. 748) and *mǣr-lēah*, " boundary-clearing " (cf. Mearley, Lancs. (Wyld, p. 188), and *mǣrmǣd*, B.C.S. 767). Phonology, § 8; App. A, § 1.

Marwood (Gainford). *c.* 1050 H.S.C. *Marawuda*; 1335 Ipm. *Marwode*; 1444 Pat. *Morwode.*
Possibly O.E. (*se*) *māra wudu*=the larger or bigger wood.

Mason (Dinnington). 1273 R.H. *Merdeffen*; 1284 Waterf. *id.*; 1296 S.R. *Merdessen*; 1336 Fine *Merdesfen*; 1479 B.B.H. *Mordesfene*; 1628 Freeh. *Mersfen*; 1663 Rental *Mairsfen, Mairson*; 1649 Comps. *Mearsfen* alias *Mearson*; 1731 Ponteland *Masson.*
Cf. the personal name *Merdo*, D.B., and place-names Marefield, Leic., D.B. *Merdefeld*, Martley, Worc., earlier *Merdeleye.* These point perhaps to a nickname from O.E. *mearð*, M.E. *merth*, " martin." Cf. a similar use of O.W.Sc.

---

[1] There is no ford near Mainsforth now, but Dr Fowler notes that in Kitchin's 18th century map of Durham the Skerne passes close by it. The course of the stream has evidently been diverted.

[2] The exact position of Marchingley is unknown, but it was probably near the March Burn (N. vi. 378).

mǫrðr. For the suffix-development cf. Hawson, Dev., F.A.
Hosefenne. Phonology, §§ 43, 53.

Matfen. 1182 Pat. *Mate(n)fen*; *c.* 1190 Godr. *Matesfen*;
1200 R.C., 1213 R.C., *c.* 1212 R.B.E. *Matefen*; 1253 Ch.
Matfen; 1278 Ass. *Materfend*; 1286 Ch. *Matfen,* 1291
Mathfen; 1298 B.B.H. *id.*; 1327 Ipm. *Matfen.*

Cf. Matson, Glouc., earlier *Mates-, Matters-, Matteres-,
Mattes-, Matre-done* or *-dune,* i.e. *Mǣðheres dūn* (Baddeley,
p. 107). In Matfen we have either this name (cf. 1278 form)
or a shortened form of it. Cf. Frisian *Mat(e), Maat, Math,
Mæt.* The change from *ð* to *t* may have been helped by the
existence of the common *Mat* for *Matthew.* (Cf. Walker,
s.n. Matlock, Derbys.). "Maeth's fen."

Maughan's House (N. Tyndale). 1279 Iter. *Mauhan.*
Pre-English.

Mayland Lea (Bedburn). 1382 Hatf. *Mayland.*

Possibly O.E. *mægðe-land*=woman's land. Cf. *mægðe
ford,* B.C.S. 906. For Maghull, Lancs., and Mayfield, Suss.,
Roberts and Wyld suggest O.E. *mǣg*=woman, virgin, but
this is a purely poetic word.

Medomsley (Lanchester). *c.* 1190 Godr. *Madmesleie*;
1207 Pap. *Madmesle*; B.B. *Medomesley*; 1303 R.P.D.
Medmesley; 1304 Cl. *id.*

Possibly "*Mǣðhelm's* clearing." Phonology, §§ 42, 53.

Meldon. 1255 Ass., 1270 Ch. *Meldon.*

Skeat takes Maulden, Beds. (p. 15), earlier *Meldone,
Maldon,* to be the same as Maldon, Ess., A.S.C. *Mǣldun,*
i.e. hill marked by a *mǣl,* i.e. a sign or cross. Phonology,
§ 21.

Melkington (Tilmouth). 1425 Raine *Millonden, Milkin-
dune,* 1636 *Melkington.*

No certainty is possible. The first element may be an
O.E. dimin. in *-ic* or *-oc,* possibly a derivative of *Mil*
(Latinised form *Milo*). Such names have their parallel
in Frisian *Myl(l)e, Milcke* (Winkler, p. 260). Phonology,
§ 10; App. A, § 1.

Melkridge (Haltwhistle). 1279 Iter. *Melkrige*; 1292
Ch. *Melkerigg*; 1479 B.B.H. *Milkrigg*; 1610 Speed *Mel-
criche*; 1663 Rental *Milkridge.*

"Milk-ridge." Cf. *meoluc-cumb*=milk-valley, B.C.S. 620. Such names are applied to rich pasturage. Phonology, §§ 27, 58.

Mereburn (Newlands). *c.* 1200 N. vi. 177 *Mereburne.* "Boundary-stream." Cf. *mærbroc,* B.C.S. 610.

Merrington, *v.* Kirk Merrington.

Mickley (Ovingham). *c.* 1190 Godr. *Michelleie*; 1255 Ass. *Mikkeleg*; 1268 Ipm. *Myckeley,* 1271 *Mickeley*; 1346 F A. *Mikkelley,* 1428 *Mykley*; 1663 Rental *Mickley.* O.E. *micela(n) leage* (dat.)=mickle or large clearing.

Middleburn (Wark-on-Tyne). 1286 Ipm. *Midelburn.*

Middleham, Bishop. *c.* 1180 D.S.T. *Midlam*; B.B. *Midelham, Midilham*; 1646 Map *Midlam*; 1715 St Mary le B. *Bishop Medlam.* **Middlehope** (Stanhope). 1418.33 *Midelhope.*

Self-explanatory. Bp. Middleham, it has been suggested, may be so called because half-way between Stockton and Auckland or Durham, these all being residences of the old Bishops of Durham. The suggestion is more ingenious than convincing.

Middlestone (Kirk Merrington). 1366 Halm. *Malderstayn, Melderstayn*; 1629 Esh. *Midleston.*

The suffix *-stayn* (O.W.Sc. *steinn,* " stone, rock ") makes it probable that this name is of Scandinavian origin. *Malder-stayn* might be from O.W.Sc. *malarsteinn,* a compound with *malar,* gen. sg. of *möl,* " pebbles." Cf. *malargrjót*=beach pebbles, *malar-kambr*=pebble-ridge. Such a compound would mean " fine pebbles or stones." For *d* cf. E.D.G. § 298. Alternatively we might connect the name with O.W.Sc. *meldr,* Scots., and Nthb. *melder,* " corn ground at one time," giving rise possibly to a compound *melder-stayn,* " grinding-stone."

Middleton (Auckland). 1104-8 S.D. *Middeltun.* (Belford) 1250 Coram. *Medelton*; 1346 F.A. *Middelton.* (Hartburn) 1346 F.A. *Middleton Morel.* (Ilderton) 1289 Ipm. *Tres Midiltonas*; 1296 S.R. *Midilest Midilton*; 1344 Sc. *Middelmast Middelton.* **Middleton-in-Teesdale.** *c.* 1200 B.M. *Midiltona*; 1271 Ch. *Middelton-super-Teisam.* **Middleton St George.** 1313 R.P.D. *Midelton Sancti Georgii.*

"Middle farm," a very common place-name, often found as *Milton*. *Morel* because held by John Morel of the Barons of Bulbeck. *St George* from the dedication of the church. There are three Middletons in Ilderton and there seems to have been a difficulty in distinguishing them. Middleton in Belford or in Ilderton may have a different history. S.D. (ii. 41, 52) speaks of *Mechil Wongtune* as the scene of the murder of Oswulf in 759. In *Libellus de primo Saxonum adventu* (ib. 376), this is called *Methel Wongtune*, and is probably the same as *Medil Wong* in the life of St Cuthbert ("Works of Bede," ed. Giles, vol. vi. p. 376). In one MS. of the *Libellus* the scribe glosses "Methel Wongtune, *id est Mitheltune.*" If *Methel Wongtune* is identical with *Medil Wong* it must be in the old diocese of Lindisfarne. Craster works out these identifications (N. x. 17) and suggests that we have here the original name of one of the Middletons in Ilderton. It might equally be the one in Belford, and the 1250 spelling rather points to the latter. If so, the name was originally *mæþelwang-tun*=farm by the place of assembly. Cf. O.E. *mæþel-stede*=meeting-place. Later this was abbreviated to *meþel-tun* and ultimately assimilated to the more usual Middleton.

Middridge (Auckland). B.B. *Midrige* (B., C. *Midderigg*); 1382 Hatf. *Midrich*.

Self-explanatory. Phonology, §§ 27, 58.

Migley (Lanchester). 1232 Ch. *Miggeleye*; B.B. *Migleia*.

"Manure-field." O.E. *micga*, North. Dial. *migg*, "manure." Cf. *micghæma gemæra*, K.C.D. 636.

Milbourne (Ponteland). 1158 Pipe *Meleburna*; 1202 Abbr. *id.*; 1255 Ass. *Melleburn*, Pat. *Milneburn*; 1263 Sc. *Melleburn*; *c.* 1250 T.N. *Milleburn, Melleburn*; 1286 Ch. *Milneburn*; 1346 F.A. *Milbo(u)rn, Milleborne*; 1428 F.A. *Milburn*; 1479 B.B.H. *Milnburn*.

Milton (Tynemouth). 1203 Ch. *Mulleton*; 1324 Inq. aqd. *Milneton*.

"Mill-stream and -farm." O.E. *mylen*=mill, *mulle* is a S. Eng. form. Phonology, §§ 10, 53.

Milkhope (Stannington). *c.* 1260 H. 3. 271 *Mylkhopeleche.*
" Hope with rich pasturage." Cf. Melkridge *supra,*
and *v. leche,* Part II.

Milkwell Burn (Ryton). 1316 Pat. *Milkewellburn.*
" Stream from the turbid spring."

Mindrum. *c.* 1050 H.S.C., 1176 Pipe *Minethrum* ;
1227 Ch. *Mindrum,* 1251 *Mundrum* ; *c.* 1250 T.N. *Mindrum* ; 1333 Ipm. *Myndrom.*

The first part of this name is cognate with the Welsh
mynydd, " a mountain," which survives in Long Mynd,
Salop, and Minton and Mindton beside it, in Stadment,
Heref., and probably in the *Minn* of Bosley Minn near
Macclesfield (T.N. *Foresta de Longe Munede).* Cf. *Munet* in
Clun, Salop (T.N.), *Dorments* farm near Minety and *Jackments* Bottom near Kemble, and other *Jackments,* Mint-
ridge, Heref., Okement Hill, Devon (=Uchmynydd), and the
many *meends* in Salop, Heref., and the Forest of Dean
(Maclure, p. 158 n. 1). The second element may be Gael.
druim, " back, ridge." Hence " hill-ridge." Cf. Mint-
ridge, Heref.

Minsteracres (Bywell St Peter). 1268 Ipm. *Mynstanes-
acres,* 1271 *Mynstanaker,* 1272 *Mynstanacres,* 1347 *Milne-
stoneacres* ; 1566 N. vi. 212 *Mynstracres* ; 1663 Rental
Minstrakers.

" Mill-stone fields," presumably from a neighbouring
quarry. Phonology, § 53. *n > r* in anticipation of following *r.*

Mitford. 1195 Pipe *Midford* ; 1229 Pat., 1255 Ass.,
1267 Ch. *id.* ; *c.* 1250 T.N. *Mitford, Midford* ; 1315 R.P.D.
Mithford ; 1489 Ipm. *Mydford, Mydforth* ; 1560 Arch.
7. 24. 119 *Mytfourth.*

" Middle ford." Cf. Midford, Som., earlier *Mitford,
Mytford.* Phonology, §§ 51, 12, 30.

Molesdon (Mitford) [mouzdən]. 1255 Ass. *Moleston* ;
c. 1250 T.N. *Molliston* ; 1269 Ch. *Molston* ; 1273 R.H.
Mollisdon, Moliston ; 1279 Anc. D. *Mulston* ; 1326 Ipm.
Molston ; 1346 F.A., 1408 Ipm., 1428 F.A. *Mollesdon* ;
1645 Map *Mosedon.*

" Moll's form or hill," *Moll* being an old Northumbrian

name. Cf. Molescroft, Yorks., earlier *Mollescroft*. Phonology, § 53. App. A, § 1.

Moneylaws (Carham). 1251 Ch. *Menilawe*; 1255 Ass. *Manilawe, Menlawe*; 1273 R.H. *Menilaw*; 1278 Ass. *Manlaus*; *c.* 1250 T.N. *Mainlawe*; 1291 Ipm. *Monilawe*; 1323 Ipm. *Monylawes*; 1428 F.A. *Monilawe*; 1480 Ipm. *Moneylawes*; 1579 Bord. *Mannylawes*.

"Many-hills." Cf. *be manige hyllan*, B.C.S. 808, Moneyhall or Moneyhull, Worc., earlier *Monhulle, Monihills*, Monyash, Derbys., and *lez Monylaws* in Heugh (1479 B.B.H.). The variant vowels are due to O.E. *manig, monig, menig*. The true Nthb. form is [moni].

Monkridge (Elsdon). *c.* 1250 T.N. *Munkerich*; 1290 Abbr. *Monkrigge*.

Monkseaton (Tynemouth). 1380 Ipm. *Seton Monachorum*.

Monkton (Jarrow). 1104-8 S.D. *Munecatun*; 1430 F.P.D. *Monketon*.

Self-explanatory. *Monk*-seaton in distinction from Seaton Delaval *infra*.

Monkshouse (Bamburgh). 1257 Raine *Broclesmouth, Brokesmuth*; 1340 Pat. *le Brokesmuthe*; 1495 N. i. 306 *le Monkeshouse ex parte boreali rivuli Broxmouth*.

First, " estuary of *Brocc* or (its dimin.) *Broccel*," later Monkshouse because used as a storehouse by the monks of Farne.

Moor, Old and New (Bothal). 1296 S.R. *Pendemor*; 1282 Newm. *Nova Pendemore*; 1346 F.A. *Mora Nova et Vetus*; 1663 Rental *Old Moor* or *Pendmoor*.

O.E. *Penda(n)-mōr*=Penda's swamp.

Moor House (Houghton-le-Spring). 1296 Halm. *Morhus*. Self-explanatory.

Moorsley (Houghton-le-Spring). *c.* 1170 Reg. Dun. *Morleslau (sic)*; *c.* 1190 Godr. *Moreslawe*; *c.* 1150 F.P.D. *Moreslau*; 1446 D.S.T. *Moreslawe*; 1539 F.P.D. *Moresley*.

"*Mōr*'s hill." Cf. *mores burh*, K.C.D. 1290. App. A, § 2.

Moralhirst (Rothbury). 1309 Ipm. *Mirihildhyrst*.

O.E. *myr(i)ge-hylde-hyrst*= pleasant-slope wood. Cf. Merril's Bridge, Notts., earlier *Miri(h)ild, Mirrihil*.

Mordon (Sedgefield). 1104-8 S.D. *Mordun.*
Cf. *Mordun,* B.C.S. 788, "swampy hill." Mordon is
" surrounded with rich low grounds verging to the marsh "
(Surtees).
Morleston (Hart). 1268 D.Ass. *Morelleston;* 1344
Ipm. *Moreliston.*
" Morel's farm." This personal name is probably of
French origin [1] (cf. Middleton *supra*), but it might be O.E.
*Mōrel, dimin. of Mōr.
Morley (Evenwood). 1312 R.P.D. *Morley.* (Hamsterley)
1382 Hatf. *Mawreley.*
The first is " swamp-clearing," the second " mower's
clearing " (cf. Fortherley *supra*), with North. M.E. *mawer*
for *mower.*
Morpeth. *c.* 1200 Joh. Hex. *Morthpath;* 1199 R.C.
Morpeth; 1210-2 R.B.E. *Morpat';* *c.* 1250 T.N. *Morpath;*
1346 F.A. *id., Morepeth,* 1428 *Morepath.*
O.E. *morð-pæð* = murder-peth (*pæð*, Part II), from some
forgotten crime. Cf. *morð-hlau,* B.C.S. 1234. Phonology,
§§ 53, 1.
Morralee (Haydon). 1279 Iter. *Moriley;* 1326 Ipm.
Moryly; 1327 Orig. *Moryleye;* 1368 Ipm. *Morele;* 1542
Bord. Surv. *Morrallee.*
O.E. *mōriga(n) lēage* (dat.)= swampy clearing. Phon-
ology, § 22.
Morton (Haughton-le-Skerne). 1278 Ipm. *Morton*
(Houghton-le-Spring). B.B. *Mortona.* (Sedgefield) 1312
R.P.D. *Morton juxta Kyllerby.* **Morton Tinmouth** (Gain-
ford). 1104-8 S.D. *Mortun;* 1271 Ch. *Mortonam in
Haliwerkesfolc.*
Morwick (Warkworth) [mɔrik]. 1171 R.B.E. *Morewic;*
1278 Ass. *Morwick;* 1628 Arch. I. 3. 94 *Morrick;* 1682
Warkw. *Morweek.*
" Farm and dwelling by the swamp." Cf. *Mortun,*
B.C.S. 565. *Tinmouth* because it once belonged to the monks
of Tynemouth. For *Haliwerkesfolc, v.* Introd. § 1. There
is also a *Morton Palms* in Haughton-le-Skerne. Surtees

---

[1] Weekley (*Romance of Names,* p. 215) takes it to be O.Fr. *morel* =
Moorish, swarthy.

K

says (3.270) that it was so called from a proprietor of late date—Bryan Palmes.

Mosscroft (Dunstan). 1269 N. ii. 186 *Musecroft*; 1323 Ipm. *Muscroft*.

" Mouse croft." Cf. *musbeorh*, B.C.S. 1242. The modern form is corrupt.

Mosswood (Shotley). 1378 Ipm. *Moseforth*; 1526 Arch. 2. 1. 136 *Mosseford*; 1569 F.F. *Mesfurthe*; 1671 Corbr. *Moswood*.

" Ford by the moss or bog." Cf. Moseley, Berks. (F.A. *Mosleye, Mesle*). Forms in *Mes-* are perhaps due to confusion of *moss* = bog, from O.E. *mōs* and the *moss*-plant, found alternatively as *mese* from O.E. *mēos*. Phonology, §§ 12, 30; App. A, § 5.

Mousen (Bamburgh). 1166 Pipe *Mulefen*, 1186 *Mulesfen*, 1195 *Mulesen*; 1255 Ass. *Mulesfen*; 1267 Ipm. *Melesfen* alias *Mulesfen*; 1428 F.A. *Mulssen*; 1538 Must. *Mowssen*; 1628 Arch. 1. 3. 95 *Moulsfen*; 1628 Freeh. *Mulsfen*.

O.E. *Mūles-fen*=Mul's fen. Phonology, §§ 53, 39.

Muggleswick. *c.* 1190, 1259 F.P.D. *Muclingwic*, 1291 *Muklyngwyk*; 1312 R.P.D. *Mukkelyngeswyk*; 1335 Ch. *Muclincgwic*; B.B. *Muglyngwyc* (B., C. *Moclyngeswyk*); 1446 D.S.T. *Mogleswike*; 1625, 1646 Stanh. *Muglesworth*.

" Dwelling of Mucel's son." Cf. *Muceling mæd*, B.C.S. 692. Phonology, § 59; App. A, § 11.

Murton (Dalton-le-Dale). 1155 F.P.D. *Mortun*. (Sedgefield). 1432.45 *Westmorton next Embleton*. (Tweedmouth) 1312 R.P D. *Morton*; 1384 Raine *Murton*. (Tynemouth) 1203 R.C. *Morton*; 1380 Ipm. *Estmureton*. Cf. Morton *supra*. *ō*> L.M.E. *ū*> *ŭ* (Phonology, § 21). Murton in Dalton is known also as Murton-in-the-Whins or Murton-juxta-Hesleden.

Nafferton (Ovingham). 1182 Pipe *Nafferton*; 1212 R.C. *id.*; 1221 Pat. *Nafretun*, 1225 *Naffreton*; 1253 Ch. *id.*; *c.* 1250 T.N. *Natferton*; 1261 Ipm., 1268 Ass., 1280 Ipm. *Nafferton*; 1263, 1289 Ipm. *Natferton*.

Lindkvist (pp. 187-8) explains this and the same name in Yorks. as O.W.Sc. *Náttfaratún*, i.e. farm of *Náttfari* or night-traveller, a nickname given by Kahle (p. 195). Cf. *Náttfaravík* (Lind. *s.n.*) Naffentorp, Skane, earlier *Natfaræthorp* (Falkman, p. 160). Phonology, § 51.

*Nakedale (S. Tyndale). 1365 Ipm. *Nakadele,* 1368 *Nakedale;* 1547 N. iv. 185 *Nakedale;* 1575 F.F. *Naketele.* Possibly " naked island " (*ele,* Part II) referring to an " eale " on the Tyne.

Nanny River (Bamburgh). 1245 Pipe *Nauny.* A Celtic river-name.

Neasham (Hurworth) [ni·səm]. *c.* 1150 S. 3. 258 *Nes(s)ham;* 1297 Pap., 1311 R.P.D., 1330 Pat. *Nesham;* 1336 Ipm. *Nessam;* 1459.35 *Neceham;* 1639 N.C.D. *Neesom;* 1671 Coniscl. *Neesam.*

M.E. *nese-ham*=homestead on the " ness " or noseshaped piece of land, *v.* N.E.D. *s.v. nese.*

Nelson (Hart). *c.* 1196 Finch. *Nelestune;* 1344 Ipm. *Neliston;* 1354 Finch. *Nelston,* 1516 *Neylson;* 1649 Comps. *Nelston.*

Possibly " Neale's farm." *Neale* is from *Nigel,* Lat. *Nigellus.* Phonology, § 53.

Nesbit[1] (Doddington). 1255 Ass. *Nesebyt, Nesebite, Nesbyte.* **Nesbitt** (Stamfordham). 1298 B.B.H. *Nesebith,* 1479 *Nesbitt;* 1709 Corbr. *Neasbitt.* (Hart) 1311 R.P.D. *Nesbitt;* 1646 Map *Nesbed.*

M.E. *nese-bit*=nose-bit, a piece of land resembling a nose in shape. Cf. Saddlebow in Wiggenhall St Mary, Norfolk.

Netherton (Alwinton). 1207 Sc. *Netterton;* *c.* 1250 T.N. *Nedderton;* 1428 F.A. *Nederton;* 1479 B.B.H. *Nethreton.* (Bedlington) *c.* 1050 H.S.C. *Nethertun.*

O.E. *neoþor-tūn*=lower-farm. Phonology, § 41.

Nether Witton, *v.* Witton, Nether.

Nettlesworth (Chester-le-Street). 1297 Pap. *Netrehworth;* 1312 R.P.D. *Netlesworth, Nettelworth.*

Cf. Nettleham and Nettleton, Lincs., D.B. *Netelham, Neteltone,* Nettlestead, Suff. (Skeat, p. 88), and Kent (B.C.S. 1322 *Netlestede*), Nettleworth, Notts. (Mutschmann, p. 96).

---

[1] Jameson gives *Nesebit, Nisbit* as a technical term for a piece of headharness. If this is the word here, the name must again have been given on the ground of some fancied resemblance.

In all these we probably have the plant-name, but Nettleton, Wilts., B.C.S. 800 *Netelingtone*, points to a personal-name. This name is, on insufficient grounds, equated with an O.E. **Nyttel*, dimin. of *Nytta*, by Ekblom (p. 130). Nettlesworth may contain the same personal-name, whatever its correct form be, or it may contain the plant-name with later pseudo-genitival *s*. For *tre v.* Zachrisson, pp. 120 ff.

Newbiggin (Blanchland). 1378 Cl. *Newbigging*. (Heighington) 1388.33 *Newbiggyng nigh Redworth* ; B.B. *New Vill next Thickley*. (Hexhamshire) 1344 Pat. *Neubiggyng*. (Middleton-in-Teesdale) *n.d.* R.P.D. *Newbygyng*. (Lanchester) 1382 Hatf. *Newbiggin*. (Newburn) *c.* 1250 T.N. *Neubiging*. (Norham) B.B. *Newbiginga* (B. *Nuburga*, C. *Neubinga*). **Newbiggin-by-the-Sea** (Woodhorn). 1268 Ipm. *Neubigging*. **Newbottle** (Houghton-le-Spring). 1197 Pipe *Newbotle*. **Newbrough** (Warden). 1203 Pipe *Nieweburc* ; 1329 Ipm. *Neuburgh* ; 1542 Bord. Surv. *Newbrough*. **Newfield** (Auckland). 1382 Hatf. *le Newfeld*. (Pelton) ib., *id.* **Newham** (Bamburgh). 1288 Ipm. *Neuham*. (Newburn) 1309 Ipm. *Neweham*. **Newhouse** (Coatham). *c.* 1090 F.P.D. *Newehusa*, 1380 *Newehous juxta Acley*. **Newland** (Bywell St Peter). 1268 Ipm. *Novalanda*, 1345 *Neulond*. **Newlands** (Bamburgh). 1318 Inq. a.q.d. *Newland*. (Hexhamshire) 1344 Pat. *Neuland*. **Newlandside** (Stanhope). 1382 Hatf. *Newlandsyde*. **Newminster**. *c.* 1200 Joh. Hex. *Novum monasterium*. **Newstead** (Bamburgh). 1377 Ipm. *Newstede*. (Ellingham) 1230 N. i. 1260 *Novum locum qui dicitur Neubigginge* ; 1377 Ipm. *Newstede*. **Newton** (Boldon). B.B. *Newtona juxta Boldonam*. (Bywell) 1346 F.A. *Neuton*. (Durham) B.B. *Newtonam juxta Dunolm*. **Newton Cap** (Auckland). *c.* 1050 H.S.C. *Neowatun* ; 1382 Hatf. *Newton capp*. **Newton Bewley** (Billingham). *c.* 1350 D.S.T. *Neuton Belu*. **Newton Hansard** (Walworth). 1362 S. 3. 88 *Newton Hansard* ; 1637 Camd. *Newton Hanset* ; 1722 Sedgf. *id.* **Newton Ketton**. 1464 F.P.D. *Newton Ketton*. **Newton-in-Coquetdale**. 1430 Ipm. *Neuton in Kokedale*. **Newton-on-the-Moor** (Shilbottle). 1346 F.A. *Newton-super-Moram*. **Newton-on-the-Sea**. 1346 F.A. *Neuton juxta* (or *super*) *mare*. **Newton Underwood** (Mitford). 1296 S.R. *Newton under Wood.* **Newtown** (Bamburgh). *c.* 1330 N. i. 196 *Nova villa super Warneth,* 1484 *New Towne juxta Bamburgh.* (Rothbury) 1248 Ipm. *Newtown,* 1309 *Le Neuton.* **Long Newton** (Teesdale). 1335 Ipm. *Langeneuton.*

The names are for the most part self-explanatory, *v.* Part II for the second elements and App. A, § 10. *Bewley* because in the manor of that name. *Cap* is possibly the same as in Capheaton *supra.* *Hansard* from the ancient lords of Walworth whose ancestor must have been " a member of one of the establishments of the German Hansa " (Forssner, p. 29) ; for *Hanset* cf. Garret Shiels *supra.* In *Newtown* an effort has been made to preserve the suffix in its fully stressed form. *Super Warneth,* i.e. on the Warren Burn (*v. infra*). *Newminster* is the new monastery founded by Ranulf de Merlay in 1139 as a colony of the Cistercian Abbey of Fountains, Yorks. *Long,* because a long, straggling village (S. 3.212). There is another Newton—Archdeacon Newton—in Darlington which Surtees (S. 3.375) says was held by lease under the Archdeacon of Durham.

Newburn-on-Tyne. Type I: *c.* 1175 S.D. *Nyweburne* ; 1203 R.C. *Neuburne.* Type II: 1204 Pipe *Nieweburc* ; 1281 Coram *Neuburgum.* Type III: 1206 Pipe *Nieweton.*

v. App. A, § 10. Probably *Newburgh* is the original form, for *new* is naturally applicable to a *burh* rather than a *burn,* but cf. Newbourne, Suff. The modern *New Burn,* a little tributary of the Tyne, may be a back-formation from the village-name.

Newcastle-upon-Tyne. The earliest name of this was *Pons Aelii.* It is found in the *Notitia Dignitatum,* and the bridge was so named after Aelius Hadrianus. The next recorded name is that found in Simeon of Durham (*Hist. Dunelm Eccl.* iii. 21) where he tells of three monks from Winchcombe who " in loco qui dicitur *Munecaceastre,* quod monachorum civitas appellatur, habitare coeperunt," and tried to revive monastic life there in the days of Bishop Walcher (1073-80). The name was perhaps only given to the site of the abortive monastery. It soon died out, and in the *Historia Regum* (vol. ii. p. 201) we are told that this

" Monkchester " is now called *Novum Castellum*. This name must have taken its origin from the castle built by Robert Curthose in 1089.

Newsham [nju·səm] (Earsdon). 1200 R.C. *Ne(h)usum*; 1207 Abbr. *Neusum*; 1461 N. ix. 208 *Newsam*; 1728 Bothal *Newsome*. (Egglescliffe) *c.* 1220 F.P.D. *Neusom*; B.B. *Newsona*; 1446 D.S.T. *Neusham*; 1652 Staindrop *Nusam*, 1734 *Nuzam*.

O.E. (*æt þæm*) *niwa(n) hūsum*=(at the) newhouses. Cf. Newsham, Lancs., Lincs., Yorks. App. A, § 6.

Ninebanks (Allendale). 1228 Gray *Ninebenkes*, 1230 *Nenbenkes*; 1296 S.R. *Nine bankes*; 1479 B.B.H. *Nynbenkys*; 1542 Bord. Surv. *Nyne Benkes*.

Probably "nine banks" on the switchback-road up the West Allen by this farm. *benk* is a M.E. variant of *bank*.

Nookton (Hunstanworth). *c.* 1190 B.B. *Knokeden*; 1649 Comps. *Knockeden*.

Possibly "valley with a knock or hill in it." Cf. *knock* (Lincs.) "a sand bank" which N.E.D. connects with Dan. dial. *knok*=little hillock, and the allied O.N. *knjúkr*, "high and steep hill of rounded form," preserved in *Knuk*, *Knyk* (Rygh. *Indledning*, p. 61). Phonology, § 21; App. A, § 1.

Norham-on-Tweed. *c.* 1050 H.S.C. *Northham*; 1097 Colding. *Northam*; *c.* 1125 F.P.D. *Nor(h)am*, *Northam*, 1273 R.H., 1340 R.P.D. *Northam*; 1430 F.P.D. *Norham*; 1584 Bord. *Norram*.

"North homestead." Cf. Northam, Dev., D.B. *Northam*, 1252 Ch. *Norham*, Hants., 1151 B.M. *Norham*. S.D. (i. 361) gives an earlier name—*Ubbanford*, i.e. ford of *Ubba*, a well-established O.E. name, probably of Frisian origin. Phonology, §§ 50, 36.

Norton (Billingham). *c.* 1000 B.C.S., 1256 *Norðtun*; B.B. *Nortona*.

"North Farm."

Nubbock (Hexhamshire). Type I: 1251 Gray *Jakele*. Type II: 1479 B.B.H. *Nobbok-scheles*; 1663 Rental *Nubbock*; 1608 Hexh. Surv. *Yokesley* or *Nubbock*.

Cf. Yoxford, Suff., and Yoxall, Staffs., which may contain O.E. *geoc*=yoke, used of a bedfellow or spouse (Skeat, p. 39). Hence " Yoke's field or clearing." The second name is a mystery.

Nunriding Hall (Mitford). 1539 Arch. 3. 4. 116 *Nuneryding.*

" Nuns' clearing," *v*. Riding *infra*. The place was *ridded* or assarted by the nuns of Holystone to whom it was given by Roger Bertram the First under the name of *Baldwineswood* (H. 2. 2. 74).

Nunstainton (Aycliffe). *c*. 1190 F.P.D. *Staynton supra Schyrnam*; 1265 F.P.D. *Staynestun*; 1387 D.S.T. *Nunstaynton*; 1719 Bp. M. *Nunstenton*.

" Stone-farm on the Skerne, belonging to the prioress and nuns of Monkton, or possibly " Stein's farm." *v*. Stainton and Stannington *infra*.

Nunwick (Simonburn). 1165 Pipe *Nunewic*.

" Nuns' dwelling," from ownership rather than residence. Cf. *nunenna beorh*, K.C.D. 623.

Oakhaugh (Brinkburn). *a*. 1201 Brkb. *Akehalgh*; 1663 Rental *Akehaugh*; 1686 N. vii. 501 *Oakhaugh*. **Oakwood** (St John Lee). *c*. 1160 Ric. Hex. *Acuudam*; 1226 B.B.H. *Acwde*, 1479 *Akwod*; 1547 Hexh. Surv. *Ackewode*, 1608 *Akewood*.

" Oak (grown) haugh " and " oak-wood." Phonology, §§ 14, 21.

Offerton (Painshaw). *c*. 1050 H.S.C., *c*. 1180 F.P.D. *Uffertun*; *c*. 1190 Godr. *id.*; 1326.45 *Ufferton*; 1552 V.N., 1637 Camd. *id.*; 1627 Houghton *Oufferton*; 1768 Map *Offerton*.

Possibly *Útfara-tún* (cf. Nafferton *supra*). The name **Út-fari* is not on record as a name in O.N., but is a possible derivative from the common *fara út*=to go (from Norway) to Iceland, also to go on a pilgrimage. Phonology, § 51.

Ogle (Whalton). 1169 Pipe *Hoggel*, 1180 *Ogle*; 1212 R.B.E. *Hoggul*; 1255 Pat. *Oggele*; 1255 Ass. *Oghyll*; *c*. 1250 T.N. *Oggill*; 1309 Ipm. *id.*; 1341 B.M. *Oggle*; 1346 F.A. *Ogle*.

Possibly O.E. *Ocga(n)-hyll*=Ocga's hill. *Ocga* was the

name of a son of Ida of Bernicia. The regular development would have been to [ɔgəl] rather than [ougəl]. Phonology, § 36.

Oldacres (Sedgefield). 1267 S. 3. 48 *Aldacres*. **Old Durham.** 1399 Acct. *Aldurham*; 1429.33 *Alderesme*. **Old Park** (Whitworth). 1382 Hatf. *Aldpark*.

Self-explanatory. Phonology, § 3.

Old Shield (Haltwhistle). Type I: Iter. *Aldithescheles, Aldichesheles*. Type II: 1268 Ass. *Aldesheles, Aldenschelys*; 1279 Ass. *Aldenescheles*; 1296 S.R. *Aldenchele*; 1298 B.B.H. *Aldschel*, 1479 *Aldscheles*.[1]

" *Ealdgyð* or *Ealdwine's* shiels." Phonology, §§ 3, 53, 49, 59.

Orchardfield (Shotley). 1378 Ipm. *Orcherfeld*; 1771 N. vi. 231 *Orchardfield*.

Self-explanatory.

Ord (Tweedmouth). 1208 R.C. *Orde*; 1539 F.P.D., 1560 Raine *Ourde*.

O.E. *ord*=point or corner of land. Cf. *to þæs hlinces orde*, B.C.S. 917. Phonology, § 12.

Ornsby Hill (Lanchester). 1408.35 *Ormysby*.

" *By* (Part II) of *Ormr*, a common Scand. name. Phonology, § 51.

Osmond Croft (Winston) [uzməncrɔft]. 1333 S. 4. 101. *Osmundcroft*; 1539 F.P.D. *Osmondecroft*; 1664 Arch. 3. 17. 124 *Osmancroft*; 1748 Gainf. *Usmancroft*.

" *Ōsmund's* croft." Phonology, §§ 21, 12, 53.

Otterburn. 1217 Pat. *Oterburn*.

" Otter-stream." Cf. *oterburna*, B.C.S. 1158, *Otterbach* and *-born*, Hesse (Sturmfels, p. 64).

Ottercops (Elsdon). 1265 Sc. *Altercopes*; 1267 Abbr. *Altercoppes*; 1273 R.H. *Antercops* (? for *Autercops*); 1306 H. 3. 2. 15 *Altercoppes*; 1586 Raine *Attercopes*; 1628 Freeh. *Ottercops*; 1635 Comm. *Attercops*.

For *Alter-* cf. Catterick *supra*. The second element is probably the pl. of *cop*=top or summit. The form has been influenced by the neighbouring Otterburn.

---

[1] The identifications made here are not always certain, and some may refer to the places mentioned under Aydon Shiel *supra*.

Ouse Burn (Newcastle-on-Tyne). 1292 Ass. *Yese*; 1671 Arch. 2. 1. 128 *Useburn*; 1732 Ponteland *Ewes Burn*.

A Celtic river-name. Initial [j] has been lost as in Earle, Easington (*supra*). Later the name was perhaps altered under the influence of the common river-name *Ouse*.

Ousterley (Lanchester). 1369.35 *Houstre*; 1382 Hatf. *Oustre(feld)*; 1391.35 *Hustre*; 1429.33 *Houstre*.

Cf. Austerfield, Yorks., earlier *O(u)strefeld*, *Austerfeld*. Moorman (p. 14) takes the first element here to be O.N. *austr*, east, but the vowel-forms and initial *h* and the absence of a second element are against this explanation for *Ousterley*. Place-names in *-tree* are fairly common (cf. Aintree, Lancs., Braintree, Ess., Picktree, Co. Durham). There is a *house-leek tree* or *tree house-leek*, a plant which grows on walls and roofs of houses. It is just possible that this may have been called, for short, *House-tree*, and the place named from it. Alternatively, we may note such compounds as *door-trees* (*v.* Potts Dultries *infra*) and *roof-tree*. There may have been a word *house-tree*, and the farm have been so called from a conspicuous piece of timbering. Phonology, § 35.

Ouston (Birtley, Co. Durham). 1328 Cl. *Ulkestan*; 1382 Hatf. *Ulleston*. (Stamfordham) 1255 Ass. *Hulkeston*, *Ulkilleston*; 1296 S.R. *Olkeston*; 1346 F.A. *Ulkeston*; 1628 Freeh. *Ulston*. (Whitfield) 1279 Iter. *Ulvestona*; 1538 Must. *Huston*; 1610 Speed *Owston*.

" Farm of *Ūlkill* and of *Ūlfr*." *Ulkill* is from O.N. *Ūlfketill*, Björkman, N.P. p. 168. Phonology, §§ 53, 59, 39.

Outchester (Bamburgh). Type I: 1236 Cl. *Ulecestr'*, 1242 *id.*; 1278 Ass. *Ulcester*; *c.* 1250 T.N. *Ulecestr'*; 1296 S.R. *Ulcester*, 1336 *Olcestre*; 1479 B.B.H. *Ulchestre*; 1577 N. i. 206 *Owlchester*; 1663 Rental *Ulchester*. Type II: 1550 H. 3. 2. 207 *Outchester*; 1579 Bord. *Utchester*.

O.E. *ūle-ceaster*=owl(haunted)chester. Phonology, § 21. Type II is corrupt.

Overacres (Elsdon). 1583 Bord. *Haveracres*; 1628 Freeh. *Overacres*, *Haueracres*; 1663 Rental *Overacris*.

" Oat-fields." North. dial. *haver* (O.N. *hafre*)=oats. Cf. *Haveracres* Halm. 1367. The modern form is corrupt.

Overgrass (Felton). 1255 Ass. *Ovegares, Oversgare* ; 1256 Brkb. *Overgares* ; *c.* 1250 T.N. *Overisgar, Overgaris* ; 1271 Ch. *Overgares* ; 1272 N. vii. 485 *Eueresgares* ; 1318 Ipm. *Overgares* ; 1346 F.A. *Overgars* ; 1638 Freeh. *Oversgrasse.*

The second element is M.E. *gares*, pl. of *gare* (S. Eng. *gore*), used of a triangular-shaped field. The first is either gen. sg. of *ofer*=shore, brink or margin, or O.E. *ufere*, upper, with pseudo-genitival *s* in certain forms. Hence " gores on or of the brink," or " upper gores," referring to the position above the valley of the Swarland Burn. Confusion of suffix is in part due to Nthb. [gars] and [gers] for *grass*. Phonology, § 54.

Ovingham-on-Tyne (ɔvindžəm]. *c.* 1200 Arch. 2. 1. 64 *Ovingeham* ; 1244 Ipm. *Ovingham* ; 1339 Perc., 1378 D.S.T. *Ovyngeham.* **Ovington** (Ovingham) [ovintən]. *c.* 1200 Arch. 2. 1. 64 *Ovintun* ; 1200 R.C. *Ovinton* ; 1255 Ass. *Ovington.* **Ovington-on-Tees** [uvintən]. *c.* 1200 Joh. Hex. *Ovendon.*

"Homestead of the sons of *Ofa*, farm and hill of *Ofa*." Cf. Ovington, Norf., Ess., Oving and Ovingdean, Suss. Phonology, § 34 ; App. A, § 1.

Owmers (Warden). 1296 S.R. *Ulmeres* ; 1298 B.B.H. *Oulemers* ; 1344 Cl. *Wolmers* ; 1364 Ipm. *Ulmers* ; 1479 B.B.H. *Olmers(se)* ; 1552 H. 2. 3. 389 *Owmers.*

O.E. *ūle-mersc*=owl-marsh. Cf. Homers Lane *supra* and Crowmarsh, Oxf., earlier *Craumares, Craumerse*. Phonology, § 39.

Owton (Seaton Carew). 1189 D.S.T. *Oveton.*

O.E. *Ofa(n)-tūn*=Ofa's farm. Cf. Owthorp, Notts., D.B. *Ovetorp*. Phonology, § 47.

Oxcleugh (Chirdon). 1279 Iter. *Oxclow.* **Oxenhall** (Darlington). 1242 D.Ass. *Oxenhale* ; B.B. *Oxenhall* (B., C. *Oxen(h)ale*) ; 1382 Hatf. *Oxenhale.* **Oxneyflat** (ib.). 1382 Hatf. *Oxenhalflat.*

"Ox-clough, oxen-haugh and *flat* (Part II) by Oxen-haugh." Cf. *oxnahealas*, B.C.S. 887 and Oxenhall, Glouc. (Baddeley, p. 118). App. A, § 6.

Painshaw or **Penshaw** (Houghton-le-Spring). *c.* 1190

B.B. *Pencher*; 1305 B.M. *id.*, 1472 *Penchare*; 1637 Camd. *Pencher*; 1649 Comps. *Pensher*; 1760 Whickh. *Painshea*; 1764 Map *Pencher*; 1803 Whickh. *Penshaw.*

A Celtic name partly anglicised. Cf. *pencersæte*, B.C.S. 455.

Pallion (Bp. Wearmouth). 1328 Arch. 3. 3. 297 *le Pavylion*; 1408.45 *Pavillion.*

Cf. Scots. *pallioun* for *pavilion.* " The summer seat and occasional residence for business or pleasure of the lords of Dalden " (S. 1. 241).

Pandon (Newcastle-on-Tyne). 1177 Pipe, 1298 Ch. *Pampeden*; 1578 Arch. 2. 1. 42 *Pandon.*

Cf. Pampisford, Cambs. Skeat inferred a nickname connected with Dan. dial. *pamper,* "short, thick-set person," Lincs. *pammy,* "thick, fat," and noted Alan Pampelin in the Ramsey Cartulary. Kahle (p. 246) confirms that by the existence of O.N. *pampi,* a nickname connected with Mod. Norw. *pampe,* " to make little halting movements." " Pampi's valley." Phonology, §§ 53, 51 ; App. A, § 1.

Park Hill (Quarrington). 1342 Hatf. *Pastura del Park.* Self-explanatory.

Parmentley (Whitfield). *c.* 1135 H. 2. 3. 18 *Parmontle*; 1279 Iter. *Permanley*; 1610 Speed *Permandley*; 1698 Whitf. *Parmaly.*

" Pearmain-clearing," pearmain being a variety of pear (N.E.D.). *parment* is due to a misunderstanding of *parmen-tree.* In the *Catholicon Anglicum* (quoted) N.E.D. we find "A Parmayn tre (*v.l.* parment tree)." Cf. Apperley *supra.*

Paston (Kirknewton) [pɔ·stən]. *c.* 1130 Perc. *Paches-tenam*; 1175 Pipe *Palestun*; 1227 Ch. *Paloxton*; 1255 Ass. *Palleston, Parleston, Palxton*; *c.* 1250 T.N. *Palwiston*; 1292 Q.W. *Palston*; 1296 S.R. *Palxston*; 1315 Inq. aqd. *Paxton*; 1334 Ipm. *Palston*; 1335 Ch. *Palkeston*; 1344 *id.*; 1441 Ipm. *Palxton*; 1542 Bord. Surv., 1855 Whellan *Pawston.*

" Pælloc's farm." **Pælloc* is dimin. of *Pælli* (L.V.D.). Phonology, §§ 59, 53.

Pauperhaugh (Rothbury) [pepəha·f]. *c.* 1120 Brkb.

Papwirthhalgh, c. 1250 *Papwurthhalgh, Papurhalgh* ; 1309
Ipm. *Pappeworthhalugh* ; 1798 Edl. *Pepperhaugh.*
Cf. Papworth, Cambs.=Pappa's enclosure (Skeat, p. 27).
The farm here must have borne the same name and
the whole name mean "haugh by Papworth." Strangely
enough there is also a *Papworthele* in Wolsingham (Hatf.
Surv.). Cf. Nthb. [pepə] for *paper.* Phonology, § 49.

Pawlaw Pike (Wolsingham). 1382 Hatf. *Pawfeld.*
Possibly O.E. *Pagan-feld*=field (Part II) of *Paga,* a very
rare O.E. name.

Pedam's Oak (Edmundbyers). *c.* 1200 F.P.D. *Pethune-
shake* ; 1364 Halm. *Pethmosake,* 1580 *Petonsake* ; 1637
Camd. *Pedumsake* ; 1764 Map *id.* ; 1804 Ebch. *Pedomsake.*
M.E. *petemos-ake*=oak by the peat-moss or bog.
Phonology, §§ 51, 14.

Pegswood (Bothal). 1258 Sc. *Peggeswurthe* ; 1261 Ipm.
Pegeswrthe, Pegiz' town ; 1663 Rental, 1750 Map, 1800
Meldon *Pegsworth.*
"Pegg's enclosure." Cf. *pecgesford* and the allied *Pæcga*
in *Pæcganham,* B.C.S. 50=Pagham, Suss. App. A, § 3.

Pelaw (Chester-le-Street). 1242 D.Ass. *Pellowe* ; 1297
Pap. *Pelawe* ; B.B. *Pelhou, Pelowe* ; 1313 R.P.D. *Pellawe,
Pelawe juxta Cestre.*

Pelton (Chester-le-Street). 1312 R.P.D. *Pelton.*
Unsolved problems. In Pelaw there would seem to
have been confusion between the suffix *hōh* (Part II) and
St. Eng. *low* for North. Eng. *law* (O.E. *hlāw*).

Pespool (Easington). 1316 Finch. *Pesepole.*
Cf. Peasmore, Berks., earlier *Pesemere, Peysmer,* which
Skeat (*s.n.*) explains as "mere near a field of peas." So,
perhaps, "pool by a field of peas." Phonology, § 21.

Philip (Kidland). 1331 Ipm. *Fulhope,* 1368 *Filhope* ;
1618 Redesd. *Filhaupe* ; 1663 Rental *Fair Philip* ; 1720
Alw. *Fill-houp,* 1729 *Philhoup.*
"Foul hope" (*hop,* Part II). Phonology, §§ 13, 36.

Picktree (Chester-le-Street). 1242 D.Ass. *Piketre* ; B.B.
Piktre.
Nthb. *picktree* for *pitchtree,* one abounding in resin.
The earliest example in N.E.D. is dated 1538.

Piercebridge (Gainford). Type I: 1104-8 S.D. *Persebrig*; 1308 Pat. *Persebrigg*; 1315 R.P.D. *Percebrig*; 1335 Ipm. *Percebrigg*; 1460 D.S.T. *Percebrig*. Type II: 1577 Barnes *Preistbrigg*.

"Piers' bridge" from its qwner or builder and, alternatively, "priest-bridge."

Pigdon (Mitford). 1226 Pipe *Pikeden*; 1255 Ass. *Pykedon, Pikeden*; 1311 Ch. *Pykeden*; 1346 F.A. *Pykdon, Pikdone*, 1428 *Pykden*; 1465 Ipm. *Pykton*. North. dial. *pike*=conical-shaped hill and *-don*. "Pigdon is picturesquely perched on the hillside which rises fairly steeply behind it" (Tomlinson, p. 273). Phonology, § 51; App. A, § 1.

Pinfold (Stanhope). 1382 Hatf. *Punfald*. O.E. *pund-fald*, with alternative *pin(d)fold, v.* N.E.D.

Pittington. *c.* 1125 P.P.D. *Pittindun*; *c.* 1180 D.S.T. *Pitindun*; 1196 Finch. *id.*; 1198 Pipe *Pitinden*; 1203 R.C. *Pittenden*; 1270 Finch. *Pytington*; 1296 Halm. *Putingd'*; 1306 R.P.D. *Pytyngden*; 1341-74 D.S.T. *Petynton*, 1391 *Pittyngton*; 1464 F.P.D. *Petyngton*.

"Hill of Pita or Pytta." Cf. *Pitanwyrð*, B.C.S. 690. For the second cf. *Pyttel* and *Putta*, which may have had an alternative form *Pytta* (cf. *Cudda* and *Cydda*). If stress is laid on the 1296 form we have *Pytta* in a South. M.E. form, due to a scribe. App. A, § 1. Phonology, § 10.

Plainfield (Flotterton). 1272 Newm. *Flaynefeld*. Possibly a scribal error. If correct it is O.W.Sc. *Fleinn* (*v.* Lind. *s.n.* and Kahle, p. 180), hence "Fleinn's field," or *fleinn*=pike, arrow, as in Flamborough Head (Lindkvist, p. 44), hence "arrow-shaped field" (*v.* Introd. p. 22). The personal name is found also in Flainville, Fleinville, Normandy (*Danske Studier*, vol. ii. p. 69).

Plawsworth (Chester-le-Street). 1297 Pap. *Plauworth*; 1312 R.P.D. *Plauseworth*; B.B. *Plausword* (B., C. *Plauseworth*); 1345 R.P.D. *Plawesworth*.

Names in *Pleg-* are fairly common in O.E., and these may have had alternative short forms, *Plega* and *Plaga*, just as *play* and *plaw* go back to W.S. *plega* and Anglian *plaga.* Hence " *Plaw's* enclosure." Alternatively we may compare *pleieswirthe,* B.C.S. 922, i.e. enclosure of play, in a late charter. There may have been a Northern English parallel form *plaw(es)-worth.*

Plenmeller (Haltwhistle). 1255 Ass. *Plenmeneure* ; 1279 Iter. *Playnmelor* ; 1302 Sc. *Playmelor* ; 1307 Pat. *Pleinmelore* ; 1663 Rental *Plenmeller.*

A Celtic name. Cf. Maylor Hund., Flints., Mellor, Lancs., and Maelor, Wales. (Morgan, p. 157.)

Plessey (Stannington). 1222 Pat. *Plesseto* ; 1255 Ass. *Pleset* ; 1257 Ch. *Plesset* ; 1328 Ipm. *Plessys,* 1335 *Plescis* ; 1491 Newm. *Plessez, Plesseto, Placeto* ; 1628 Freeh. *Plessy.*

Cf. Plessy, Herts, and Pleshy, Ess., named perhaps from one of the numerous *Plessis* in France, or perhaps directly from N.Fr. *plessis,* " terrain enclos de haies entrelacées " (Bescherelle, *Nouv. Dict. Nationale*), LL. *pleisseicium, plessetum, plassetum,* " sylvula, seu parcus undique clausus " (Ducange), a derivative of *plectere,* to weave.

Plundenburn (Alnwick). *c.* 1220 Tate ii. 386 *Plundenburne.*

Possibly " plum-valley, -stream." Cf. Plumptree, Notts., D.B. *Pluntre,* Plungar, Leic., F.A. *Plomgarthe.* Phonology, § 51.

Podge Hole (Bedburn). 1382 Hatf. *Poydeshole.*

Possibly " Poid's hole." Cf. M.Sc. *poid,* " a vile person." Phonology, § 31.

Pokerly (Lanchester) [pɔkəli]. 1242 D.Ass. *Pokerlege* ; 1277 Pat. *Pokrely* ; B.B. *Pokerleia* ; 1636 St Mary le B. *Pockerly.*

" Goblin-field." Cf. *poker,* " hobgoblin, bugbear, demon," once common in England but now more common in America (N.E.D. *s.v.*). It is the same as Dan. *pokker,* Swed. *pocker,* " devil."

Polam (Darlington). 1382 Hatf. *Polumpole.*

Probably " pool-homestead." Cf. Poolham, Yorks.

Pollard's Lands (Auckland). 1382 Hatf. *Pollarden* ; 1435.33 *Pollardene.*

The *dene* was held by John Pollard in 1382.

Poltross Burn (Irthing, R.). 1279 Iter. *Poltroske*; 1637 Camd. *Poltrosc.*

A Celtic river-name.

Pont, R. (Ponteland). 1268 Ass. *Ponte.* **Pont Burn** (Pontop). 1153-9 Newm. *Pont.*

O.E. *Panta,* the original name of the Upper Blackwater in Essex may be the same, but the forms *Punt-* and *Pount-* (*v.* Ponteland *infra*) would then be difficult to explain unless these are due to attempts to connect it with M.E. *pounte,* a bridge.

Ponteland. 1248 Newm. *Eland*; 1255 Ass. *Elaund*; 1268 H. 3. 2. 110 *Punteylond*; 1278 Ass. *Eylaund*; 1291 Tax. *Pount Eland*; 1292 Q.W. *Punteylond*; 1295 Ipm. *Pont Eyland*; 1312 R.P.D. *Ponteland*; 1346, 1428 F.A. *Eland*; 1663 Rental *Pont Island.*

The *eland* formed by the Pont. O.E. *īegland* and *ēaland,* M.E. *e(y)lond,* are used of land surrounded by marshes as well as of an island. Cf. Elland, Yorks., and Ealand, Lincs.

Pontop (Lanchester). 1240-9 F.P.D. *Pontehope.*

" Hope by the Pont Burn (*v. supra*).

Pooltree (Lynesack). 1431.34 *Pultre.*

" Pool-tree," i.e. tree by the pool. Cf. Polstead, Suff. (Skeat, p. 27), Polehanger, Beds., Polam, *supra,* and Poolham, Lincs.

Portgate (St John Lee) [puˑrtgət]. 1278 Ass. *Portyate*; *c.* 1356 B.M. *Portchet*; 1382 Pat. *Porteyete*; 1663 Rental *Portgate.*

The second element is O.E. *geat*=gate, Mod. dial. *yet* or *yat,* the form *gate* being a modern substitution, and it probably refers to some opening in the Roman wall. The first is O.E. *port,* " a town," or *port,* " a gate." Craster (N. x. 35) takes it to be the former and interprets *Port-* as " market-town," a fair having been at one time held here. *gate* he takes to mean " way " (the Scand. *gate*=road or way), but the M.E. forms forbid this. Alternatively he takes the name to be the equivalent of O.E. *burh-geat* and to have had the same meaning as Ger. *burg-gasse,* " market-place." This is more than doubtful. It is dangerous solely on the ground of etymological identity to give a new meaning to O.E. *burh-geat*, and even the identity is doubtful (*v.* Kluge, *s.v.*).

Possibly the name is best explained by assuming that the name was originally *æt porte*, i.e. at gate, and that later the name was explained by adding the English *geat* (cf. Kirkley *supra*). O.E. *port* did not survive in M.E. with this sense, and such an addition might well be thought necessary. For *Portchet*, cf. *orchard*<O.E. *ortgeard*.

Potts Dultries (Otterburn). 1275 Pat. *Dortrees*; 1276 De Banco *Durtrees*; 1663 Rental *Potts Durtrees*.

" The *door-trees* or door-posts of *Potts.*" A family of this name once lived here (Arch. 1. 2. 330). Phonology, § 12.

*Pounteys Bridge. 1345 R.P.D. *Pounteys*; 1446 D.S.T. *Poyntesse*.

M.E. *pount*, " bridge," and *Teys*, Tees. The bridge no longer exists.

*Powtreuet (Falstone). 1325 Ipm. *Poltrerneth*, 1329 *Poltrevet*; 1330 Cl. *Poltrerneth*; 1370, 1376 Cl. *Peltreuerot*, *Poltreuerot*; *c.* 1590 Map *Powtreuet*.

A Celtic name. Cf. Powter How, Cumb., earlier *Poltraghaue* and *Polterheued* in the Lanercost Foundation Charter. (Sedgefield, p. 89.) Phonology, § 39.

Prendwick (Alnham). 1255 Ass. *Pridewyk, Prandewick*; 1275 Perc. *Prendewyk*; *c.* 1250 T.N. *Preudewic* (*sic*); 1428 F.A. *Prendwyke*; 1542 Bord. Surv. *Prendyke*.

" Dwelling of *Prende* (?)." Cf. *Prendestreteland* in Corbridge (N. x. 97) and the place-names *Prandingea, Prandinghe* in Winkler (p. 295). Phonology, § 49.

Pressen (Carham). 1176 Pipe *Prestfen*; 1251 Ch. *Pressen*; 1255 Ass. *Prestfen*, 1278 *Pressefen*; 1309 Ipm. *Presfen*; 1428 F.A. *Pressen*. **Preston** (Ellingham). 1288 Ipm. *Preston*. (Jarrow) 1104-8 S.D. *Preostun*. (Tynemouth) 1200 R.C. *Preston*. **Preston-le-Skerne.** 1091 F.P.D. *Prestetona*; 1384 B.M. *Preston super Skiryn*. **Preston-on-Tees,** B.B. *Prestona*; 1402.33 *Preston-upon-Teas*. **Prestwick** (Ponteland) [prestik]. *c.* 1250 T.N. *Prestwic*; 1428 Freeh. *Prestick*.

"Priests' fen, farm, and dwelling," probably from possession. Phonology, §§ 53, 51, 49.

Prudhoe-on-Tyne [prudə]. 1173 Pipe *Prudho*; 1217 Pat. *Prudhou*; *c.* 1250 T.N. *Prudehou*; 1307 Ipm. *Prodhow*; 1416 Inq. aqd. *Prudhowe*; 1479 B.B.H. *Proudehowe*; 1539 F.P.D. *Prowdhow*; 1642 Ryton *Priddowe*.
"Pruda's *hōh* (Part II), cf. *Pruda* (L.V.D.). Alternatively, the first element might be L.O.E. *prūd*<O.Fr. *prūd, prōd*, "proud," "gallant," descriptive of its proud position above the Tyne. Phonology, § 21.

Puncherton (Kidland). *c.* 1250 H. 3. 2. 43 *Pun(t)-chardon*; 1296 S.R. *Punchardon*; 1760 Alnham *Pungherton*.
Puncherton is so called after a Norman owner named *Punchardon*, from Pontchardon in Normandy. Cf. Heanton Punchardon, Dev. The family is often mentioned in early Nthb. records.

Pye Close (Frosterley). 1382 Hatf. *Piotland*.
"Land infested by the *piot*," North Country dim. of *pie*=magpie.

Quarrington. *c.* 1190 Godr. *Querendun*; *c.* 1150 F.P.D. *Querindone*; 1299 Acct. *Queringd'*; B.B. *Querindune* (B., C. *Queryngdon*); 1382 Hatf. *Queringdon*; 1443 Acct. *id.*; 1457.34 *Wharyngdon*; 1500.36 *Queryngton*; 1649 Comp. *Wharrington*.
Probably O.E. *cweorn-dūn*=quern-hill, i.e. one where stones for querns were found or prepared. Cf. *cweornclifu, cweornwelle*, B.C.S. 887, 1129, Quarrendon, Bucks., D.B. *Querendone*, Quorndon, Leic., Ch. Hy. II. *Querendona*, 1316 F.A. *Querndon*. App. A, § 1. Phonology, § 28. Cf. Wharmley *infra*.

Raby (Staindrop). *c.* 1050 H.S.C. *Raby*.
"Town by the land-mark (O.W.Sc. *rá*.)." (Lindkvist, p. 188.)

Raceby (Garmondsway). 1344.45 *Raceby*.
"The *by* of *Hreiðr*." *Hreiðr* is a hypothetical short form of the common O.W.Sc. name *Hreiðúlfr*. *v.* Lindkvist on Raysdale, Yorks., earlier *Reythesdale* (p. 75). Phonology, § 53.

Rackwood (Bedburn). 1382 Hatf. *Rakwod*.

L

Cf. Rackham and Racton, Suss. Roberts (p. 124) assumes from O.E. *Raculf* and *Raculfesceaster* a name *(H)raca*, but this is impossible as *Raculf* is simply a respelling of Romano-British *Regulbium*. No explanation can be offered.

Rainton (Houghton-le-Spring).   *c.* 1125   F.P.D. *Reinuntun, Re(n)ingtun, c.* 1150 *Raintonam, c.* 1190 *Reiningtone,* 1185 *Re(i)nintun,* 1203 *Reynton,* 1228 *Reiningtone;* 1253 Ch. *Reignton;* 1260 D.S.T. *Estringtona;* 1296 Halm. *Reynton;* 1311 Finch. *Estreynington;* 1430 F.P.D. *Raynton,* 1539 *Rauntone;* 1793 St Mary le B. *Renton.*

Probably the same as Rennington *infra*, though Lindkvist (p. 75), not knowing the history of that name, takes the first element to be a patronymic from O.N. *Hreinn.* Phonology, § 59.

Ramshaw (Evenwood).   1382 Hatf. *Ramsale;* 1747 Staindrop *Ramsey.* (Haltwhistle) 1312 Ipm. *Ramschawes;* 1372 Swinb. *Rampeshawe;* 1726 Whitf. *Ramsey Rigg.* **Ramshaw Well** (Windyside Fell). 1458.35 *Ramshawewell.* **Ramshope** (Elsdon).   *c.* 1230 H. 2. 16 *Rammeshope; c.* 1320 B.M. *Rameshopp;* 1542 Bord. Surv. *Rampshepp-head;* 1663 Rental *Ramshope.*

" Raven's haugh, wood, and hope." O.E. *hræfnes>* *hremnes, hramnes>rams.* The references may be to a bird or to a man of that name. Phonology, § 55; App. A, § 7.

Raredean (Cornsay). 1382 Hatf. *Rewardon;* 1688 Lanch. *Rardon,* 1715 *Rareton,* 1740 *Reardown,* 1750 *Raredane.* Unexplained.

Ratchwood (Bamburgh).   1279 N. i. 119 *Wrethewode;* 1620 N. i. 256 *Wretchwood;* 1663 Rental *Rateswood.*

O.E. *ureccea(n)-wudu* = outlaw(s)-wood. Nthb. [ratʃ] for *wretch.* Cf. Wretchwick, Oxon. (Alexander, p. 228). Phonology, § 40.

Ratton Row (Haydon). 1257 Ch. *Ratuneraw;* 1268 Ass. *Ratunrowe.*

Cf. Rattenraw in Redesdale, *Ratten Rawe* in Durham (1306), *Ratonraw* in Bamburgh (1430), quoted by Tate

(ii. 387). All mean "rat-row," with M.E. *ratoun* (O.Fr. *raton*)=rat. This name is fairly common in North. Eng., and it was probably used in contempt of a row of houses so wretched that they might be imagined to be given up to the rats alone.

Ravensfield (Stanhope). 1382 Hatf. *Ravenfeld.* **Ravens-flat** (Belmont). 1346 Halm. *Ravenflat.* **Ravensheugh** (Wark-on-Tyne). 1354 Pat. *Ravenshugh.* **Ravenside** (Chopwell). *c.* 1315 Newm. *Ravenside.* **Ravensworth** (Lamesley). 1104-8 S.D. *Rœveneswurthe.*

"Raven's field, flat, *hōh*, hill and enclosure," *Raven* being either the bird or a personal name. The *Raven-* names are probably of younger formation than the *Ram-* ones given above.

Ray (Kirkwhelpington). *c.* 1300 Abbr. *Raye*; 1542 Bord. Surv. *Reye*; 1663 Rental *Rais.*

Possibly so called from *ray* or darnel (cf. Friars Goose *supra*), or from Dial. *wray*=landmark, of which Lindkvist (p. 188) believes the more correct form to be *ray* (O.W.Sc. *rá*). Cf. Raby *supra*.

Raylees (Elsdon). 1377 Swinb. *Raleys*, 1409 *Ralees*; 1579 Bord. *Releas*; 1663 Rental *Reelees*; 1673 Elsd. *Reelees, Reallees.*

"Roe(deer)-clearings." Cf. *rahgelega, rahslede*, B.C.S. 455, 564. *re* is the common North. and Scots for *ra* or *ray*.

Reaveley (Ingram). 1268 Ipm., *c.* 1250 T.N. *Reveley*; 1663 Rental *Reavley.*

"The reeve's clearing." Cf. *Essays and Studies, u.s* vol. iv. pp. 64-5, and Raveley, Hants. (Skeat, p. 334).

Redburn (Haltwhistle). 1255 Ass. *Redburn.* (Rookhope) 1382 Hatf. *id.*

"Red-stream" from its peat-stained waters or "reed-stream." For the latter cf. *hreodburna*, B.C.S. 983 and Redbourn, Herts., K.C.D. 962 *Reodburne* (Skeat, p. 15).

Redesdale [ridzdəl]. 1075 H. 2. 3. 3 *Redesdale*; 1203 R.C. *Riddesdale*; 1274 Arch. 3. 3. 189 *Redisdale*; 1320 Ipm. *Redesdale*; 1327 *id.*; 1337 F.P.D., 1446 D.S.T. *Riddesdale*; 1542 Bord. Surv. *Ryddesdayle.*

"Valley of the Rede." Phonology, §§ 21, 7.

Redeswood. 1255 Ass. *Rode-, Rede-wode* ; 1663 Rental *Reedswood.*

Probably "Rede-wood," i.e. by the Rede. The possessive form may be due to the neighbouring Reedsmouth and Redesdale.

Redford (Hamsterley). 1314 R.P.D. *Le Roteford* ; 1342 Ipm. *Rotiford* ; 1369.45 *Rutynford* ; 1382 Hatf. *Ridforth.*

"Rotten-ford." M.E. *roten, rotin* is often applied to ground which is very soft or yielding, e.g. a "rotten" bog. The modern form is corrupt.

Redheugh (Gateshead). 1290 F.P.D. *Redhoghe.* (Thorneyburn) 1290 Ipm. *Le Redehouef* ; 1663 Rental *Reedhaugh.*

"Red" or "reed heugh" (*hōh*, Part II). Cf. Redburn *supra.* App. A, § 6.

Redhills (Durham). 1438 Acct. *Redehylles.*

"Red hills" from the colour of the soil or, possibly, "cleared" hills. Cf. *reda,* "to clear up" (Heslop).

Redmarshall (nr. Stockton). 1260 Pat. *Redmerhill* ; 1311 R.P.D. *Redmeshill,* 1314 *Redemershill,* 1345 *Redmershill* ; 1372 Pat. *Ridmershale, Ridmershill* ; 1400 D.S.T. *Redmershyll,* 1507 *Redmersell.*

O.E. *hrēod* (or *rēad*)-*meres hyll* = hill of the reed (or red)-mere, cf. *hreodmeresheafod,* B.C.S. 725 or, less probably, *hrēod* (or *rēad*) *mersc hyll* = reed (or red) marsh hill. Cf. Surtees (3.76) : "Its tower and tufted trees are seen . . . over a level district of loam and red clay, where the floods of winter would formerly collect and rest on the tenacious soil in a broad discoloured pool or mere, and hence most literally the name 'the hill of the Red Mere.'" For *s* and *h* >[ʃ] cf. Evesham [i·vʃəm] from Eves-ham. App. A, § 6.

Redmires (Wolsingham). 1382 Hatf. *le Redmyres.* **Redpeth** (Haltwhistle). 1255 Ass. *Redepeth.* **Redworth** (Heighington) B.B. *Redwortha.*

"Red or reed swamps (*mȳrr*, Part II), path and enclosure." The latter might also be *Rǣda's* enclosure. Phonology, § 21.

Relley (St Oswalds, Durham). c. 1210 Finch. *Rilli* ; 1310 R.P.D. *Rilley* ; 1637 Camd. *Relley.*

Perhaps for earlier *Ridley*, *v. infra* and Phonology, § 51.
Cf. Strelley, Notts., earlier *Stratlega*, *Stretlee*.

Rennington (Embleton). 1104-8 S.D. *Reiningtun*; 1175
Pipe *Renninton*; 1255 Ass. *Renington*; 1256 Ch. *Renigton*;
1266 Ipm., *c.* 1250 T.N. *id.*; 1307 Ch. *Renington*; 1538
Must. *Rynington*; 1579 Bord. *Rynnengton*.

Cf. Simeon of Durham (i. 80) who tells us of one Franco,
one of the bearers of the body of St Cuthbert in its wander-
ings, whose father was Reingualdus, " a quo illa quam
condiderat villa Reiningtun est appellata." Reingwaldus =
O.E. *Rægenweald* from O.N. *Rögnvaldr*. Hence Rægenweald's farm, *v.* Introd. p. xxvi. *Rægen* or *Rein* would be
a shortened form of it. Phonology, § 7.

Rickleton (Chester-le-Street). 1339 F.P.D. *Rykeling-
den*; 1421.45 *Riklinden*; 1649 Comps. *Rickleden*.

Cf. *Ricola*, A.S.C. and *Ricula* in Schönfeld. " Valley
of *Ricel* or *Ricola* and his sons." App. A, § 1. Phonology, § 59.

Ricknall (Aycliffe). 1091 F.P.D. *Richenehalla*; B.B.
Rikenhall; 1307 R.P.D. *Rikenhale*, 1311 *Rikehale*,
Richale, *Rykehal(l)e*.

Possibly " Ricwine's haugh," though the name is continental rather than native. App. A, § 6. Phonology, § 49.

Riddlehamhope (Hexhamshire). *a.* 1214 Dugd. vi. ii.
886 *Redeleme*; 1338 N. iv. 70 *Ridlam*, 1333 *Redelem*; 1547
Hexh. Surv. *Ridelamehoppe*; 1663 Rental *Ridlamhope*.

Probably " hope by the ridded or cleared *ham*." *v.*
Leam *supra*, and Ridley *infra*.

Riddyng House (Rogerley). 1382 Hatf. *le Ryddyng*.
Riding Lee and Mill (Shotley). 1262 Ipm. *Ryding*; 1298
Arch. 3. 2. 3 *le Ruddyng*; 1312 Ipm. *Ryddyng*, 1323 *La
Lye*, 1335 *La Riddyng*; 1575 N. vi. 270 *Rydinge mylne*;
1428 F.A. *Rydyng le Lee*; 1454 Pat. *Redyng*; 1526 Arch.
2. 1. 136 *Riddinge*.

" Rid(d)ing " is a common term for a clearing (O.E.
hryding). Heslop (*s.v.*) quotes B.B. for an example of this
term in the sense of *assart*, and Hodgson (2. 1. 94) shows that
its Latin equivalent was *incrementum*, i.e. a place taken in
or enclosed from a common or lord's waste. The vowel

should be short. A similar change of vowel has taken place in the Yorks. and Lincs. Ridings, which are, of course, of totally different origin. "Clearing and mill by the ridding."

Ridlees (Alwinton). *c.* 1320 B.M. *Reddeleys*; 1720 Alw. *Redlees*.

Possibly from North. dial. *redd*, "to clear, prepare," with later assimilation to the more usual type (*v.* Ridley *infra*) or *Reddeleys* may be for *Riddeleys*. Phonology, § 10.

Ridley (Bywell St Peters). 1268 Ipm. *Ryddeley*. (Haltwhistle) 1279 Iter. *Rideley*.

"Cleared clearing." *rydd* is pp. of *rid*, "to clear," from O.N. *ryðja*.

Rift Dean Burn (Heddon). 1288 De Banco *Rysdenburn*.
If this form is correct and not an error of transcription, the first element must be O.E. *hrysc-denu*=rush valley (cf. Roseden *infra*) and the modern form be corrupt. Otherwise no suggestion can be offered.

Rimside Moor (Eglingham). 1268 Pat. *Rimescid*, 1472 *Rymessid*.

Possibly "shore-, edge- or bank-side or-hill." Cf. O.E. *rima*=shore, edge, and Rimpton, Som., B.C.S. 931 *rimtun*. *boscus de Remelde* in the Assize Rolls (1278) seems to be identical with the place, and should possibly be *rim-hylde*=edge-slope. Alternatively, we may have gen. sg. of O.E. *Rim*, a personal name.

Riplington (Whalton). 1251 Sc. *Riplingtone*; 1255 Ass. *Ripplinton*; *c.* 1250 T.N. *Riplingdon*; 1298 B.B.H. *Riplengton*; 1309 Ipm. *Ripplinton*.

"Farm of Rippel or his sons." *Rippel* (cf. Ripplesmere Hund. Berks., Skeat, p. 80, and Riplingham, Yorks.) is a dimin. of *Rippa* (cf. *rippanleah*, K.C.D. 1361). App. A, § 1.

Risebridge (nr. Durham). 1311 R.P.D. *Rysebrigge*.
"Hrisa's bridge." Cf. Risbridge, Suff. (Skeat, p. 10), and Risborough, Bucks., earlier *Hrisanbyrg*. So also Riseley, Beds. (p. 38), though Skeat explains it differently.

Ritton (Netherwitton). Type I: 1135-54 Perc. *Rittona*; 1139 Newm. *Rittun*; 1290 Abbr. *Ritton*. Type II: 1208 Perc. *Westrington*, 1225 *id.*, 1268 *Esttrington, Westtrinton*.

A difficult name, possibly from O.E. *Ridda(n)-tun*=
Ridda's farm. Type II may show an alternative develop-
ment from *Riddington* (i.e. Ridda's farm) to *Rington.*
Phonology, §§ 51, 59.

Rivergreen (Meldon). 1268 Ass. *Reshon* (*sic*); 1277
Ch. *Revehou*; 1590 Anc. D. *Reffho(we)*; 1663 Rental *River-
green.*
" The reeve's *hōh* of land." Cf. Reaveley *supra* and
Ryhope *infra.* The modern form is corrupt.

Rock. 1164 Pipe *Roch*; *c.* 1250 T.N. *Rok*; 1314 Ipm.
Rokk.
The limestone here is very near the surface, cropping
out in various places. The forms go back to O.Fr. *roche*
and *roke,* and carry the history of the word a good deal
further back than N.E.D., cf. Roch(e), Yorks, and Pembr.

Roddam (Ilderton). 1135-54 Perc. *Roden*; 1203 Pipe
Rodun; 1207 Perc. *Rodenham*; 1222 Pipe *Rodon*; 1230
Cl. *Rodun*; *c.* 1250 T.N. *Rodum*; 1278 Ass., 1289 Ipm. *id.*;
1307 Ch. *Rodom*; 1308 Ipm. *id.*; 1542 Bord. Surv. *Rod-
dome*; 1663 Rental *Rodham.*
Cf. Roade, Northts., D.B. *Rode,* Road, Som., D.B. *Rode,*
Rothe End, Ess., D.B. *Roda,* Odd Rode, Chesh., D.B. *Rodo,*
Rhodes or Royds in Rothwell, Yorks., 1283 Ch. *Rodes,*
Royd in Soyland, ib. 1297 *Rode,* and possibly Rowden,
Yorks., D.B. *Rodun,* also the common suffixes *-royd* and *-rod*
in Lancs. and Yorks. For *-rod,* Wyld (p. 377) suggests O.N.
rjóðr, " a clearing," with an intermediate form *rōd,* but
O.N. *rjóðr* would give M.E. *rethe,* and probably survives in
Reeth, Yorks., D.B. *Rie.* The distribution of this element
suggests rather a native word, and there is evidence for the
existence of such in O.E. itself. Cf. B.C.S. 208 *andlang rode,*
1230 *id.,* 1129 *andlang ðære bradan rode,* 419 *on norðan siolta
roda oð ða eastroda* and *rodstubban* (Earle, p. 393).[1] It is
possible that these are in some cases from *rōd*=rood, measure
of land or " strip of cultivated land," but the latter sense
is very doubtful, and *rōd* does not explain the phonological
development to forms like *Roade* given above. More

---

[1] Ambiguous examples, in which *rod* might be O.E. *rōd*=cross, have
been omitted.

probably we have an Eng. suffix cognate with Scand. *rud*, Germ. *rod*, *rot*, *rad*=clearing, elements which are very common in place-names. -*rod(e)* is one of the commonest of place-name suffixes both in Germany and the Low Countries. A full discussion of the suffix will be found in *Nomina Geographica Neerlandica*, Part II, pp. 32-45, with lists of names (pp. 46-78), and Jellinghaus (p. 112) gives full examples from Westphalia, and Sturmfels (p. 69) from Hesse. It is unlikely that an element so common in the other Germanic dialects should have left no trace in English. The oblique case form *rode* would give Mod. Eng. [roud]. Yorks. and Lancs. *Royd* show a local sound development of *ō* to *oy* (cf. Wright, *Windhill Dialect*, § 109). Roddam may be for O.E. **Rodham*=homestead by the clearing, cf. Rodheim, Hesse (*loc. cit.*), earlier *Rodeheim*, or from dat. pl. *rodum*=(at the) clearings, cf. Ober-, and Nieder-roden, Hesse, Roden, Holland. The suffix -*rod(e)s* is fairly common in Nthb. field-names, cf. *le Smalrodes*, *Hudesrodes*, *Lamerodes* in B.B.H., and *Summerods*, *Oxenrods* in Hexh. Surv.

*Rodestane (Tynemouth). 1320 N. ix. 34 *Rodestane*. Cf. *rōde-stan*, B.C.S. 1127, "rood-stone." Possibly identical with the Holy Stone, the socket of a cross near Backworth (N. viii. 413 n.).

Rogerley (Stanhope). B.B. *Rogerleia*. " Roger's clearing."

Rookhope (Stanhope). c. 1190 B.B. *Rokehope*; 1323.45 *Rukhop*; 1338 Acct. *Rokop*, 1339 *Rukehop*. "The hope infested by rooks or belonging to Rooke." Cf. O.E. *hrocanleah*, B.C.S. 1047. Phonology, § 18.

Rosebrough (Bamburgh). 1252 Pipe *Osberwick*; 1278 Ipm. *Osburwick*; 1346 F.A. *Osborwyk*. "Dwelling of *Ōsburh* (f.)." This identification (N. ii. 225) may be correct for the site, but the names are not connected.

Roseden (Ilderton). 1255 Ass., c. 1250 T.N., 1307 Ch. *Russeden*; 1346 F.A. *Russhden*, *Russeden*, *Rosden*, 1428 *Rusden*; 1580 Bord. *Rossedoun*; 1663 Rental *Rosdon*; 1712 Egling. *Rosden*; 1754 Chatton *id.* "Rush-valley," cf. *riscdene*, B.C.S. 945 and Rushden,

Northts. *Ros*(*h*)- shows the same phonological development as the 16th and 17th c. forms *rossh, roche* given in N.E.D. for the independent word. The modern form with *ō* is corrupt. App. A, § 1.

Ross (Belford). 1249 Ipm. *Ross.*
Cf. Ross, Heref., Roos, Yorks, D.B. *Rosse.* A name of Celtic origin. Cf. Ir. *ros*, Welsh *rhos*, promontory, moor, waste, highland.

Rothbury [rɔtbari). c. 1100 Hexh. Pr. *Routhebiria*; 1166 Pipe *Roebi*, 1176 *Robirei, Roberi*; 1200, 1203 R.C. *Robery*, 1204 *Rodbery*; 1210-2 R.B.E. *Roburiam*; 1212 R.C. *Roubir*; 1219 Pat. *Roobiry*; 1228 Cl. *Robir*; Pat. *Rothebiry*, 1235 *Robery*; 1248 Ipm. *Roubiri*; 1255 Ass. *Roubir, Rowebyr*; 1258 Newm. *Routhbiry*; 1271 Ch. *Rodebir, Robery*; 1278 Ass. *Rothbyry*; 1290 Ch. *Rothebiri, Roubiri*; 1291 Tax. *Routhebyr*; 1331 Perc. *Routhebiry*; 1340 F.A. *Rothebury, Routhbery*; 1722 Houghton *Rodbury*; 1733 Ponteland *Rodberry.*
Lindkvist (pp. 158-9) takes this to mean " at the red fort," from O.W.Sc. *rauðr*, red+-*bury*, but *v.* Introd., p. xxii. for the improbability of such hybrids, and further, there is, so far as we can see now, no justification for calling Rothbury " red." Rather we must take the first element to be O.W.Sc. *rauði*, " red," used as a nickname (cf. Eng. *Routh*) and interpret the name as " Red one's *burh.*"

Rothley (Hartburn). 1233 Pipe *Rotheley.*
" *Hroða*'s clearing," *Hroða* being short for a name in *Hroð-.*

Roughley Wood (Edlingham). 1296 N. vii. 105 *Ruely*; 1396 Ipm. *Ruthle*, 1402 *Roghle.* **Roughside** (Edmundbyers). 1382 Hatf. *Rughside.* **Roughside Moor** (Falstone). 1357 Pat. *Rughside.* **Rowhope** (Kidland). 1233 Newm. *Ruhope*; 1304 Pat. *Rughope*; 1542 Bord. Surv. *Rowehoope*; 1773 Alw. *Roeup.* **Rowley** (Hexhamshire). 1226 B.B.H. *Ruley*; 1295 S.R. *Rouley*; 1298 B.B.H. *id.*, 1479 *Roulye.* (Mugglewick) R.P.D. *Rouley.* (Norham) 1228 F.P.D. *Ruleya.* **Ruchester** (Chollerton). 1348 N. iv. 333 *Rowchestre.*
In all alike the first element is the adj. " rough," either from the Nom. *rūh* >M.E. *rogh*, or the oblique wk. form *rūga*(*n*) > M.E. *rowe.*

Rowley Burn (Hexhamshire). An earlier form of this name is found in Bede (iii. 1), viz., *Denisesburna*, id est rivus Denisi.

Rudchester (Ovingham). *c.* 1250 T.N. *Rucestre* ; 1251 Pat. *Rodecastre* ; 1255 Ass. *Rucestre*, 1268 *Rouecestre* ; 1296 S.R. *Roucestre* ; 1324 Ipm. *Rouschestre, Roucestre* ; 1346 F.A. *Rouchestre*, 1428 *id.* ; 1663 Rental *Routchester* ; 1683 Ovingham *Rouchester*.

Possibly " Red-one's chester." Cf. Rothbury *supra*, but the early forms are difficult.

Rugley (Shilbottle). 1255 Ass. *Rogeley* ; 1267 Ch. *Rugeley* ; *c.* 1280 Perc. *Rogele* ; 1307 Ch. *Rugeley* ; 1333 Ipm. *Ruggeley* ; 1346 F.A. *Roughle* ; 1348 B.M. *Reuclay* ; 1428 F.A. *Rugley*.

" Rugga's clearing." Cf. *ruggan sloh*, K.C.D. 667, but it is just possible it may be " rough clearing," for northern forms, *roge* and *rug*, of this word occur.

Rumby Hill (Newton Cap). 1382 Hatf. *Ronundby*.

Probably the early form should be *Romundby*, i.e. Hrómundr's *by*. Cf. Romanby, Yorks., D.B. *Romundrebi.* Phonology, §§ 59, 51.

Rushyford (Windlestone). 1242 D.Ass. *Risseforthe ;* *p*. 1336 Robt. de Greyst. *vadum cirporum* ; 1316 R.P.D. *Ryssheford*.

O.E. *hrysca-ford*=ford of the rushes. Phonology, § 30.

Ryal (Sedgefield). 1382 Hatf. *Ryghill*. (Stamfordham) 1255 Ass. *Ryhull* ; 1268 Ipm. *Rihill* ; 1346 F.A. *Riell* ; 1663 Rental *Ryall*. **Ryle, Great and Little.**[1] (Whittingham) 1176 Pipe *Rihul* ; 1428 F.A. *Ryle*. **Ryton-on-Tyne.** *c.* 1190 Godr. *Ritun* ; 1242 D.Ass. *Rieton* ; 1307 R.P.D. *Ryton.* **Ryton Woodside,** 1493.36. *Wodsid nigh Ryton.*

" Rye-hill and -farm."

Ryhope (Bp. Wearmouth). *c.* 1050 H.S.C. *duas Reofhoppas* ; *c.* 1190 Godr. *Refhope* ; 1197 Pipe *Riefhope* ; B.B. *Refhope* (B. *Resehoppe*, C. *Roshepp*); 1327 Pat. *Revehop* ; 1335.45 *Reffhop* ; 1384.45 *Revehop* ; 1764 Map *Ri(veh)op*.

---

[1] Chastellain (*Chronique des derniers Ducs de Bourgogne*, ed Lettenhove, iv., 278), speaking of Queen Margaret's Nthb. expedition of 1463, mentions a retreat before *Rel*. Bates (*Border Holds*, p. 438) takes this to be Gt. Ryle, others identify it with Rye Hill in Slaley.

"The reeve's *hop* or enclosure." Perhaps there were two such originally, *v. Essays and Studies*, u.s. pp. 64 ff.

Sacriston Heugh (Witton Gilbert). 1312 R.P.D. *Segrysteynhogh*; 1536 Acct. *clivus Sacristae*; 1637 Camd. *Segerstonhough*; 1577 N.C.W. *Sackerston Heughe.*
"The heugh of land where the sacrist of Durham had his country estate." For *segrystein* < A.F. cf. *prebenda sacristæ*=Segerston prebend at Southwell.

Sadberge (Haughton-le-Skerne). *c.* 1150 Finch *Satberga*; 1189 D.S.T. *Sadberg*; *c.* 1190 Godr. *Sedberuie*; 1176 Pipe *Sethberga*; *p.* 1214 Geoffr. de Cold. *Sathbergia*; 1234 Pat. *Sedberg*; 1238 Cl. *Sedberue*; 1307 R.P.D. *Sadberg*; 1318 Ch. *Se(d)berge*; 1435 Pat. *Sadberg*; 1535 Finch. *Sadbury*; 1584 Arch. 3. 1. 25. *id.*
The vowel of the first element of this name is uncertain. In Sedbergh, Yorks., *e*-forms predominate, and Moorman (p. 165) explains it as from O.N. *set-berg*, "hill whose top suggests a seat by its shape." Cf. N.G. xi. 32 *Setberg*, earlier *Sedberge, Settberg*. There may have been a variant form in *a*. Cf. Norw. *sete* and *sate*, alike used of a little flat place on a rock or hill-top, and Sedbury, Yorks., 1283 Kirkb. Inq. *Sadbergh*. App. A, § 12.

St John Lee (nr. Hexham). 1310 B.B.H. *Capella Beati Johannis de Lega*; 1310 Pat. *Eccl. Sancti Johannis de Leye.*
"Church of St John in the clearing," St John being St John of Beverley, whose hermitage was close by.

St John's Chapel (Weardale). 1335 Ch. *Eccl. S. Johannis cum villa sua.* Self-explanatory.

Salt Holme (Cowpen Bewley). 1338 Acct. *le Holme.*
Saltwell (Gateshead). *c.* 1190 B.B. *Saltewelmedewe.* **Saltwick** (Stannington). 1268 Ass. *Saltwyk*; 1676 Mitford *Saltik.*
The *holm* (Part II) and dwelling where salt was once worked or sold, the salt-spring. Cf. *sealtwelle*, B.C.S. 240, *in wico emporio salis quam nos Saltwich vocamus*, B.C.S. 130, *Sealtham* 734, *Sealtleah* 540.

Sandoe (St John Lee) [sanda]. 1225 Gray, 1232 Ch. *Sandho*; 1328 B.B.H. *Sandhou,* 1479 *Sandow*; 1663

Rental *Sandhoe*; 1724 Corbr. *Sandy*. **Sandyford** (New-castle-on-Tyne). 1384 Ipm. *Sandeforthflat*; 1556 Arch. 2. 1. 32 *Sandeford Deane*.

" Sand-*hōh* (Part II) and ford." Phonology, § 36.

Satley (Lanchester). 1228 F.P.D. *Sateley*; 1304 Cl. *Satley*; 1311 F.P.D. *Satteley*; 1312 R.P.D. *Satley*.

A difficult name. The first element might be the same as in Sadberge *supra*, cf. Norw. *Saatvet*<*Satapveit*= thwaite on the flat hill top (N.G. v. 397), or possibly O.N. *saata*, "haystack," which Rygh (N.G. v. 276) finds in several place-names, hence "field by the hay-stack," but as there is no evidence that these words were ever naturalised in England, it is highly improbable that either suggestion solves the problem.

School Aycliffe, *v*. Aycliffe, School.

Scots House (Boldon). 1382 Hatf. *Scothous*.

Probably named from Galfridus Scot, who held land in Newton-by-Boldon (*Hatf. Surv.*, p. 98, S. 2. 59). Similarly Dendy has shown that Scotswood (Newcastle-on-Tyne) was so called from its one-time owner.

Scrainwood (Alnham) [ska·nwud]. 1255 Ass. *Scrawene-wude*; 1288 Ipm. *Scranewod*; 1324 Perc. *id.*, *Scravenwod*; 1318 Inq. a.q.d. *Scranewod*; 1346 F.A. *Skranewyk*; 1421 Ipm. *Screnwode*; 1428 F.A. *Scranewod*; 1542 Bord. Surv. *Skreynwood, Skrenwood*; 1580 Bord. *Screanewood*; 1663 Rental *Scarnwood*.

Initial [sk] points to a personal name of Scand. origin. Possibly it is O.W. *skraffinnr*=chatterer, used as a nick-name. No certainty is possible, but this would explain the early forms. Phonology, § 54.

Scremerston (Ancroft) [skraməsən]. *c*. 1130 Perc. *Scrimestan*; 1228 F.P.D. *Scremerestone*; 1237 Cl. *Scremeston*; 1248 Sc. *Skremerstone*; 1539 F.P.D. *Screymerston*; 1542 Bord. Surv. *Scrymmerstone*.

A difficult name. The first element may be the name *Skirmer, Skurmer*<O.F. *escrimeur*, "fencer" (Weekley, p. 112). The second is probably O.E. *stān*=rock or stone, hence "Skrimer's boundary-stone." If it were *tūn*, it would be "Skrimer's farm." The [a] of the pronunciation

in our days and in those of Raine (p. 235) is difficult. Phonology, § 10; App. A, § 7.

Seaham. *c.* 1050 H.S.C. *Seham.* **Seaton** (Lesbury). 1280 Ch. *Seyton.* (Seaham) *c.* 1190 Godr. *Sethune.* (Woodhorn) 1268 Ipm. *Seton.* **Seaton Carew.** 1345 R.P.D. *Seton Carrowe.* **Seaton Delaval** (Earsdon). 1200 Ch. *Seton;* 1270 Ch. *Seton de la Val, Seyton.*

"Homestead and farm by the sea." *Carew* because in the hands of Petrus Carou in 1189 (D.S.T. lx.). *Delaval* from the family of that name (N. ix. 135), who took their name from the castle of La Val in the Lower Marne Valley.

Sedgefield. *c.* 1050 H.S.C. *Ceddesfeld;* *c.* 1190 Godr. *Segesfeld, Seggesfelde;* 1307 R.P.D. *Seggefeld,* 1311 *Seggesfeld;* 1507 D.S.T. *Segefeld.*

Apart from the form in H.S.C. the name would clearly be O.E. *Secgesfeld*=Secg's field. Cf. Sedgeberrow, Worc., B.C.S. 964 *secgesbearwe.* If, however, the identification is correct the history is different. O.E. *c* (palatal) occasionally becomes *s* or *c* (=*s*) under A.N. influence. Zachrisson (pp. 19-20) gives examples, Cerne, Cerney, Cippenham, Cirencester, and Baddeley notes further Sezincote, Glouc., earlier *Cheisnecote* (p. 137). *ds >ge* as in Hedgeley *supra* (Phonology, § 31). Hence the name is "Cedd's field."

Seghill (Earsdon). 1271 Ch. *Sihala, Syghal;* 1295 Ty. *Seyhale;* 1296 S.R. *id.;* 1318 Inq. a.q.d. *Syhale, Sikhale,* 1336 *Sighale;* 1363 N. ix. 14 *Seighale;* 1392 Pat. *Seghall;* 1428 F.A. *Syghale;* 1542 Bord. Surv. *Syghell;* 1596 N. ix. 69 *Sighell;* 1663 Rental *Sighill;* 1727 N. ix. 71 *Seghill;* 1855 Whellan *Sighill, Seghill, Sedgehill.*

Names in *Sige-* are very common in O.E. and would seem to have had alternative pet forms *Sigga* and **Siga.* The former is found as an alternative name for *Sigefrith,* Bp. of Selsey. These names would give M.E. *Sigge* and *Seye,* and the wide variety of M.E. forms is probably due to alternative forms "*Sigga's healh*" and "*Siga's healh*" (Part II). The final predominance of *Sig-* forms may in part be due to antiquarian influence. Camden (p. 811, Holland's tr.) says, "Verily *Segedunum* is all one with Seghill in English." The identification is wrong, but it has doubtless done its work in moulding both the first and second elements in the name. App. A, § 6. Phonology, § 10.

Selaby (Gainford). 1197 Pipe *Selebi* ; 1317 Cl. *Seletby* ; 1322 Pat. *Seleteby* ; 1335 Ipm. *Seletby* ; 1460 Pat. *Seleby* ; 1480.35 *Seletby* ; 1558 V.N. *Selletbye* ; 1601 Wills *Sel(a)bye.*
" The *by* of **Sǽ-liði.*" This name is not actually found in O.W.Sc., but cf. *Haf-liði*=ocean-traveller. It is perhaps worth noting that in the *Lay of Maldon* the O.E. poet speaks of the Vikings as *sǽ-lida(n)*, the English equivalent of *sǽ-liði.* Cf. Follingsby *supra.*

*Sessinghope (Blanchland). 1336 Ipm. *Sessynghop,* 1364 *Sessinghope* ; 1425 Pat. *id.* ; 1538 N. vi. 232 *Cissenhope* alias *Cisseyhope,* 1595 *id. Cessinghope.*
Cf. Sessay, Yorks., D.B. *Sezai,* Kirkby's Inq. *Cessay.* Possibly from O.E. *Cissa,* a name found in L.V.D., with the same development of palatal *c* as in Sedgefield *supra,* hence "Cissa's hope."

Settling Stones (Newbrough). 1255 Ass. *Sadelingstan, Sadelestanes* ; 1298 B.B.H. *Sadelingstanes* ; 1452 Ipm. *Sadelyngstanes* ; 1542 Bord. Surv. *Satlyngestones* ; 1663 Rental *Satlingstones.*
settling- and *saddling-stone* are terms for a whetstone (Heslop, *s.v.*). The phonology is difficult. The vb. *settle* has M.E. forms *settle* and *sattle*<O.E. *setlan* and *sætlan.* The noun *settle* goes back to W.S. *setl,* with Anglian forms *seðl, sedl* (cf. Budle *supra*), Mod. North. dial. *seddle* and *saddle.* In M.E. it may well have been the case that on the analogy of sb. *settle,* vb. *settle* or *sattle,* there arose a series —sb. *seddle,* vb. *seddle* or *saddle.* This would explain all the forms given above. Phonology, § 14.

Sewing Shields (Haltwhistle) [sjuˑiŋʃiˑlz]. 1279 Iter. *Swyinscheles, Sywinescheles* ; 1286 Ipm. *Schiwynscheles, Siwinshell* ; 1296 Ch. *Sewynsheles* ; 1407 B.M. *Swynscheleys* ; 1479 B.B.H. *Sewyngshelez* ; 1610 Speed *Sewenshield* ; 1663 Rental *Sueingsheels* ; 1711 N.C.D. *Sewen Shields.*
" Shiels of *Sigewine*" (D.B. *Siwinus,* L.V.D. *Siwine, Sewin*).

Shadfen (Morpeth). 1257 Ch. *Shaldefen* ; 1270 Ipm.

Schaldefen. **Shadforth** (Pittington). *c.* 1190 Godr. *Schelde-ford*; B.B. *Shadeford* (B., C. *Shaldeforth*).

" Shallow fen and ford." O.E. *sceald*=shallow is discussed by Stevenson in *Philol. Soc. Trans.*, pp. 532-6. Cf. Shadwell, Norf. and Middx. The rivulet at Shadforth is called the *Shald*, probably an early back-formation. For a variant form, *v.* Shilford *infra.*

Shaftoe (Hartburn). 1230 Sc. *Shatpho (sic)*; 1255 Ass. *Shafhou, Schafthowe, Shaftho*; *c.* 1250 T.N. *Schafhou*; 1346 F.A. *Schafthow, Schaffhow.*

Probably O.E. *sceaft-hōh*=" shaft-shaped *hōh* " (Part II) or " *hōh* by or with the shaft-shaped crag," referring to one of the bold crags of Shaftoe. There is also a name *Sceaft(a)* in O.E., cf. Shaftenhoe, Herts. (Skeat, p. 36) and *sceafteshangra*, B.C.S. 629, which might be the first element.

Sharperton (Alwinton). *c.* 1250 T.N. *Scharberton*; 1296 S.R. *Scharperton*; 1303 Pat. *Sharberton*; 1307 Ipm. *Schar-berton*; 1313 Perc. *Skarberton*; 1314 Ipm. *Scharperton*, 1326 *id.*; 1346 F.A. *id., Scharpton.*

O.E. *scearda-beorg tūn*=farm by the notched hill or hill with a gap in it. Cf. *to ðæm sceardan beorge*, B.C.S. 978. The change to *Sharper-* is probably due to association with the common word *sharp.*

Shawdon (Whittingham). 1232 Pipe *Schaheden*; *c.* 1250 T.N. *Schauden*; 1428 F.A. *Shaweden*; 1542 Bord. Surv. *Shawdon.*

O.E. *sceaga-denu*=wood-valley. App. A, § 1.

Sheddon's Hill (Birtley, Co. Durham). 1382 Hatf. *Shedneslawe.*

Possibly " Sceldwine's hill." The name is not found in O.E., but is a possible formation. For loss of *l*, cf. Shadfen and Shadforth *supra.* Phonology, § 49.

Sheepwash (Bedlingtonshire). 1177 Pipe *Sepewas*; 1296 S.R. *Schipwas*; 1379 H. 3. 2. 68 *Shepwassh*; 1577 Barnes *Schipwesshe*; *c.* 1750 Wallis *Shipwasshe.*

" Place for washing sheep," with North. *ship* for *sheep.*

Shellbraes (Bingfield). 1479 B.B.H. *le Schellawe.*

" Hill with a shiel." Later the suffix was changed and the vowel shortened.

Shelley (Netherwitton). 1290 De Banco *Shelyngley*; 1292 Ass. *Shelingley*; 1663 Rental *Shelley*.

" Clearing with a ' shieling ' on it." Cf. Sheilleys, in Galloway (Maxwell, p. 285). Phonology, §§ 22, 59.

Sheraton (Monk Heselden). *c.* 1050 H.S.C. *Scurufatun*; *c.* 1190 Godr. *Scurvertune*; *c.* 1250 F.P.D. *Surueton*; B.B. *Shurutona* (B., C. *Surueton*); 1307 R.P.D. *Schurueton*; 1395.35 *Shorowton*; 1499.44 *Sherowton*; 1580 Halm. *Sherifton*; 1649 Comps. *Sheraton*.

A difficult name, probably from O.E. *Scurfan-tūn.* There is an O.E. name *Sceorf* found in *sceorfes stede*, B.C.S. 339, *sceorfes mor*, K.C.D. 650. This name seems to be identical with O.E. *sceorf*=scurf, and must have been given as a nickname. Of *sceorf* there was a L.O.E. form *scurf*, due to Scand. influence (*v.* N.E.D.). This in its turn is identical with *Scurfa*, the name of a Danish jarl (cf. Dan. *skurv*=scurf). Sheraton shows the initial cons. of *Sceorf* (*sc*=sh) and the vowel of *Scurfa*. Such a hybrid might well arise when O.E. *sceorf* and *scurf* existed side by side. There is evidence for similar sound-substitution in other words even when English and Scandinavian pairs did not exist, e.g. O.E. *sciftan*, M.E. *shiften*< O.N. *skipta*, and *v.* Snook Bank *infra*. The later development was influenced by the common word *sheriff*. Cf. Shurton, Som., earlier *Schurreveton*, *Shereveton*, and Scruton, Yorks., D.B. *Scurueton*.

Sherburn (Pittington). *c.* 1190 B.B. *Scireburne*; 1311 R.P.D. *Scherborn Balyen*; 1391 D.S.T. *Schirborn Balyen*.

" Sheer or clear stream." Cf. *scirburna*, B.C.S. 455.

Shiel Hall (Slaley). 1296 S.R. *Schelis*. **Shieldfield** (Newcastle-on-Tyne). 1255 Ass. *Schenefeud*; 1259 Ipm. *Selingfeld*; 1378 Pat. *Schelesfeld*; 1399 Ipm. *Schelefeld*. **Shields, North.** 1267 Ipm. *Chelis*; 1273 R.H. *Nortschelis*; 1291 Ty. non fuerunt ibi nisi tres *sciales* tantum;[1] 1445 Inq. a.q.d. *Seles*; 1663 Rental *North Sheeles*; 1607 Tyn. *Sheilds*. **South Shields.** 1235 F.P.D. *Scheles*.

" Shiels " and " field with shiels or shieling on it " (*scheles*, Part II).

---

[1] This quotation is due to N.E.D. (*s.v. shiels*), which also gives, from Bulleyn's *Book of Simples* (1562), " the Sheles by Tinmouth Castle."

Shield Dykes (Alnwick). 1288 Ipm. *Swynleys*; 1314 Ipm. *Swynleysheles*; 1538 Must. *Schelldyke.*
" Swine-clearing, shiels by the same, dyke by the shiel."

Shilbottle. 1228 F.P.D. *Siplibotle*; 1237 Cl. *Schiplibotle*, 1238 *Shimplingbot*; 1256 Ch. *Sheplengbotle*; 1266 Ipm. *Syplingbotill*; 1278 Ass. *Schepelingbotel*; 1288 Brkb. *Schiplingbotil*; · c. 1250 T.N. *Shipplingbothill*; 1291 Tax. *Schiplinbotel, Schiplebodil*; 1296 S.R. *Schiplingbotill*; 1311 R.P.D. *Shuplingbotill*, 1312 *Shypbotill*; 1314 Ipm. *Schippelyngbotell, Schiplyngbodel*; 1336 S.R. *Shilbotill*; 1346 F.A. *Schilbotel.*
" Shimpel's building." Cf. Shimpling, Norf. and Suff. Skeat (p. 74) takes this to be a patronymic from an unrecorded name *Scimpel*, a nickname by origin, meaning "jester." Cf. Mod. Du. *schimpen*, " to scoff at." Phonology, §§ 53, 59.

Shilburnhaugh (N. Tynedale). 1329 Ipm. *Shovel-, Sholeburn*; 1330 Cl. *Shouelburn*; 1637 Camd. *Shilbornhaugh.*
" Haugh by Scufel's stream." *Scufel* is dimin. of *Scufa*. Cf. *scufan beorh*, B.C.S. 457. Phonology, § 45.

Shildon (Auckland). 1214 Pipe *Sciluedon*. (Blanchland) 1269 N. vi. 303 *Silvedene*; 1475 N. vi. 340 *Shyldeyn*.
O.E. *scylf-dūn* and *-denu*=shelving-hill and valley. Cf. *scylf-hrycg*, B.C.S. 547.

Shildon (Bywell St Peter). 1240 Cl. *Silvingdon*; c. 1250 N. vi. 250 *Schilyngdon*; 1255 Ass. *Shilvesdon*; 1526 Arch. 2. 1. 133 *Sheldon Moore.*
" Hill of Scylf(a) or his sons." Cf *scylfes wille*, B.C.S. 197 and *Scylfingas* in *Beowulf*, the name of the Swedish Royal House.

Shilford (Styford). 1262 Ipm. *S(y)eldeford*; 1297 Cl. *Shelforth*; 1377 Ipm. *Sheldeforth*, 1421 *Sheldeford*; 1453 Pat. *Shilforth*; 1663 Rental *Shilford.*
M.E. *schelde-ford*=shallow-ford. Cf. Shelford, Cambs. (p. 63) with *scheld*, a mutated form of *schald* (O.E. *sceald*), found in Shadforth *supra*.

Shilmore (Kidland). 1292 Ass. *Shouelmore*; 1380 Ipm. *Sholemorelaw*; 1380 Ass. *Shelmerlaw*; 1642 Arch. 3. 4. 120 *Shillmore.*

M

" Scufel's swamp." Cf. Shilburnhaugh *supra.*

Shilvington (Morpeth). *c.* 1250 T.N. *Schullington*; 1316 Ipm. *Schillington, Shilvington,* 1323 *Shilvyntoune.*

"Farm of Scylf(a) or his sons. *v.* Shildon in Bywell *supra.*

Shincliffe [ʃiŋkli]. *c.* 1125 F.P.D. *Sinneclif, Scinneclif*; *c.* 1180 D.S.T. *Sineclive*; 1203 R.C. *Sinecliue*; 1304 Ch. *Shynecliue,* 1335 *Sinecliue*; 1383 Halm. *Shenclyf*; 1450 D.S.T. *Shynclyff*; 1467 Acct. *Shyncley*; 1646 Map *Shinkley.*
A doubtful name. Possibly the first element is O.E. *scinna*=demon, spectre, hence " ghost's cliff," or it may be O.E. *Scyne* as in *Scynesweorð,* B.C.S. 820, hence " Scyne's cliff." Phonology, §§ 10, 56; App. A, § 7.

Shipley (Bedburn). 1349.35 *Shepley*; 1382 Hatf. *Shipley.* (Ellingham) 1247 Sc. *Scepley*; 1252 Pipe *Scippele*; *c.* 1250 T.N. *Schipley, Schepley*; 1346, 1428 F.A. *S(c)hipley.*
" Sheep-clearing." Cf. Shipley, Yorks., Shipmeadow, Suff., Shipton, Ox.

*Shirmonden. *c.* 1250 T.N. *Chirmundesden*; 1324 Ipm. *Schirmundesdene, Shirmunden,* 1386 *Shirmounden.*
" Scirmund's valley." **Scīrmund* is a possible O.E. name.

Shitlington (Wark). *c.* 1240 Swinb. *Sutlingtun*; 1279 Iter. *S(c)hutelington*; 1358 Pat. *Shutlyngton*; 1663 Rental *Shitlington.*
" Farm of Scyttel or his sons." **Scyttel* is a dimin. of *Scytta* inferred from *scyttan-dun, -mere,* B.C.S. 216. Cf. Shitlington, Yorks., Shillington, Beds., earlier *Shitlington,* and Chesters *supra.* The *Scyttel* of Chesters and Shitlington, Nthb., were quite possibly the same man.

Shittleheugh (Otterburn). 1378 Ipm. *Shotelhough*; 1618 Redesd. *Shittelhaughe*; 1663 Rental *Shittleheugh.* **Shittle hope** (Stanhope). 1382 Hatf. *Shuttilhopfeld.*
Cf. Shutlanger, Northts., earlier *Schutel(h)anger.* All alike are probably so named from some fancied resemblance to a " shuttle " (M.E. *shittle, shotel, shuttle*) though they may contain the name found in Shitlington *supra.* App. A, § 6.

Shorden Brae (Corbridge). *c.* 1290 Perc. *Schortedene*; 1761 Corb. *Shorden bray,* 1771 *Shorden brea.*

" Short valley." Cf. *to scortandene*, B.C.S. 1125.
Shoreston (Bamburgh) alias **Shoston** (Whellan, p. 567).
Type I : 1176 Pipe *Schoteston*, 1187 *Stotesdona*, 1189
Stodeden, 1191 *Shotesdon* ; 1236 Cl. *Shoston* ; 1253 Pipe
Shocton, 1257 *Soteston* ; 1255 Ass. *Socheston, Scoteston* ;
1273 R.H. *Socston* ; 1296 S.R. *Scoston* ; 1335 Pat. *Schet-
tesdon*, 1373 *Shosseton* ; 1628 Arch. 1. 3. 95 *Shosten* ; 1663
Rental *Shotton*. Type II : 1245 Ipm. *Shorstone* ; 1579 Bord.
Shorestoune.
" Scot's Hill " (*sc*=[ʃ]) as in *scotteshealh*, B.C.S. 240,
Shottisham, Suff., Shottesbrook, Berks., Shotswell, Worc.
Phonology, § 53. The M.E. forms show common editorial
confusion of *t* and *c*. Type II is probably due to an attempt
to associate the name with the " shore " near which it stands.
App. A, § 1.
Shoresworth (Norham). *c.* 1125 F.P.D. *Scoreswurthe,
Schoresurtha*, 1203 *Sorwurth* ; 1331 Bury *Schoresworth* ;
1539 F.P.D. *Shoreswod* ; 1730 Tweedm. *Choswood* ; 1778
Lowick *Shosewood*.
Cf. Shoresworth, Lancs., earlier *S(c)horesw(o)rth, Sorisurth,
Schereswurth, Sheresworth*. Sephton (p. 212) derived this
from a name *Scorra*,[1] inferred from *Scorranstan*, B.C.S.
574, but this leaves the forms with *e* unexplained. Sherston,
Wilts., is perhaps identical with *Sc(e)orstan*, A.S.C., and is
found as *Soristone, Scorestan, Sorestan, Schor(e)stan* from
D.B. to 1252, and as *Sherston, Sereston, Sharston* from 1250-
1428. Ekblom (p. 147) takes this to be " shore-stone,"
explaining the change from *o* to *e* as due to a dissimilatory
process when *Shorestan* had become *Shoreston*. No parallel
for such a change can, however, be found. Shoresworth,
Nthb., can only be explained as " Scorra's enclosure " if we
dissociate it from the other names. App. A, § 3.
Shortflatt (Bolam). 1284 Ipm. *le Scortflat*.
" Short flat." *v. flat*, Part II.
Shotley. 1255 Ass. *Scotelye* ; *c.* 1250 T.N. *Schotley*.
Shotton (Easington). *c.* 1050 H.S.C. *Sceottun* ; B.B. *Siotona*
(B., C. *Shotton*). (Staindrop) *c.* 1050 H.S.C. *Scottun* ; 1428.33
Shotton nigh Raby. (Stannington) 1270 Ch. *Shotton*.

[1] Björkman (N.P. p. 124) takes this to be an anglicising of O.N. *skorri*.

Shotton-in-Glendale. *c.* 1050 H.S.C. *Scotadun*; 1284 Sc. *Shottone.*

Cf. Shotley, Suff., which Skeat (*s.n.*) takes to be O.E. *scota-leage* on the analogy of *scotta þæð*, B.C.S. 1282, *scottarið*, Earle, p. 310, all from *scota* gen. pl. of *scot*=small building or hut. This word is inferred from O.E. *ge-sceot*, once used of an inner-room, *sele-scot* (=tabernaculum) in the Rushworth Gloss. (Mercian), *sele-gescot*, *-gesceot* with the same sense in various renderings of the psalms and also in *Christ* and *Exodus*. If this is correct Shotley is " clearing with the huts," Shotton-in-Glendale is " hill of the huts," and the other Shottons might be " hut-farm." Shotton in Stannington, and possibly the one in Staindrop, might equally well contain the personal name *Scott(a)*. *v.* Shoreston *infra.*

Silksworth (Bp. Wearmouth). *c.* 1050 H.S.C. *Sylceswurðe*; *c.* 1180 F.P.D. *Sylkeswrtha*; 1203 Pipe *Selkesurch* (*sic*); 1322 Inq. a.q.d. *Silkesworthe.*

" Silk's enclosure." Cf. O.N. *Silki* used as a nickname (Jónsson, p. 345). O.E. *Seolca* (m.), *Seoloce* (f.) are late and may be anglicisings of the Scandinavian name. Cf. Silkstone, Yorks., D.B. *Silchestone.*

Sills (Redesdale). 1324 Ipm. *Suleshop*; 1723 Alw. *Sils.* " Syla's (*hop*) " (Part II). For this name cf. Redin, p. 79. For the loss of the second element cf. Bellshiel *supra.*

Simonburn (N. Tyndale). 1230 Ch. *Simundeburn*; 1291 Tax. *Symmundburn*; 1596 Bord., 1809 Stanh. *Simmonburn.* **Simonside** (Monkwearmouth). 1276 F.P.D. *Symondset*; 1335 Ch. *Simondesete*; 1539 F.P.D. *Symon(d)syd(e)*. (Rothbury) 1273 R.H. *Simonseth*; 1278 Ass. *Simundessete.*

" Sigemund's burn and seat " (*sæte*, Part II). For the vowel shortening at one time found in Simonburn, cf. Symonds Yat on the Wye, and for other names with the same first element, *v.* Moorman in *Essays and Studies, u.s.* vol. 4, pp. 84-103. Phonology, § 22; App. A, § 8.

Sipton Shiel (Allendale). 1491.36 *Shipstane shele*; 1547 Hex. Surv. *Siptenshel.*

Possibly " sheep-stone shiel," (cf. Sheepwash *supra*), but why so called it is impossible to say. The modern form is due to a process of dissimilation *sh——sh > s——sh.*

Skerne, R. 1381.32 *Skyren*; 1402 F.P.D. *Skyryn*, 1430 *Skeryn*.

Cf. Skerne, Yorks., D.B. *Schirne*, later *Skiren*, *Skyryn*. We may compare Norw. *Skirna*, which Rygh (*Norske Elven-avne*, p. 217) connects with O.N. *skirr*=clear, bright, *skirna* =to clear up, and the farm name *Skjern*, which he says is taken from a stream close at hand. Similarly Skerne, Yorks. is probably so named from Skerne Beck. Hence " clear, bright stream."

Skirningham. *c.* 1090 Hist. de Obsid. Dunelm *Skirn-ingheim*, *Skerningeim*; 1135-54 F.P.D. *Schirningaham*; 1203 R.C. *Skirningeham*.

A purely Scandinavian name. " Homestead (O.W.Sc. *heimr*) by the Skerne (*v. supra*) *ings* or meadows. *v.* Introd., p. xxvii.

Slaggyford (Knaresdale). 1218 Pipe *Chaggeford*; 1257 Swinb. *Slagingford*, 1267 *Slaggingford*, 1335 *Slaggiford*, 1353 *Slaggyford*.

Possibly the first element is dialectal *slag*, as in *Promptorium Parvulorum*, " slag or fowle way . . . *lubricus, lutosus, limosus*," and still used in Scots dialect (E.D.D.). If so, the name may be " ford by the muddy *ings* " (Introd., p. xxvii.)

Slaley. 1166 R.B.E. *Slaveleia*; 1170 Pipe *Slaulea*; 1255 Ass. *Slaveleia*; *c.* 1250 T.N. *Slaveley*; 1262 Ipm. *Slaueley*; 1332 Ch. *id.*; 1428 F.A. *Slauley*; 1479 B.B.H. *Sclavelye*, 1507 D.S.T. *Slaveley*; 1526 Arch. 2. 1.137 *Slaveley, Slalee*; 1538 Must. *Slale*.

Possibly the first element is the common word *slave*, and the clearing may be so called because cultivated by serfs. No example of *slave* is given before 1290 in N.E.D. Phonology, § 46.

Slatyford (Stanhope). 1382 Hatf. *Slaterforth*.

" Slater's ford," Slater being used as a personal name. Phonology, § 30.

Sledwick (Whorlton). *c.* 1050 H.S.C. *Sliddeuesse*; 1104-8 S.D. *id.*; 1306, 1316 (R.P.D.) *Sledwys*; 1336 Ipm. *Sledewys*; 1487 Pat. *Sledwys, Seldwise*; 1592 Wills *Sledwish*.

" Sledda's meadow," *v. wisce*, Part ii. Phonology,
§ 7 ; App. A, § 8.

Sleekburn (Bedlingtonshire). *c.* 1050 H.S.C. *Sliceburne ;*
1181 Pipe *Slickeburn ; c.* 1190 Godr. *Slikesburne ;* 1225 Sc.
Slikeburn ; 1236 Newm. *id.* ; B.B. *Slik(e)burna ;* 1610
Speed *Slekbornes.*[1]
" Sleek, smooth-flowing stream." M.E. *slike*, " smooth,"
Mod. Eng. *sleek, slick,* and *sleck,* the last two in dialect
only.

Slingley (Seaham). 1155 F.P.D. *Slingelawe ;* 1422.45
Slynglawe.
Cf. Slingsby, Yorks., earlier *Slengesby,* which Björkman
(Z.E.N. p. 77) takes to contain the Norse nickname
*Sløngr or **Slengi.* Cf. Norw. dial. *sleng,* " a growing
youth, an idler," and North. dial. vb. *sling,* " to go about
idling." " Sleng's Hill." App. A, § 2.

Smales (Greystead). 1279 Iter. *Smale ;* 1329 Ipm.
hopa q.v. Smale.
" *Smala's* hope," spoken of for short as " Smale's (cf.
Kirkharle *supra*), or, less probably, "small hope" (cf.
Smailholm, Roxburghshire), *smale* being later used alone,
and given pseudo-genitival suffix.

Smallhope Burn (Lanchester). 1382 Hatf. *Smalhop-
ford ;* 1479 B.B.H. *Smalhopburne.* Self-explanatory.

Snabdaugh (Greystead) [snapduf]. 1325 Ipm. *Snabo-
thalgh ;* 1663 Rental *Snabdaugh.*
snab=projecting part of a hill or rock, a rough point
or steep place, the brow of a steep ascent (Heslop). *-ot* is
perhaps the diminutive suffix. If so, the name is " haugh
by the little rock or hill." Phonology, § 50.

Snape Gate (S. Bedburn). 1382 Hatf. *Snaypesgest.* (Stanhope) ib. *Snaypgest.* Cf. *Snaypgest* in Newton by Durham,
and *Snapgest* in *Quarrington* (Hatf. Surv., and 1453.34).
A personal name is out of the question, as we cannot
believe that four *Snapes* happened to possess a *gest,* whatever
that might be. There is a North. M.E., and Mod. Eng. dial.
sneip, snayp, snape (<O.N. *sneypa*), meaning " to be hard

---

[1] In the Bedlington Parish registers the 17th c. form is *Slikbury ;* 18th
c., *Sligburn.* (Information kindly given by the Rev. A. C. Fraser.)

on, rebuke, or snub," and the suggestion may be hazarded that a piece of land which made no response to cultivation, or a farm which was notoriously inhospitable, might be dubbed " Snape-gest." Cf. Unthank *infra*.

Snipe House (Alnwick). *c.* 1290 Perc. *Swinleysnepe* ; 1663 Rental *Snipe House.*

The modern form is clearly corrupt. *Snepe* is found elsewhere as *Sneap*, in " the sneap," the name given to a well-known horseshoe-bend on the Derwent, in Sneap Plantation, only two miles from Snipe House, also in " The Sneap," the name of a house standing on high ground between the Tarset Burn and the Tarret Burn in North Tynedale. Earlier forms of this are *Snepe* and *Snipe*,[1] and there is little doubt that all these may be explained by connexion with *sneap* vb. and sb., check or rebuke. Cf. Norw. *snøypa*, to withdraw, pinch, M.Sw. and Sw. *snöpa*, to castrate. Perhaps this " swine clearing " was called " sneap " from a sharp bend in the neighbouring stream.

Snitter (Rothbury). 1175 Pipe *Snitere* ; 1176 Pipe, *c.* 1250 T.N. *Snitter(a)* ; 1248 Ipm. *Snither* ; 1309 Ipm., 1334 Perc. *Snytir.*

Cf. Snetterton, Norf., D.B. *Snetretuna*, F.A. *Sniterton*, Snitterby, Lincs., D.B. *Esnetrebi*, Lincs. Surv. *Snitrebi*, Snitterton, Derbys., D.B. *Sinitretone*, 1287 Ipm. *Sneterton*, Snitterley, Norf., 1317 Ch. *Snyterle*, and unidentified *Snitertun*, D.B. Yorks (*Yorks. Arch. Journ.*, xiv. p. 419).

The variety of the second element makes it likely that the first is a personal name, and the distribution of the name makes a Scandinavian name likely. Cf. also Nétreville, in Normandy, earlier *Esne(u)treville*, which Fabricius (*Danske Minder i Normandiet*, p. 263), and Jakobsen (*Danske Studier*, 1911, pp. 59-84) agree in taking to be from O.N. *Snørtr*, gen. sg. *Snartar*. This seems a little doubtful in face of the entire absence of forms in *rt*, for such loss of *r* has few parallels. The only ones noted in Björkman (N.P. and Z.E.N.) are O.Sw. *Anger* for *Arnger*, Norw. *Andorr*=O.N. *Arnþorr*, *Suatricus* for *Suartricus*, and rare *Tochil, Toustain* (N.F.), *Suætbrand* (L.V.D.) for more common *Torchil,*

---

[1] Information kindly given by the owner, J. H. Holmes, Esq.

Torstein, Swartbrand. Walker (p. 223) suggests an O.E.
name **Snythere,* a variant of *Snothere,* given by Searle, but
this latter is only Searle's conjectural restoration of a late
O.E. name *Snoter.* This is more probably O.E. *snotter*=
wise, used as a nickname. If so, *snytre* might be a variant
showing mutation (cf. *snytre* found once in O.E. poetry).
Early nicknames are of Scandinavian rather than English
origin, and that might account for the local distribution of
this name. Whatever the name, *Snitter* must, if associated
with these other place-names in which the personal name
remains, be one of those names in which the suffix has been
lost. Cf. Kirkharle *supra.*

If dissociated from these names, there is another possible
explanation. There is a Sw. dialectal *snyte,* " corner of a
field, angle," Norw. *snytt,* " point, top," and *Snyta* is a
common Norwegian mountain-name (N.G. vi. 189, xv. 72).
Possibly the English settlement was named after some
Scandinavian one in which this word was used in the
plural form, hence " corners or angles of land."

Snotterton (Staindrop). 1411 S. 4. 140 *Snotterton.*
" Snotter's farm." Cf. Snitter *supra* for this name.

Snook Bank (Long Framlington). 1264 Brkb. *Schakel-
zerdesnoke*; 1273 R.H. *Skalkelyerdesnoke*; 1702 Long Frame
Snukbank.

For *snoke, v.* Blyth *supra.* For the first cf. Shacklecross,
Derbys., 1235 Ch. *Shakelcros,* Shackleford, Surrey, 1355
Pat. *Shakelford,* and possibly Shackerstone, Leic., earlier
Schakeliston, Schakereston. No such personal name is
known in O.E., but cf. O.N. *Skökull,* which is clearly its
cognate. This is a nickname from *skökull*=pole of a cart
or carriage, and is found in Yorkshire Scackleton, D.B.
Scacheldene, Scagglethorpe, E.R., D.B. *Scachetorp,* W.R., D.B.
Scachertorp. An English name *Shakel* may have been
formed on the analogy of Anglo-Scand. *Skakel* (cf. Sheraton
supra). The name would then mean " snook by Shackle's
yard " or " Shackleyard's snook," with early use of a place
as a personal name.

Alternatively we may note dialectal *shackle* with various
meanings (E.D.D.), with a possible compound *shackleyard*=

yard where cattle are " shackled " or chained up. If so, the name means " snook by the shackle-yard."

Snope (Knaresdale). 1325 Ipm. *Suanhope (sic)* ; 1695 Knaresdale *Snowup*, 1710 *Snoap*.

The same name as Snowhope *infra*.

Snowhope (Stanhope). 1382 Hatf. *Snawhopkerr*. " (Marsh by) the hope where the snow lies long." Phonology, § 16.

Sockburn-on-Tees. A.S.C. *Soccaburh* ; 1104-8 S.D. *Socceburg* ; 1268 D.Ass. *Sockeburne* ; 1380 Pat. *Sokeburne*. A difficult name. There is no O.E. name *Socca*. There is an O.N. *Sokki*, probably a nickname by origin, but this could hardly be found in a name in an entry dated 780. The name *Soca*, found in Notts. in 958 (B.C.S. 1044) may well be the Scand. name. App, A, § 10.

Softley (Auckland). *c.* 1200 Finch. *Softe-lawe, c.* 1280 *Softeley.* (Knaresdale) 1277 Swinb. *Softeley.* " Soft or spongy hill and clearing." App. A, § 2.

Soppit (Otterburn). 1292 Ass. *Sokepeth* ; 1323 Perc. *Sokpeth, Soppeth* ; 1333 *id.*, 1338 *Sokpeth* ; 1586 Raine *Sopoth* ; 1618 Redesd. *Soppat* ; 1663 Rental *Soppet(h)*. *v. pæð*, Part II. The first element may be dial. *sock*=wet or moisture collecting in or percolating through a hill, drainage of a dunghill. Hence " *peth* along which drainage runs." Cf. Middendorf (p. 120) on O.E. *soces-seað*, which he takes to mean " pool of drainage," and to be equivalent to Mod. dial. *sock-pit.* Phonology, § 51.

Southwick (Monkwearmouth) [sudik]. 1104-8 S.D. *Suthewic* ; 1580 Halm. *Suddick.* " South dwelling," Cf. Sud-bourne and -bury, Suff. Phonology, § 21.

Sowerhopeshill (Cheviot). *c.* 1050 H.S.C. *Suggariple.* The identification is uncertain, the meaning still more so.

Spain's Field (Stanhope). 1382 Hatf. *Spaynesfeld* ; 1420.45 *Spanesfold.* Cf. Spain's Hall, Finchingfield, Ess., so called because held in D.B. by Henry de Ispania. Hence the first element is probably a personal name.

Spartylea and **Spartywell** (Allendale). 1547 Hexh. Surv. *Sperterley, Spertewell.*

If *Sperter-* is due to a copyist's error, the first element may be *spart* = dwarf-rushes or coarse, rushy grass, a North. and Scots dialect word. Spargrave, Som., 1262 Ch. *Spertegrave,* would suggest that in M.E. this word was used in other parts of the country, or else that we have to do with a lost personal name, cf. *Sparteswelle Mor,* K.C.D. 1367 (late copy).

Spen (Chopwell). 1312 R.P.D. *le Spen.*

Cf. Newm., p. 24, le *Spen.* No definite solution can be offered. There is a word *spine, spen,* or *spend* (O.E. *spind*) = greensward, turf, but there is no evidence that the word was ever used in the North.

Spennymoor (Whitworth). *p.* 1336 Robt. de Greyst. *Spendingmor;* 1381 Pat. *Spennyngmore;* 1446 D.S.T., 1539 F.P.D. *id.*

Cf. Spennithorne, Yorks., earlier *Spenningthorne.* No. O.E. name of this form is known, but cf. *Johannes Speninc* (Socin, *M.H.G. Wörterbuch*) and Förstemann's *Spani, Spaneldis, Spenneol,* which he associates with O.H.G. *spanan,* to entice (cf. O.E. *sponnan*).

Spindleston (Bamburgh). 1165 Pipe *Spilestan,* 1176 *Spinestan,* 1186 *Spindlestan;* 1255 Ass. *Spinelstan;* 1428 F.A. *Spyndelestane.*

O.E. *spinel-stān* = spindle-rock, so called from a detached upstanding pillar of whinstone (Tomlinson, p. 440). Jakobsen noted a similar use of "spindle" in Shetland (p. 149). Cf. also "Spindle Rock," St Andrews. App. A, § 7.

Spithope (nr. Catcleugh). 1324 Ipm. *Spithope.*

" Spit-shaped hope " possibly, though *spit,* meaning " tongue of land," is first recorded in the 16th cent. in N.E.D.

*****Spredden** [1] (Styford). 1262 Ipm. *Spyriden;* 1273 R.H. *Spiridon;* 1280 Ipm. *Spyrindene;* 1313 Fine *Spiryden;* 1318 Inq. a.q.d. *Spiredene.*

Spurlswood (Evenwood). *c.* 1280 Finch. *Spirleswod.*

No suggestion can be offered for these names.

---

[1] There are two fields of this name on the farm of Brocksbushes (N. vi. 234).

Stagshaw (Corbridge) [stadži] and [stein∫ə]. Type I:
1296 S.R. *Stagschaue*; 1315 R.P.D. *Staggeshaghe.* Type II:
c. 1340 N. x. 434 *Stainscau.*
Type I is "stag-wood," Type II is "wood by the stain
or rock" (M.E. *stain* < O.N. *steinn*).
Staindrop (nr. Raby). 1131 F.P.D. *Standrop, c.* 1150
Steindrope; 1253 Pap. *Stentrop*; 1311 R.P.D. *Stayndrop*;
1507 D.S.T. *Standropp*; 1539 F.P.D. *Standrop*; 1748
Coniscl. *Stainthrope.*
The first element is O.N. *steinn*, "rock." -*drop* in Bur-
drop, Oxon., and Souldrop, Beds., is a variant form of *thorpe*
Part II), but the early and uniform use of *drop* makes this
explanation impossible for Staindrop. Lindkvist (p. 84
n. 4) suggests O.N. *dropi*, drop, or O.W.Sc. *drop*, " a drop-
ping or dripping." Later, the second element was interpreted
as *thorpe*. Cf. Camden (p. 737), who speaks of Staindrop,
which is also called *Stainthorpe*, " stony village." ⸴

Stainton, Great and Little or **Stainton-in-the-Street.** 1091
F.P.D. *Staninctona, c.* 1250 *Steinintune*; 1284 Finch.
Staynton, n.d. Steintona; 1312 R.P.D. *Staynton in Strata.*
v. Stannington *infra.* It stands on "an ancient Roman
cross-road running in almost a direct line from Old Durham
and Mainsforth through Bradbury" (S. 3. 61). Phonology,
§ 95.
Stamford (Embleton). 1244 Ipm. *Staunford*; 1257 Ch.
Stanford. **Stamfordham** [stanətən]. 1187 Pipe *Stanford-
ham*; 1246 Ch. *Staunfordham*; 1249 Ch. *Stamfordeham*;
1270 Ipm. *Stanfordham* alias *Stamfordham*; 1409 Swinb.
Stanerdame; 1428 F.A. *Stanfordham*; 1460 H. 3. 1. 29
Stanwardham; 1559 F.F. *Stanerden*; 1717 Elsdon *Stanerton.*
" Stony-ford and homestead by the same." Phonology,
§ 51; App. A, § 7.
Stanhope. B.B. *Stanhopa.* **Stanley.** 1297 Pap. *Stanley*;
1340 R.P.D. *Stanlawe.* **Stanton** (Longhorsley). 1200 R.C.
Stantuna; 1379 Ipm. *Staynton*, 1480 *Staunton.*
" Stone or rocky- hope-, clearing or hill and -farm."
Phonology, §§ 14, 21.
For *Staynton, v.* Nunstainton *supra.*
Stannington. Type I: 1255 Ass. *Steynington*; 1270 Ipm.

Stayngton; *c.* 1250 T.N. *Staungton*; 1303 Var. *Stainton*. Type II: 1257 Ch. *Staningion*; 1271 Ch. *Stanigton*; 1312 R.P.D. *Staungton*; 1346 F.A. *Stanyngton*. "Farm of Steinn or his sons." For this name *v.* Björkman, N.P., p. 130. Type II is an anglicising of the Norse name. Phonology, §§ 59, 22.

Stawàrd (Haydon). 1271 Sc. *Staworthe*; 1279 Iter. *id.*; 1290 Ipm. *Stannord*; 1326 Pat. *Staward*, 1373 *Staworth*; 1542 Bord. Surv. *Stawarde*.

O.E. *stānweorþ*=stone-enclosure. Cf. O.E. *stān-wielle* > Stawell and Stowell, Glouc. (Baddeley, pp. 147-8), Stowell, Som.

Steel (Hexhamshire). 1268 Ass., 1298 B.B.H., 1308 Cl., 1479 B.B.H. le *Stele*. (Chesterhope) 1359 Cl., 1395 Ipm. *id.* Cf. also *le Stele*, in Benfieldside (Hatf. Surv.), *Bromhoppe cum Stele* (Coram 1291), Hawksteel (Hexh. Surv. *Haukestele*). Here and in Todburn Steel, Steel in Lilswood, Steel Cleugh in Ridley, Steel Rigg on the Wall, we have the word *steel* used in Scots dialect of (1) a wooden cleugh or precipice, (2) a ridge projecting from a hill, and found also as the name for long lines of rocks projecting into the sea, e.g. Long Houghton Steel, Whitburn Steel.

The Steel in Hexhamshire is the name given to the long point or tongue of land formed by the junction of the Rowley Burn with the Devil's Water. This was once known as *Ruleystal* (Gray 1233). Whether this is the correct early form of *steel* is unknown, for the history of *steel* in this sense is not known.

Stella (Ryton). B.B. *Stelyngleye*; 1382 Hatf. *Stelley*; 1438 Acct. *id.*; 1635 Comm. *Stelhoe*; 1663 Ryton *Stellay*, 1698 *Stella*.

Stelling (Bywell St Peter). *c.* 1250 T.N. *Stellyng*. Cf. *stelling*=cattle-fold (Heslop). The first name is "clearing with a cattle-fold." Cf. Shelley *supra*. App. A, § 7.

Steward Shiel (Muggleswick). 1382 Hatf. *Stewardhall*, *Stewardshell*.

Eggleston (p. 145) says that this was a residence of the steward of the Bishop of Durham.

Stickley (Horton). 1203 R.C. *Stikelawe* ; 1255 Ass. *id.* ; 1270 Ch. *Stickelawe* ; 1533 N. ix. 134 *Styklaye.*

Cf. Stickford, Lincs., D.B. *Stichesforde*, Stickney, ib., D.B. *Stichenai*, and Winterbourne Stickland, Dors., F.A. *Wynterburne Stikeland.* O.E. *sticca*, " stick, peg," does not seem to have been used in place-names, and would here give no satisfactory sense. There is no O.E. name *Sticca* and no O.N. one is recorded, but it may be that in *Stykkis-eyjar*, *-hólmr, -völlr* in Iceland (Kålund, *op. cit.* vol. i., pp. 541, 444, 63), we have such a name. Cleasby-Vigfusson (*s.v.*) takes the meaning of the middle name to be " island of the piece," but this does not seem very probable. If **Stykki* (a nick-name derived from *stykki*) was in use, it might be expected in Lincs., and is quite possible in Nthb. Stickland, Dors., may contain a M.E. derivative of this name. The *n* in Stickney is a difficulty unless it develops from a weak form already in use in O.E. Possibly in this case the name may be *Sticwine*, found once in O.E.

Stillington (Redmarshall). *c.* 1190 Godr. *Stillingtune.*

Cf. Stillingfleet and Stillington, Yorks., D.B. *Steflinghefed*, Kirkb. Inq. *Stivelingflete*, D.B. *Stivelinctun.* These point to an O.E. name **Styfel*, a dimin. of **Styfa*, a name found in *Stifingehæme*, B.C.S. 1142, Steeton in Sherburn, Yorks. (*c.* 1030 Yorks. Ch. *Styfetun, Styfingtun*), and in Bolton Percy, D.B. *Stivetone.* Another dimin. is **Styfic* or **Styfeca*, which Skeat finds in Stetchworth, Cambs. (p. 27), and Stukeley, Hunts. (p. 335). *Styfa* is allied to *Stybba* and *Stuf*, recorded by Searle. Phonology, § 51.

Sting Head (Elsdon). *a.* 1226 Newm. *Steng* ; 1536 Arch. 3. 8. 20 *The Stinge.*

O.E. *steng*=pole. Cf. *stenges healh*, B.C.S. 890, which Middendorf (*s.v.*) takes to mean " haugh of the pole." Phonology, § 7.

Stirkscleugh (Hesleyside). 1279 Iter. *Strikeliscloyche.*

" Styrcol's clough." *Styrcol* is a L.O.E. name of Scand. origin. (Björkman, N.P., pp. 132-3). Phonology, § 54.

Stobbilee (Lanchester). 1292 Pat. *Stubbiley.* **Stobs House** (Dipton). 1347.31 *le Stobbes.* **Stobswood** (nr.

Chevington). 1252 Pat. *Stubbes*; 1255 Ass. *Stobbeswude*; 1297 Newm. *Stobbeswood*; 1723 Bothal *Stobesworth*. **Stubb House** (Whorlton). 1333 S. 4. 101 *Stubhous*.

O.E. *stubb*=tree-stump, with adj. *stubby*, covered with such. *stobb* is a common dialectal variant. App. A, § 3.

Stockerley (Iveston). 1382 Hatf. *Stokerley*.

" Stocker's field." *Stocker*=one who fells or grubs up stumps of trees. Cf. Stockerton, Galloway (Maxwell, p. 296), Fortherley, and Morley *supra*.

Stockley (Brancepeth). *c.* 1200 B.M. *Stocheleya*. **Stocksfield** (Bywell St Andrew). 1244 Cl. *Stokesfeud*; 1255 Ass. *Stokesfeld*. **Stocksfield Burn.** *c.* 1220 N. vi. 254 *Stochisburne*. **Stockton-on-Tees.** 1228 F.P.D. *Stoketone*; 1249 Ch. *Stocton*; 1311 R.P.D. *Stok(e)ton*. **Stokoe** (Greystead). 1279 Iter *Stokhalche*; 1330 Orig. *Stokehalgh*; 1663 Rental *Stokoe*.

The first element in Stockley and Stockton is probably O.E. *stocc*=stock or post. Cf. *stoc-tun*, B.C.S. 1007 meaning "enclosure formed by stocks or posts." *Stockley* is the clearing marked or enclosed by such. Stocksfield and its burn are apparently "field and stream by (or marked by) the posts." The long vowel of Stokoe furnishes a difficulty. Ekblom (*Place-Names of Wiltshire*, p. 21) shows that O.E. *stōc* is a ghost-word and that *Stoke* in place-names is dat. sg. of O.E. *stoc*, with lengthening of vowel in the open syllable, the word *stoc* seeming to have no definite meaning beyond that of " place." Such a form could hardly be found in the first half of a place-name, and perhaps the first part of this one is as corrupt as the second (App. A, § 6) and the place really means " haugh marked by a stock or post."

Stonecroft (Newbrough). Type I: 1175 Pipe, 12th c. B.B.H. *Stancroft*; 1327 Cl. *Stauncroft*. Type II: 1262 Ch. *Stuincroft*; 1298 B.B.H., 1325 Ipm. *Stayncroft*. Type III: 1663 Rental *Stonecroft*.

Cf. *Stanecroft* in Warkworth (iv. v. 13). Self-explanatory. Type I is North. Eng., II shows Scand. influence, III is due to Standard English.

Stoney Burn (Riding Mill). *c.* 1275 N. v. 377 *Stainesden Burn*.

"Burn in Steinn's valley." Cf. Stannington *supra*. The modern form is anglicised.

Stotfield Burn (Stanhope). 1382 Hatf. *Stotfeld*; 1580 Halm. *Stotfolde Burne*. **Stotfold** (Elwick). *a.* 1244 B.M. *Stotfald, Stodfald*.

The second name is O.E. *stōd-fald*=stud-enclosure. Stotfield may be the same or possibly it is a compound of O.E. *stot*, "horse," M.E. *stott*, "ox, steer," and *feld*, "field." Phonology, § 51 ; App. A, §

Stotgate (Bear Park). 1380 Acct. *Stottesȝate*, 1438 *Stotyate*; 1446 D.S.T. *le Stotyate*.

stott is North. Eng. for "steer" and also for a "heifer," and this may be "steer's gate." It was sometimes used as a nickname. Cf. the personal name *Stott*. *v. geat*, Part II.

Stranton (nr. W. Hartlepool). *c.* 1130 Ch. *Strantun*; *c.* 1190 Godr. *Straintune*; 1158 Pipe *Stranton*; 1451 D.S.T. *Straunton*, 1507 *Stranton*.

"Strand-farm," as suggested by Surtees (3. 121). Cf. *Stranda-tún* in Iceland (Jónsson, p. 469). Phonology, §§ 51, 5.

Streatlam (Barnard Castle). *c.* 1050 H.S.C. *Stretlea*; 1316 Cl. *Stret(e)lam*, 1317 *Stretlem*; 1336 Ipm. *Stretlom*; 1656 Staindrop *Streatenam*, 1659 *Streatnam*.

Cf. B.C.S. 625 *strætlea*. "Clearing by the Roman road," and later, "homestead by the same," with loss of unstressed *h*. Phonology, § 21.

Strother (Boldon). *c.* 1190 F.P.D. *Estrother*. (Haughton) 1273 Swinb. *Haluton Strothir*; 1279 Iter. *Halchtona Struther*; 1663 Rental *Strudder*. *v. strother*, Part II. Phonology, §§ 12, 41.

Stubb House (Whorlton). *v.* Stobbs *supra*.

Sturton (Warkworth). *c.* 1220 Newm. *Strattona*; 1241-8 *Stretton*.

O.E. *stræt-tūn* (O. North. *strēt-tūn*)=farm by the "street" or paved road. Cf. Sturton, Lincs., Notts., Yorks., Stirton, Yorks., and numerous Strettons and Strattons. Phonology, §§ 21, 54.

Styford (Bywell St Andrew). 1210-2 R.B.E. *Styfford* ;
1262 Ipm. *Stiford* ; *c.* 1250 T.N. *Stifford* ; 1273 R.H., 1278
Ass., 1312 Ipm. *Stiford,* 1316 *Styford* ; 1346 F.A. *Stifford* ;
1425 Ipm. *Styford.*
O.E. *stīg-ford*=ford by the *stīg* or path. Cf. Stifford,
Ess., D.B. *Stiforda,* and Parford, Dev., B.C.S. 1331 *pathford.*

Summerhouse (Gainford). *c.* 1200 B.M. *Smuhusum*
(*sic*) ; 1207 F.P.D. *Sumirhusum* ; 1316 R.P.D. *Somerhouse.*
O.E. (*æt þæm*) *sumor-hūsum*=(at the) summer-houses.
Cf. N.E.D. which gives an early quotation from a custumal
of Newington by Sittingbourne, in Kent, which tells us that
the men living in the weald have to provide a "*domus
aestivalis* quae Anglice dicitur *Sumer-hus.*" "Summer-
residence in the country."

Sunday Burn (N. Tyndale). 1291 Ipm. *Sunday-burn.*

Sundaysight (N. Tyndale). 1325 Ipm. *Sundayheugh.*

Sunderland-by-the-Sea. *c.* 1168 F.P.D. *Sunderland.*

Sunderland Bridge. 1163-80 F.P.D. *Sunderland* ; 1383.32
S. nigh Durham. **Sunderland** (Stanhope). 1457.35 *Sunder-
land-shele.*

Cf. also *Sunderland,* B.C.S. 1298 and *Sunderland* in
Warkworth (N. v. 113) from O.E. *sunder-land*=land set
apart for some special purpose, private land. Plummer
(*Bedae Opera Historica,* Introd., p. ix.) suggests that when
Bede says (Eccl. Hist. v. 24) he was born *in territorio* of
the monastery of Wearmouth and Jarrow, he is really
referring to Sunderland-by-the-Sea, for the O.E. Bede (v. 23)
renders this phrase *on sundurlonde.*

Sunderland, North. 1176 Pipe *Suðlanda ;* *n.d.* Nost.
Cart. *Sutherlannland* [1] ; 1187 Pipe *Sunderland* ; 1236 Cl.,
1248 Ipm., 1278 Ass. *id.*

The earliest forms suggest O.N. *suðr-land*=south-land,
identical with Sutherland in Scotland. The form in the
Nostell Cartulary shows a curious doubling of the suffix.
Later the name was assimilated to a more common type.

Sunniside (Lamesley). 1322 Cl. *Sonnyside* ; 1342 Ipm. *id.*
"Sunny-hill."

Sunnyside (Wolsingham). 1382 Hatf. *Sonnyngside.*

[1] This reference is due to N. i. 306.

"Hill of Sunna or his sons." Cf. *Sunnandun*, K.C.D. 920 and Sunningwell, Berks., *sunningauuille*, B.C.S. 366.

Swainston (Elwick). 1351 B.M. *Swayneston.*
"Sveinn's farm." *Sveinn* (O.E. *Swegen*) is a common Scand. name.

Swalwell (Whickham). B.B. *Swalwels.*
"Swallow wells or springs." Cf. Hawkwell *supra.* *Swallow* is not the first element in Swalecliffe, Kent, as is often asserted on the authority of B.C.S. 756, which speaks of "nomen . . . rupis irundinis, id est swealewan clif." This is only an early piece of etymologising, for Swalecliffe is on the Swale, which in B.C.S. 341 is called *suueluue flumen.* Cf. also Swale, Yorks., Bede *Sualua.*

Swarden Burn (Eachwick). 1479 B.B.H. *Swardonsyde.*
Swarland (Felton). 1255 Ass. *Swarla(u)nd, Swarelaund* ; 1278 N. vii. 387 *Swerlaund* ; *c.* 1250 T.N. *Swarland* ; 1310 Sc. *Swareland* ; 1707 Ford *Swarlin.*

O.E. *swāre-* or *swāre-land=*"heavy, sluggish land." *Swarden-syde* probably describes a hill with similar soil. Phonology, § 56. App. A, § 1.

Sweethope (Thockrington). 1280 Wickw. *Suethoppe* ; 1663 Rental *Sweetup.*

So called probably from the quality of the land or pasture.

Swinburn (Chollerton). *c.* 1250 T.N. *Swineburn* ; 1346 F.A. *Swymburn, Swynbourn.* **Swinhoe** (Bamburgh). *c.* 1250 T.N. *Swinhou* ; 1280 Ch. *Swyneho* ; 1315 Ipm. *Swynowe.* **Swinhope** (Weardale). 1313 R.P.D. *Swynhopelawe.*

"The burn, *hōh* (Part II), and hope haunted by the wild boar." Phonology, §§ 51, 36.

Tanfield (Beamish). *c.* 1190 Godr. *Tainefeld* (*sic*) ; *c.* 1175 Joh. Hex. *Tamefeld* ; *c.* 1300 Lewes *Taundfeld* ; 1297 Pap. *Taunfeldleye* ; 1312 R.P.D. *Taunfeld* ; 1382 Hatf. *Ta(u)mfeld* ; 1483.35 *Taundfeld.*

"Field by the Team (earlier *Tame*), R. Phonology, §§ 52, 55, 5.

Tarsett (Thorneyburn). 1269 Pat. *Tyrsete* ; 1279 Iter. *Tyrset* ; 1329 Ipm. *Tirset* ; 1542 Bord. Surv. *Tarsett.*

O.E. *Tira(n)- sǣte=*Tir's farm, *Tir(a)* being short for such a name as O.E. *Tir-weald* or *-wulf.* Phonology, § 8.

N

Team, R. 1277 Pat. *Thame*; 1349.45 *Tame.*

Cf. Thame and Thames, Oxf., Tame, Staffs., Teme, Worc., as river-names.

Tecket (Simonburn). 1279 Iter. *Teket*; 1663 Rental *Teckett.*

A Celtic name.

Tedcastle (Haydon). 1364 Ipm. *Tadecastell*; 1671 Arch. 2. I. 127 *Teadcastle.*

"Tada's Castle." Cf. Tadcaster, Yorks. (Moorman, p. 180), Tadlow, Cambs., Tadley, Hants., B.C.S. 1152 *Tadanleage.* The variant vowels may be due to association with *toad* (O.E. *tadige, tadde*), in the North. dial. forms *ted* and *tead.*

Tees, R. 1104-8 S.D. *Teisa.*

Temple Heap (Thirlwall). 1479 B.B.H. *le Temelhope.*

Tepper Moor (Simonburn). 1479 B.B.H. *Tepermore.*

No explanation of these names can be offered. For the last, cf. Teppermuir, Perthshire.

Thackmire (Castle Eden). *n.d.* F.P.D. *Thacmere.*

Cf. Thakeham, Suss., which Roberts (p. 156) takes to be from O.E. **þaca*=thatcher (cf. *þacian*, to thatch). The second is *mere* (*v.* Part II). Hence "Thatcher's pool or boundary."

Thickley (Redworth). *c.* 1050 H.S.C. *Thiccelea*; 1104-8 S.D. *Ticcelea*; 1312 R.P.D. *Thikeley*; 1331 B.M. *Thickley.*

"Clearing in or by the thicket (O.E. *þicca*)."

Thirlwall (Haltwhistle). 1255 Ass. *Thurlewall*; 1279 Iter. *Thirlewalle*; 1479 B.B.H. *Thrilwall.*

Fordun's *Scotichronicon*, II. vii.; III. x., xliii., says that *Thirlwall* was the name given to the wall which the Romans drew across Britain from sea to sea in order to keep back the attacks of the Scots, and that this name means *Thirlitwall* or *murus perforatus*, because, with the aid of the country folk, they *thirled* or pierced it in many different places so that they might always be able to pass to and fro through it. The name was certainly never applied to the wall as a whole, but certain gaps, of which Thirlwall was one, may have been so called. *þȳrel* is used in O.E. as an adj.

meaning "pierced." (Cf. Middendorf, p. 141). Phonology, § 54.

Thirston (Felton) [θrustən, θristən]. 1257 Newm. *Thrasterston*; Ipm. *id.*; 1278 Ass. *Traterston*; *c.* 1250 T.N. *Th(r)a(s)friston*; 1298 Ipm. *Traustreston*; *n.d.* Newm. *Thrastreston, Thresterston*; 1332 Fine *Thrastreston, Thracheston, Thareston*; 1346 F.A. *Trasterton, Thartreston*; 1388 Ipm. *Thristerton*, 1417 *Thresterton*; 1428 F.A. *Thersterton*; 1580 Bord. *Thrustoun*; 1628 Freeh. *Thriston*. The first element is M.E. **thrastere, *threstere*, an agent noun from O.E. *þræstan*, M.E. *þreste, þraste, þarste*, "to push, stab, thrust." It must have been used as a nickname, perhaps in the sense of a pushful person, a "thruster." This would give Mod. Eng. *Threston, Thraston, Tharston*. The modern pronunciations are due to associations with the vb. *thrust* (North. dial. *thrist*), a vb. with which *þræstan* has been confused throughout its history.

Thockrington. 1274 Giff. *Thokerington*. A difficult name, but there is little doubt that the first element is a personal name and should be associated with O.E. *þocerian*, "to move to and fro, run up and down," or with O.N. *þoka*, "to move," with agent noun **þokari*, used as a nickname. Hence "farm of Thocker or his sons."

Thornbrough (Corbridge). 1255 Ass. *Thorneburg, Thorneburn'*; 1262 Ipm. *Thornbg'* alias *Thorneburi*; *c.* 1250 T.N. *Thorneburg*; 1682 Arch. 2. 1. 106 *Thorbrough*. **Thornhope Beck.** *c.* 1150 F.P.D. *Thornhopeburn*. **Thornhope** (Knaresdale). 1279 Iter. *Thornhoppe*; 1855 Whellan *Thornup*. **Thornley** (Wolsingham). 1382 Hatf. *Thornley*. (Kelloe) 1104-8 S.D. *Tornalau*; 1460 Pat. *Thornelawe*. **Thornton** (Hartburn). 1249 Ipm. *Thurneton*; 1479 B.B.H. *Temple Thornton*. (Norham) B.B. *Tornet', Torent*. (Tyndale) 1262 Ch. *Thornton*; 1316 Ipm. *Therntoun*.

"The *burh*, hope, clearing, hill and farm by the thorn bushes," or, in the last case, perhaps, "enclosure made of thorn-bushes." Cf. Thornbury, Glouc., *þorn-leah*, B.C.S. 1282, *þorntun*, B.C.S. 1033. *Thern-* and *Thurn-* point to O.E. *þyrne* rather than *þorn* (cf. Farnham *supra*). App. A, § 10; Introd., p. xix.; Phonology, §§ 54, 36. There are

three Thorntons in Hartburn. Temple Thornton belonged
to the Knights Templars, another was known as *Thornton
Giffard* (Pat. 1358).

Thorneyburn (N. Tyndale). 1325 Ipm. *Thorny-
bourne.* **Thornyhaugh** (Brinkburn). 1309 Ipm. *Thorni-
halugh.*

"Stream and haugh overgrown with thorn-bushes."

Thorngrafton (Haltwhistle). *c.* 1150 H. 2. 3. 383 *Thor-
graveston*; 1175 Pipe *Thorgrafton, Thoringraston*; 1279 Iter.
Thorngarstona; 1298 B.B.H. *Thorngraffton.*

O.E. *þorn-grāf-tūn*=farm by the thorn-copse, with
pseudo-genitival *s* in some forms.

Thorpe (Easington). *c.* 1050 H.S.C. *Thorep*; 1197 Pipe
Torp; 1539 F.P.D. *Thropp juxta Esyngtoune.* **Thorpe
Bulmer** (Hart). 1312 . R.P.D. *Thorpebulmer.* **Thorpe
Thewles** (Grindon). 1265 Finch. *Thorpp Thewles*, 1402
Thropthewlesse.

v. þorp, Part II. *Bulmer* because granted by Bp.
Kellaw to Ralph de Bulmer (S. i. 61). *Thewless*, i.e. with-
out morals. Cf. Wicked Widford, Herts., Drunken Thoresby,
Lincs.

Threepwood (Haydon). 1308 Arch. 2. 17. 43 *Trepwoode*;
1364 Ipm. *Threpwode.*

"Wood of disputed ownership." Cf. Nthb. *threaplands*
(Heslop, *s.v.*), Threapwood, Chesh., Threapland, Yorks.
and Cumb.

Thrislington (Bp. Middleham). 1300 F.P.D. *Thur-
staneston*, 1309 *id.*; 1382 Hatf. *Thrustanton*; 1475 Finch.
Thrustyngton, 1478 *Thurstyngton*, 1511 *Trystillyngton*;
1637 Camd. *Thruslington.*

"Thorsteinn's farm," found also as Thurstaston, Chesh.,
Thurston, Suff., Thruxton, Norf., Thrussington, Leic.
The intrusive *l* and the change from *u* to *i* may be due to
confusion with North. dial. *thristle*, used for both *thistle* and
throstle (*v.* Heslop). Phonology, §§ 54, 13.

Throckley (Newburn). 1160 Pipe *Trocchelai*, 1176
Trokelawa; 1210-2 R.B.E., 1255 Ass. *id.*; *c.* 1250 T.N.
Throkelawe; 1309 Ipm. *Throckelawe*; 1479 B.B.H.
Throkelaw.

" Throc's hill." Cf. *þroc-brig and -mere* (B.C.S. 391, 508), Throcking, Herts. (Skeat, p. 38), Throckmorton, Worc. (Duignan, p. 162). App. A, § 2.

Throp Hill (Mitford). 1166 R.B.E. *Trophil*; *c.* 1250 T.N. *Throphill*; 1273 R.H. *Troppil'*; 1322 Ipm. *Throppell*; 1346 F.A. *Tropphil*; 1421 Ipm. *Thropell*; 1663 Rental, 1807 Meldon *Thropple*. **Thropton** (Rothbury). 1176 Pipe *Tropton*; 1334 Perc. *Thorpton*.
" Hill and farm by the *þorp* " (Part II). Cf. Dunthrop, Heythrop and Thrup, Oxf. Phonology, §§ 54, 36.

Throston (Hart). *n.d.* Lewes *Thoreston*; 1344 Ipm. *Thorston*; 1475.35 *Thirston*, 1480 *Thruston*.
" Thor's farm." For this name *v.* Björkman, N.P., pp. 146-7. It is probably not to be taken as from the god of that name. Phonology, § 54.

Thrundle (Chilton). 1392 F.P.D. *Thurnedale*. **Thrunton** (Whittingham). *c.* 1180 Newm. *Trowentona*; 1199 Pipe *Torhenton*; 1253 Ipm. *Throunton* alias *Trowynton*; 1258 Newm. *Thrownton*; 1260 Ipm. *Trovinton, Thowerton,* 1265 *Throwinton,* 1266 *Trowinton*; *c.* 1250 T.N. *Throingtun*; 1278 Ass. *Thorowinton, Trowenton*; 1312 Inq. a.q.d., 1320, 1422 Ipm. *Throunton*; 1649 Arch. 2. 1. 55 *Thrunton*; 1650 Comps. *Throunton*.
" Thurwine's dale and farm." Cf. *Thruwin*, L.V.D., and *Thurwineholm*, K.C.D. 566. Björkman (N.P. p. 164) explains this name as of hybrid origin from O.N. *þor* or *þur* and O.E. *winé*. Phonology, §§ 49, 54.

Till, R. *c.* 1050 H.S.C. *Till*; 1255 Ass. *Tylle, Tilne*; 1560 Raine *Tilne*.
A Celtic river-name. Phonology, § 56.

Tillmouth. 1104-8 S.D. *Tillemuthe*; *c.* 1250 T.N. *Tillemue*; B.B. *Tilmouth*. Self-explanatory.

Tinely (Ellingham). 1278 Ipm. *Tyndeley*; 1663 Rental *Tyneley*.
Cf. *le tyndlaw* in Southwick (Halm. 1380). Both alike are probably named from some fancied resemblance to the projections on a harrow or fork (O.E. *tind*, later *tine*).

Titlington (Eglingham). *c.* 1150 Perc. *Thitelittonam*; 1166 Pipe *Tithlington,* 1197 *Titlinton,* 1252 *Titlington*;

1268 Ass. *Tyttelington* ; *c.* 1250 T.N. *Titlington* ; 1320 Pat. *Tidilyngton* ; 1336 Ch. *Tedlintone, Titlingtona.*
" Farm of *Titel* or *Tyttla* (*Tytel*) or his sons." For the first, cf. Bede's *Titillus, Titelescumb,* B.C.S. 1191, *Titlesham,* B.C.S. 198, and *Titlandun,* B.C.S. 667. It is a dimin. of Titta. For the second *v.* Searle. There is yet a third possibility, viz., that it is the O.W.Sc. *titlingr,* a nickname meaning " sparrow " (Jónsson, p. 310). For *d, v.* Zachrisson, p. 43 n.

Todburn (Longhorsley). 1434 R.C. *Totborne* ; 1663 Rental *Todbourne.*
Hodgson (2. 2. 206) is probably correct in associating this name with *tod*=fox. It might, however, be from the personal name *Tota.* Phonology, § 57. " Fox-stream " or " Tota's stream."

Todhill (Haltwhistle). 1312 Ipm. *Todholes.*
" Tod or fox holes." Cf. Foxhall, Suff., D.B. *Foxehola* and Foxholes, Yorks.

Todridge (Bingfield). 1479 B.B.H. *Todrige* ; 1663 Rental *Todrish.*
" Fox-ridge." Cf. Todburn *supra.* Phonology, § 58.

Toft House (Elsdon). 1397 Pat. *Toft* ; 1663 Rental *Tofthouse.*
" House by the clearing." *v. toft,* Part II.

Togston (Warkworth). 1129 Pipe *Toggesdena,* 1176 *id.,* *Tockisdena* ; 1248 Ipm. *Togesdene* ; 1255 Ass. *Tokesden, Togesden* ; *c.* 1275 Newm. *Toggesden* ; 1307 Ipm. *Tokisdene* ; *c.* 1250 T.N. *Tog(g)isden, Toggesden* ; 1346 F.A. *Tog(g)esdon,* 1425 *Toggesden* ; 1638 Freeh. *Toggesdon* ; 1663 Rental *Togston.*
" Tocg's valley." Cf. O.E. *Tocga.* Unvoicing of *g* to *k* may have been assisted by association with O.E. *Tocca* and O.N. *Tóki,* which is common in L.O.E. as *Tokig* and *Tochi.* Phonology, §§ 50, 51 ; App. A, § 1.

Tone (Birtley). *a.* 1182 Newm. *Tolland* ; 1296 S.R. *id.* ; 1568 N. iv. 297 *Tonande,* 1592 *Towlands* ; 1663 Rental *Tone House* ; 1693 N. iv. 297 *Towlands* alias *Tone House.*
A difficult name. Alternative suggestions may be offered :—(1) *toll-land,* i.e. land on which toll is paid, though

no such compound is on record. (2) O.E. *Tollan-land* (cf. *tollandene*, B.C.S. 689), i.e. Tolla's land or "land of *Toli*," a Scand. name common in England. (3) Cf. S.Sw. *toland=* tow or flax land (Lindroth, p. 48). [1]

Tosson (Rothbury). 1203 Pipe *Thosan*; 1229 Pat. *Thossan*; 1240 Newm. *Tossen*, 1245 *Tossan*; *c.* 1250 T.N. *Tossen*; 1265 Ass. *Tosham*, 1278 *Tossen*; 1280 Ipm. *Tossan*; 1331 Inq. a.q.d. *Tossam*; 1346 F.A. *Tosson*, 1428 *id.*; 1542 Bord. Surv. *Tosson*.

Two Scandinavian parallels offer themselves for this difficult name. (1) Norw. *Taasen* (N.G. ii. 102)<*Tossini*, which is possibly a compound of O.N. *vin*, "grass-land." (2) *Tossene* in Bohuslän, earlier *Tossini*, which Lindroth (p. 48) explains as *Tos-vin*, i.e. field of tow or flax. In either case the name must have been imported as a whole, for the suffix *-vin* was no longer a living one in the Viking Age.

Tow Law (Wolsingham). 1423.33 *Tollawe*.
Possibly "hill of *Tolla* or *Toli*," *v.* Tone *supra.*

Town Green (Knaresdale). *c.* 1235 H. 2. 3. 18 *Townegreene*.
" Green by the town or farm."

Tranwell (Morpeth). 1267 Ipm. *Trennewell*, 1270 *Trenwell*, 1288 *Tranewell*; 1296 S.R. *Tranwell*; 1310 Ch., 1316 Ipm. *id.*, 1323 *Tranewell, Trenwell*; 1356 Cl. *Tranewell*; 1428 F.A. *Trenwell*.

Cf. Tranby, Yorks., Tranmere, Chesh., Trenholme, Yorks., from O.N. *trani* = crane, here used as a nickname.

Trefford (Egglescliff). 1189 D.S.T. *Treiford*; 1649 Comps. *Trafford*. (Coatham Mundeville) 1268 D.Ass. *Tre(f)ford*; 1382 Hatf. *Trefforth*.

Probably the same as Treyford, Suss. [tri·fəd, trefəd], earlier *Treverde, Triferd, Tre(u)ford*, which Roberts (*s.n.*) explains as " tree-ford," i.e. one marked by a tree or made of timber, but the phonological development is difficult.

Trewhitt (Rothbury) [trufit]. 1229 Pat. *Tyrewyt*; 1255

[1] Tolland, Som., D.B. *Talanda*, 1334 Ch. *Taland*, 1266 Pat. *Tolaunde*, must be an entirely different name.

Ass. *Tyr(e)wyt*; 1296 S.R. *Tirwyth*; 1327 Inq. a.q.d. *Tirwhite*; 1346 F.A. *Tirwith*; 1356 Newm. *Tirwhit*; 1428 F.A. *id.*; 1436 Ipm. *Tyrwhitte*; 1542 Bord. Surv. *Trewhytt*.

An unsolved problem. It is impossible to say whether the name has anything to do with Dial. *tirwhit* = lapwing. This is probably the source of the surname *Tyrwhitt*.

Trewick (Bolam). *c.* 1250 T.N. *Trewick*; 1638 Freeh. *Truick*.

O.E. *trēo-wic*=dwelling by the tree. Cf. Treeton, Yorks.

Trewitley (nr. Hebron). 1255 Ass. *Thurwyteley*; 1314 Ipm. *Tirwhitley*; 1663 Rental *Trewhitley sheels*.

"The clearing of *Dorviðr.*" *v.* Lind. *s.n.*, who gives other forms (*Truth, Trwd, Toruid, Toruit, Torved*) which help to explain the later developments. The name may also have been influenced by the not very distant Trewhitt (*v. supra*).

Tribley (Chester-le-Street). 1242 D.Ass. *Tribelege*; B.B. *Tribleia*.

An unsolved problem.

Trickley (Wooler). 1177 Pipe *Trikelton*; *c.* 1250 T.N. *Trikilton*; 1387 Ipm. *Trikulton*.

Possibly "sheep-dung farm," from the rare English word *trickle*=sheep's dung. The later development is without parallel.

Trimdon (Sedgefield). 1197 Pipe *Tremeldon*; B.B. *Tremeduna*; 1262 B.M. *Tremedon*; 1312 R.P.D. *Tremdon*; 1400 D.S.T. *Trimdon*.

Cf. Trimley, Suff., D.B. *Tremelaia* and D.B. *Treme(s)lau*, the name of a Warwickshire Hundred, pointing to a personal name *Trem(a)*. Phonology, § 10.

Tritlington (Hebron). 1210 Pipe *Tirlington*, 1212 *Tierclinton*, 1252 *Tirtlinton*; *c.* 1250 T.N. *Tirtlington*; 1255 Ass. *Tritlinton, Tyrtlington*; 1346 F.A. *Tyrtelyngton*, 1428 *Trytlyngton*.

"Farm of *Tyrhtel* or his sons." Phonology, § 54.

Troughburn (Heathpool). 1352 Cl., 1359 Sc. *Trollop*; 1367 Pat. *Trolhop*; 1542 Bord. Surv. *Trohope*; 1593 F.F. *Trowupp*.

"Troll-hope." *troll* is common in Scand. place-names.

There is no other evidence for its use at such an early date in England as this name would suggest, but it is possible it may be this word. For the phonetic development cf. *trow* (<O.N. *troll*) in Shetland and Orkney dialect. The " burn " lies in the " hope."

Troughend (Otterburn) [trufend]. 1279 Iter. *Trequenne* ; 1290 Abbr. *Troquenne* ; 1292 Q.W. *Troghwen, Trehquen* ; 1327 Orig. *Torquen* ; 1331 Ipm. *Troghwenn*, 1399 *Troughwen* ; 1460 H. 3. 1. 29 *Trowhen* ; 1618 Redesd. *Troughwen* ; 1663 Rental *Trough End* ; 1612 Elsd. *Troughen*, 1692 *Troughend*.

A Celtic name. Phonology, § 54.

Trows (Kidland). *a.* 1197 Newm. *Wytetrowes* ; 1227 Ch. *Whytecrowes*, 1271 *Wytetrowes* ; 1542 Bord. Surv. *The Trowes.*

" White troughs," with Nthb. [trou] for *trough*, used either in its ordinary sense or of a " dish or depression in stratified rocks." Cf. also *trow-stones* used at Houghton-le-Spring (*c.* 1860) for stone mortars used in preparing frumenty (Dr Fowler).

Tudhoe (Brancepeth). 1279 S. 3. 297 *Tudhow* ; 1296 Halm. *Tudhowe* ; 1684-6 Houghton *Tudda, Tuddy.*

" Tudda's *hōh* " (Part II).

Tughall (Bamburgh). 1104-8 S.D. *Tughala* ; *c.* 1175 Hist. Reg. *Tuggahala* ; 1251 Sc. *Tugehale* ; 1255 Ass. *Tuchehal* ; 1297 Ch. *Tughale* ; 1538 Must. *Tugell* ; 1663 Rental *Tuggell.*

" Tugga's *healh*" (Part II). Cf. Tugford, Salop, earlier *Tugaford, Tuggeford, Toggeford. Tugga* or *Tucga* is a variant of *Tocga* (cf. Togston *supra*), and has its parallel in such pairs as O.E. *Tocca* and *Tucca, Dodda* and *Dudda* (cf. Hudspeth *supra*), and in the history of Mod. Eng. *tug*, M.E. *toggen.* App. A, § 6.

Tunstall (Bp. Wearmouth). 1197 Pipe *Dunstall* ; B.B. *Tunstall.* (Stranton) 1475.35 *Tu(n)stall-by-Stranton.*

O.E. *tūn-steall*=farm-stead, farm-yard. Cf. Tunstall, Staffs., Kent, Lancs., Norf., Suff., and Dunstall, Lincs., D.B. *Tonestale*, Lincs. Surv. *Tunstal, Dunestal.* The interchange of *t* and *d* is unexplained.

Turret Burn (Redesdale). 1325 Ipm. *Trivetbourne*; 1769 Alw. *Truereghet*.

A doubtful identification and an insoluble problem.

Tursdale (Kelloe). *c.* 1150 F.P.D. *Trellesden, c.* 1200 *Trillesdene*; 1340 R.P.D. *id.*; 1337 R.P.D. *Trollesdale, Trullesdale*; 1432.45 *Tirlesden*; 1649 Comps. *Tursdaile*.

" Thrall's valley." *thrall* is from M.E. *threl, threlle*, and *thrill* (O.N. *þrǽll*). With metathesis this gives *thirl* or *thurle* (cf. *thirl* sb.[2] N.E.D.). With fresh metathesis we get *Trull-*. Change from initial *th* to *t* is very common in place-names of Scand. origin, and is common also in the Scandinavian dialects themselves. App. A, § 11.

Tweed, R. *c.* 750 Bede *Tuidi*; *c.* 1050 H.S.C. *Tweoda*; *c.* 1125 F.P.D. *Tweodam*, 1430 *Twede*. A Celtic river-name.

Tweedmouth [twedməθ]. *c.* 1180 D.S.T. *Toedmuthe*; *c.* 1250 T.N. *Tvedemue*; 1539 F.P.D. *Twedmouth*. Self-explanatory. Phonology, § 21.

Twizel (Norhamshire). *c.* 1250 T.N. *Tvisele*; B.B. *Tuisill* (B., C. *Twisele*); 1560 Raine *Twizell*. **Twizell** (Chester-le-Street). 1328 Cl. *Twysilles*; B.B. *Tuisela*. **Twizle** (Morpeth). *c.* 1050 H.S.C. *Twisle*; 1663 Rental *Twizle*.

O.E. (*æt þæm*) *twislan*=at the fork or junction of two streams. *v. twisla*, Part II.

Tyne, R. Ptolemy Τύνα; Ravenna Geogr. *Tinea*; *c.* 750 Bede *Tinus, Tina*. A Celtic river-name. Cf. Tyne, R., in Haddingtonshire (Johnston, p. 292).

Tynemouth [tinməθ]. A.S.C. *Tinanmuþ, Tinemuþa*; *c.* 1125 F.P.D. *Tinemuthe*; 1260 Finch. *Tynemue*; 1485 Pat. *Tynnemouth*; 1637 Camd. *Tinmouth*.

For the local pronunciation, *v.* Phonology, § 21, and note Morton Tinmouth *supra*. The place had an earlier Celtic name. Cf. Leland (*Collectanea*, ed. Hearne, 1774, vol. iv. p. 43), " Locus ubi nunc coenobium Tinemuthense est antiquitus a Saxonibus dicebatur Benebalcrag," and Camden (p. 811), " Yet some there be who think that the rampire and not the wall, went as farre as to the very mouth of Tine, which is called *Tinmouth*, and stifly affirm that it was called *Pen-bal-crag*, that is, *the head of the rampire in the rocke."* Phonology, § 21.

Ulgham (Morpeth) [ufəm). 1139 Newm. *Wlacam*; 1226 Pipe, 1251 Ch. *Ulcham*, 1290 *Ulgham*; 1296 S.R. *Ulweham*; 1316 Ipm. *Ulcham, Ulghham, Ulougham*; 1570 N.C.W. *Howgham*; 1663 Rental *Ougham*; 1812 Corbr. *Uffham*. **Ulwham** (Featherstone). 1479 B.B.H. *Ulg(he)ham*; 1745 Lambley *Ulpham*.

Cf. Ufton, Warw., early forms *Ulfetune (c.* 1100), D.B. *Ulchetone*, 13th c. *Ulston, Oluston, Olufton.* Duignan (pp. 114-5) gives these forms, and takes it to be " *Ulf* or *Wulf's* farm," but *Ulfetone*, though nominally from a charter of Earl Leofric (dated 1043), printed in Dugdale's *Antiquities of Warwickshire*, is really a 13th c. form. *v.* Charter Rolls, *s.a.*, 12 Hy. 3, where it is clear that all the names have been given M.E. forms prevailing at the time of the *inspeximus.* *Ulf* is simply a later development of *Ulche.* The latter is found in D.B. *Ulchenol* (once Cheshire, now Flintshire). *Ulche* might go back to O.E. *Ulca, Ulga,* or *Ulha*, but no such names are known, though *Ull(o)ca* is a possible dimin. from O.E. *Ulla.* Alternatively there are forms *Ulchel, Ulchil, Ulchet* for O.W.Sc. *Úlfketill* (cf. Ouston *supra*), which might give *Ulch(el)ham, Ulch(el)ton.* Loss of *el* in such consonantal combinations would not be surprising.

Ulnaby Hall (Coniscliffe). *n.d.* Newm. *Vluenebi*; 1314 R.P.D. *Ulneby*, 1340 *Ulmeby*; 1366.32 *Oulneby*; 1595 Coniscl. *Ounbie*, 1777 *Ulmby.*

" The by of *Úlfheðinn.*" Lind. *s.n.* gives a late Norw. form *Vlfuen* for this name. Phonology, §§ 59, 39, 51.

Unthank (Alnham). *c.* 1250 T.N. *Unthank.* (Bywell) *c.* 1200 Abbr. *Unthanc.* (Stanhope) 1416.33 *Unthank.*

Cf. also *Hunthank* in W. Auckland (Hatf. Surv.), and in Shotley (Ipm. 1262), and *Unthank* in Plenmeller, and near Tweedmouth on the modern map. The name must have been given to a piece of land whose soil was particularly stubborn and " ungrateful."

Urpeth (Chester-le-Street). 1297 Pap. *Urpath*; B.B. *id.*; 1307 R.P.D. *Urpeth*; 1382 Hatf. *Urpath.*

The first element is probably a personal name *Ur(a)*. Such is not recorded, but cf. *Uro, Urard, Urold, Urolf* on the Continent, and one O.E. compound with *Ur-*, viz., *Urbaldus* (Searle). There is also a very doubtful O.N. name *Uri* (Björkman, N.P., p. 171, Z.E.N., p. 92). The same name is probably found in Urlay Nook, Co. Durham. Hence "*Ur(a)* or *Uri's* path" (*þæð*, Part II).

Ushaw (Esh). *a.* 1196 Finch. *Ulveskahe*; 1312 R.P.D. *Uuesshawe*; 1382 Hatf. *Ulleschawe*; 1393.35 *Ulshaw*.

"Ulf's wood." *Ulf* being the common O.W.Sc. *Úlfr. skahe* is due to confusion of English *sceaga* and Scand. *skógr*, wood.

Usway Burn (Kidland). *n.d.* Newm. *Osweiburne*; 1743 Ilderton *Useyfoord*. **Usworth** (Washington). *c.* 1190 Godr. *Osurde*; *c.* 1190 F.P.D. (*H*)*oswrth*; B.B. *Useworth* (B., C. *Osseworth*); 1312 R.P.D., 1326 Pat. *Oseworth*; 1353 F.P.D. *Useworth*, 1354 *Osworth*; 1560 V.N. *Usworth*.

"Burn by Osa's road, enclosure of the same." *Ōsa* is a shortened form of one of the numerous O.E. names in *Ōs-*. Phonology, §§ 18, 21.

Vauce (Haydon). 1329 Pat. *Vaus*; 1421 Inq. a.q.d. *le Vaux*; 1655 Haydon *Voase*; 1663 Rental *The Vawse*.

Named after or by some Norman lord. *Vaux* (pl. of *val*, valley) is common in Fr. place-names.

Wackerfield (Staindrop). *c.* 1050 H.S.C. *Wacarfeld*; 1268, 1310 Pat. *Wakerfeld*; 1686 Staindrop *Wackerfeild*.

The name *Wacer* is found twice in L.O.E., one example coming from Swaffham, Cambs., so it is probably from O.N. *Vakr*, at least in these cases. The name is fairly common in Icelandic place-names (Lind. *s.n.*). It may, however, have existed as a purely English one, for we seem to have a patronymic formation from it in Wakering, Ess., D.B. *Wacheringa*, Ch. *Wakeringes*, where Scand. influence is unlikely. The name is certainly a very old Teutonic one, cf. *Vaccarus* (*Wacar*), the name of a 6th cent. king of the Warni. As the second element in a name, it is found in O.E. *E(a)dwæcer* (11th c.), and D.B. *Aluuacre* for *Aelf-wæcer*. It is identical with O.E. *wacor*=wakeful, watching.

Wadley (Bedburn). 1382 Hatf. *Wadley.*
" Clearing where woad grows." Cf. *wadleah,* B.C.S.
1222, *wadlond* 356.

Wainhope (Plashetts). 1279 Iter. *Waynhoppe*; 1325
Ipm. *Waynhop*; 1376 Cl. *Wayneshopp.*
O.E. *wægen-hop*=wain or waggon-hope (with pseudo-
genitival *s* in one form), but this does not sound very prob-
able. *Vagn* is fairly common as a personal name in
Denmark, and is found in late O.E., as *Wagen(e),
Wagan* (Björkman, N.P., p. 172), but this should give
Mod. Eng. *Wawn,* not *Wain.* (Cf. M.E. *Wawan* quoted by
Björkman, and Wawne, Yorks., D.B. *Waghene.*)

Waldridge (Chester-le-street). 1297 Pap. *Walrigg*; 1345
R.P.D. *Walrygge*; 1382 Hatf. *Walrig*; 1636 Witton
Warrish.
Possibly O.E. *Wēala-hrycg*=ridge of the foreigners or
Britons or *Wala(n)-hrycg*=Wala's ridge. Cf. D.B. *Wala.*
Phonology, §§ 58, 27.

Walker (Newcastle-on-Tyne). 1267 Ipm. *Walkyr*; 1346
F.A. *Walker, Walcar.* **Wall** (N. Tyndale). 1165 Pipe
Wal; 1296 S.R. *Walle.* **Wallbottle** (Newburn). 1176 Pipe
Walbotle; *c.* 1250 T.N. *Walbothhill*; 1428 F.A. *Walbotell*;
1610 Speed *Wawbottle.* **Wallsend-on-Tyne.** *c.* 1125 F.P.D.
Wallesende, Waleshende; 1279 Ass. *Wallesende*; 1464, 1539
F.P.D. *Walleshend.* **Walltown** (Haltwhistle). 1279 Iter.
Waltona; 1542 Bord. Surv. *Wawetoune.* **Walwick** (Haydon)
[wɔlik]. 1262 Ch. *Wallewik*; 1390 Sc. *Walwik*; 1542 Bord.
Surv. *Wallyk.*
" Marsh (*kiarr,* Part II) by the Wall, Wall, Wall-building
(*botl,* Part II), Wall's end, Wall-farm, Wall-building," all so
called because on the line of the Roman Wall. *Wallbottle*
stands on the line of the Roman Wall, and is possibly
identical with the *vicus regis . . . qui vocatur* " *ad murum* "
of Bede. For -*hende, v.* Phonology, § 38. With reference
to the spelling *Wawetoune,* Bates (*Border Holds,* p. 48 n.)
said that Dr Lyon of Hexham could tell from what town-
ship along the Wall any man came by hearing him pronounce
" Wall." Some said Wa', some Wo', some Wael', etc., etc.,
the only thing none of them said was " Wall."

Wallington (Hartburn). 1255 Ass. *Warlington*; 1262 Ipm. *Walington*; 1346 F.A. *id.*, *Waungton*. Cf. Wallingford, Berks., K.C.D. 716 *Wealingaford* and Wallington, Norf., Herts., Surr. *Wealing* is a patronymic from O.E. *wealh*, "foreigner, Briton," hence "farm of the sons of the Briton."

Walworth (Heighington). 1207 F.P.D. *Walewrth*; 1291 *Waleworth*; 1313 R.P.D. *Walworth*, 1345 *Walleworth*. O.E. *wēala-weorþ*=enclosure of the foreigners or Britons. Cf. the numerous Waltons in England.

Wansbeck, R. 1139 Newm. *Wenespic*; 1255 Ass. *Wanespik*; 1271 Ch. *Wanspic*; 1436 Ipm. *Wanspyke*; 1552 Bord. Laws. *Wandesbeck*; 1610 Speed *Wanspek*.

Heslop suggested that the river-name was derived from Wanny's Crags in which it rises, its name being originally "Wanny's Crags (or Pike) Water." The suggestion is just possible, provided it is quite certain that *Wanny's Pike* is not itself an invention (or distortion of some earlier name), due to an earlier antiquarian, who knew the old forms of Wansbeck. There is a *Vanspeck* in Skåne, which Falkman (pp. 37, 94) takes to be for earlier **Vatnsbekkr*, "stream of water," but the order of development of the Nthb. forms seems to make any relationship between these names impossible.

Warcarr (Thirlwall). 1479 B.B.H. *Wyrthkeryne*. Unexplained.

Warden-on-Tyne. *c.* 1175 Joh. Hex. *Waredun*; 1205 Pipe *Wardon*; 1296 S.R. *Wardun*; 1542 Bord. Surv. *Warden*.

Skeat explained Warden, Beds., as O.E. *weard-dūn* (cf. B.C.S. 1176), "watch-hill," and it is probable that this is the meaning of the name of this hill which dominates the junction of the N. and S. Tyne. App. A, § 1.

Warden Law (Houghton-le-Spring). *c.* 1104-8 S.D. *Wrdelau*[1]; B.B. *Wardona*.

[1] Dr Fowler says that this probably refers, not to Warden Law, but to the hill just east of Durham City, now known as Nine-tree Hill. It was also known as Munjey or Mountjoy Hill, and tradition has it that it was so called because here pilgrims from the south got their first view of Durham.

" Watch-hill," first with alternative, and later with combined suffixes. Cf. *Warde-knolle*, Halm, 1345. App. A, § 11.

Wardley (Jarrow). 1260 Finch. *Wardeley*. Cf. *Weardan dun*, B.C.S. 789. " Wearda's clearing."

Wardrew (nr. Gilsland). 1479 B.B.H. *Wardrew*. Probably Celtic. Cf. Cumrew, Cumb., earlier *Comreu*, *Cumreu, Cumrewe* (Sedgefield, *s.n.*). Ekwall (p. 111) identifies the suffix with Welsh *rhiw*, " hill, ascent."

Warenford (Bamburgh). *c.* 1200 N. i. 306 *Warendforthe*; 1234 Ch. *Warneford*; 1313 Ipm. *Warenford*; 1628 Freeh. *Warneford*. **Warenton** (Bamburgh) [waˑntən]. 1208 Pipe, 1243 Ipm. *Warnetham*; 1255 Ass. *Warendeham*; *c.* 1250 T.N. *Warneth'm*; 1296 S.R. *Waryndham*; 1297 Ipm. *Warendham*; 1305 Inq. a.q.d., 1330 Ch. *Warndham*; 1346 F.A. *Warndam*; 1663 Rental *Warndon*. ***Warnmouth.** *c.* 1050 H.S.C. *Warnamuthe*. **Warren Burn** [weˑrən]. *c.* 1025 H.S.C. *Warned*; 1560 Raine *Warne*.

" Ford on, homestead by, mouth of the Warren Burn." App. A, § 7. The river-name is pre-English.

Wark-on-Tweed [waˑk]. 1157 Pipe *Werch*. **Wark-on-Tyne.** 1279 Iter. *Werke*; 1294 Ch. *Wark*. O.E. (*ge*)*weorc*=fortifications. For pronunciation, *v.* E.D.G. p. 686.

Warks Burn (N. Tyndale). 1293 Ass. *Werkesburn*. " Wark's stream," named from Wark-on-Tyne.

Warkworth [wɔˑkwəθ]. *c.* 1120 Hexh. Pr. *Wercheorda*; 1104-8 S.D. *Werceworde*; 1160 Pipe *Wercwurda*, 1162 *Werchesurda*; 1199 R.C. *Werkwurth*; 1291 Tax. *Werkesworth*; 1428 F.A. *Warkeworth*.

Not " wark-worth," i.e. fortified enclosure, for there is no evidence for the possibility of such a compound in O.E. Rather we have O.E. *Werce*, a woman's name, found as *Verca* in Bede's *Life of St Cuthbert*, c. 35, or the name *Weorc* found in *Weorcesmere*, B.C.S. 782, Worsborough, Worsall, Wortley, Yorks., Warkton, Northants, Worksop, Notts., Workington, Cumb.[1]

---

[1] Warkworth is probably to be identified with the *Wyrcesford* of the *Historia Sancti Cuthberti*.

Warland (Lanchester). 1311 R.P.D. *Warlandes*; 1382 Hatf. *Warlandfeld.* **Warton** (Rothbury). *c.* 1250 T.N. *Warton.*

Possibly these names contain a name **Wǣra*, a shortened form of one of the numerous O.E. names in *Wǣr-*. *Ware* is found once as the name of a moneyer to Cnut, and there is a patronymic *Warincus* (= Waring) in D.B.

Washington. 1197 Pipe *Wessinton*, 1211 *Wassinton*; 1280 F.P.D. *Quessington*; B.B. *Wassyngtona* (B., C. *Wessington*); 1311 R.P.D. *Wessington*, 1314 *Wasshington*, 1340 *Wessington*; 1400 D.S.T. *id.*, 1507 *Weshington*; 1747 Houghton *id.*

Cf. Washingley, Hunts., earlier *Wasingelei*, *Wassinglei*, Washingford, Suff., D.B. *Wasingaford*, Washingborough, Lincs., D.B. *Washingeburc*, Lincs. Surv., *Wassingburgh*, Washington, Suss., B.C.S. 834 *Wasingatun*. Skeat (p. 335) and Roberts (p. 169) takes *Washing-* to be a patronymic from *Wassa*, found in *wassanburna*, B.C.S. 236. The change from *ss* to *sh* is noteworthy, and Roberts suggests it may be due to the influence of the common word *wash*. The latter certainly accounts for *Wesh-* forms, for [weʃ] is Nthb. and Dur. for *wash*.

Waskerley (Shotley). 1262 Ipm. *Waskerley*; 1312 Q.W. *Waskreley*; 1663 Rental *Warscally*.

This name is discussed by the present writer in *Essays and Studies, u.s.* vol. iv. p. 69. It is there suggested that the first element may be either O.N. *vatnskjarr* (later *Vatskiær*, *Wazkere*)=marsh of water, or O.N. *váskjarr*=wet marsh. Cf. *Vasakärr* in Skåne [1] (Falkmann, p. 96). "Clearing by or at the marsh."

Waskerley Park and Beck (Stanhope). 1242 D.Ass. *Walkeropburne* (sic); 1311 R.P.D. *Wascroppeheued*; 1373.32 *Park of Wastrepp*; 1446 D.S.T. *Wascroppheued*; 1464 F.P.D. *id.*; 1637 Camd. *Wascrop Burn*; 1768 Map *Wes-*

---

[1] Falkmann connects this with O.N. *veisa*=standing pool, but this would give O.Dan. *vese* rather than *vase*. It may be noted that the cognate O.E. *wāse* gives St. Eng. *ooze*, but the Nth. form would be *wase*, and it is just possible that *Wasker* is a hybrid compound of this and the anglicised *ker.*

crow River. **Waskrow Bridge** (Wolsingham). 1382 Hatf. *Westcropbrig.*

Difficult names, but it may be suggested that *Wascropp* is for *Wask(e)r-hopp*, with the same first element as in Waskerley *supra.* The *heued* is the head of the " hope " where the burn rises. When the *p* of *Wascrop* and the *b* of *burn* and *bridge* came together *Wascrowburne* and *Wascrobrig* were misunderstood and a river-name *Wascrow* was formed. In modern times *Wascrop* and *Wascrow* have, in the case of the park and burn, been assimilated to the not very distant Waskerley discussed above. Introd., p. xix.

Waterfalls (Thockrington). 1296 S.R. *Waterfelles.*

The falls at the head of the Dry or Swin-burn. *Fells* is now replaced by *falls*, a fresh formation from the verb.

Wear, R. *c.* 750 Bede *Uiuri* ; De situ Dunelm (12th c. MS.) *Weor* ; *c.* 1200 Finch. *Wyry.* **Wearhead.** 1372.32 *Wereheved.*

A Celtic river-name. Chadwick (*Essays and Studies presented to Wm. Ridgeway,* p. 319) suggests that it is the same as that found in Weaver, R. (Chester), Waver, R., Cumb., and is ultimately identical with Germano-Celtic *Weser.*

Wearmouth, Bp. and Monk. *c.* 750 Bede *Uiuræmuda* ; 1104-8 S.D. *Guiramuthe* ; *c.* 1125 F.P.D. *Wiramutham* ; 1306 R.P.D. *Wermouth Episcopi* and *id. Monachorum* ; 1438 Misc. *Warmouth* ; 1539 F.P.D. *Wermoth* ; 1631 Whitb. *Warmouth* ; 1723 Castle E. *Warmoth* ; 1733 Ingram *Warmouth.*

Wearmouth of the Bishop and the Monks of Durham respectively. For pronunciation, *v.* Phonology, § 8, and cf. 1733 Corb. *Wardale,* 1799 Warkw. *Wardell,* and the personal names *Wardale, Wardle, Wardell,* all for *Weardale.*

Weedslade (Long Benton). 1196 Pipe *Wideslad* ; 1203 Coram *Witheslad* ; 1209 Sc. *Widdeslade* ; 1255 Ass. *Wydeslade, Wyteslade* ; *c.* 1250 T.N. *Wydeslad* ; 1315 R.P.D. *Wyteslade* ; 1346 F.A. *Wedslad, Whitslad* ; 1360 Sc. *Weteslade* ; 1460 H. 3. 1. 30 *Witteslade* ; 1663 Rental *Weatslett.*

O.E. *wipig-slæd*=withy or willow-valley (B.C.S. 158,

o

550). Cf. *Crawford Charters* (p. 2) where O.E. *wiþig-slæd*
has given M.E. (15th cent.) *Wydeslade*, and note also Widford, Glouc., Widley, Hants., B.C.S. 142 *withiglea*, Widdial,
Herts., D.B. *Widehale*. Phonology, § 42. The later vowel
development is difficult. *i* was lowered to *e* (ib. § 10) and
then apparently was lengthened, perhaps under the influence of dialectal *weet* for St. Eng. *wet*. Cf. Weeton, Yorks.,
earlier *Widetune*, *Witheton*, *Wieton*, *Witon*, and Lancs.,
earlier *Withetun*, *Wetheton*, *Weeton* (Sephton, p. 192).
Moorman (p. 202) derives Weeton, Yorks., from O.N. *viða-
tún*, but the phonetic development is no easier, and for the
compound here suggested we may compare *wiðigham*,
B.C.S. 1307, *wiðigwic*, ib. 702.

Weetwood (Chatton). 1196 Pipe *Wetewude* ; 1255 Ass.
Wetwod' ; 1262 Ipm. *Wethwde*, *Wettwod*, 1314 *Wetewod*,
Wytewod ; 1542 Bord. Surv. *Wetewod* ; 1579 Bord. *Wheitt-
wod* ; 1628 Freeh. *Weetwood*.

O.E (*se*)*wǣta wudu*=wet wood, with the same fluctuation between *wet* and *weet* that we find in modern dialect.

Weldon (Longhorsley). *c.* 1250 Newm. *Welden* ; 1421
Ipm. *id.*

O.E. *wielle-denu*=spring-valley.

Wellhaugh (Falstone). 1303 Pat. *Wellehawe* ; 1663
Rental *Wellhaugh*.

This identification is made by the editor of the Patent
Roll, but it is very doubtful if it is correct. There is no
particular reason apparent for associating *Wellehawe* with
Nthb. at all, and as Wellhaugh stands on a very clearly
marked "haugh" of the North Tyne, it is pretty certain
that *haugh* and not *hawe* is the original suffix in the name.
For *Wellehawe*, cf. Wellow, Notts., earlier *Welhagh*, *Wel-
hawe*, *Wellaw* (Mutschmann, p. 148).[1]

Welton (Ovingham). 1203 R.C. *Waltenden* ; *c.* 1250
T.N. *Weltedene* ; 1271 Ch. *Waltedene* ; 1292 Ass. *Weltes-
dene* ; 1307 Ch. *Welteden* ; 1346 F.A. *Welldon* ; 1638 Freeh.
Welton.

[1] In the Fine Rolls (1316) *Weelhall* or *Welhall*, belonging to the bishopric
of Durham, is identified with Wellhaugh, but reference to R.P.D. shows
that this place was in the diocese of York, and not in Nthb. at all.

" Wealt(a)'s valley." Cf. Waltham, Herts., for which Skeat (p. 32) suggests a name *Wealta* derived from O.E. *wealt*, unsteady. Nicknames in O.E. are doubtful unless formed quite late, but such a name may have been coined under Scand. influences. For *Walt-* and *Welt*, cf. *walter* and *welter*, variant derivatives of M.E. *walten*, to roll. App. A, § 1.

Westernhope (Stanhope). 1418.33 *Whestanhope*; 1457.35 *Westanburnshele*.

Cf. *le whystan*, nr. Fontburn, Ass. 1292. Both alike are probably from O.E. *hwæt-stān*=whet-stone, hence " hope where whet-stones are found."

Westgate (Stanhope). 1457.35 *Westyatshele*. v. *geat*, Part II.

Westoe (Jarrow). *c.* 1125 F.P.D. *Wiuestoue*, 1228 *Wiuestowe*; 1446 D.S.T. *Wyvestowe*; 1539 F.P.D. *Westowe*.

" Wifa's place." Cf. *wifanstoc*, B.C.S. 624, and *stōw*, Part II. Phonology, §§ 53, 10.

Westwick (Barnard Castle). 1091 F.P.D. *Westewic*. Self-explanatory.

Whaggs (Whickham). 1382 Hatf. *le Whag*.

Probably a dialectal form of *quag*, " a bog." Phonology, § 28.

Whalton [wa·tən]. 1203 Pipe *Walton*; 1205 Perc. *Whalton*; 1218 Pipe *Wauton*; 1241 Cl. *Whauton*; 1250 Ipm., *c.* 1250 T.N. *Walton*; 1268 Ass. *Hwalton*; 1271 Ch., 1291 Tax. *Walton*; 1298 B.B.H. *Whalton*;· 1312 R.P.D. *Qualton*; 1317 Ch. *Whalton*; 1333 Newm. *id.*; 1424 Pat. *Qwalton*; 1638 Freeh. *Whawton*.

Cf. Whalley, Lancs., and Whalley, Derbs., which Wyld (p. 262) and Walker (p. 261) agree in connecting with O.N. *hvàll*=hill, but as this word was never naturalised in England, such a compound as O.N. *hvàll+-ley* is very unlikely, and, at least in the case of the Lancs. name, is impossible, for A.S.C. (*sub anno* 798) gives forms *Hweallege, Hwællæge*. The MSS. (D and E) date from *c.* 1100, but the names are no doubt as old as the entries, so Scand. influence is out of the question. There is an O.E. adj. *hweall, hwal, hwæl*, " bold, impudent," but it is difficult to see how this could

be used of a farm or clearing. There is also a very rare
O.E. *Hwala* [1] which might have given rise to Whalton, but
it can hardly be found in Whalley.

Wham (Lynesack). 1315 R.P.D. *Quwam, Qwhom. v.*
hvammr, Part II.

Wharmley (Newbrough). 1279 Iter. *Quarenley* ; 1289
Sc. *id.* ; 1325 Ipm. *Quarneley* ; 1392 Sc. *Quarnele* ; 1663
Rental *Wharnley.* **Wharnley Burn** (Healeyfield). 1399
Accts. *Wharnowe* ; 1792 Muggles. *Wharnayebourne,* 1801
Wharneyburne.

O.E. *cweorn-lēage* and *hōh*=mill-clearing and heugh
of land. Cf. Quarrington *supra* and Quarmby, Yorks.,
for which Goodall (p. 234) notes late forms *Wherneby,*
Wharneby. Note also *wherne-house*=mill-house, in a
Southwell visitation (N.E.D.). The same phonetic change
is found in the Shetlands in *Hwern-bregg* and *Hwern-gert*
from O.N. *kvern* (Jakobsen, p. 179). App. A, § 7.

Wheatley (Lanchester). 1311 R.P.D. *Wetley* ; 1382
Hatf. *Whetlay.* (Kelloe) *c.* 1190 Finch.*Wuetlawe* ; 1335
R.P.D. *Quetelawe.*

"Wheat-clearing and hill." App. A, § 2. Cf. *huæta*
leage, hwætedun, B.C.S. 204, 183.

Wheatridge (Earsdon). 1296 S.R. *Whytrig* ; 1579 N.
ix. 96 *Whitriche* ; 1855 Whellan *Whitridge.*

O.E. *hwīt-hrycg*=white-ridge. The modern form is
corrupt. Phonology, §§ 21, 27, 58.

Whessoe (Haughton-le-Skerne). 1304 Pat. *Wessehou* ;
1307 R.P.D. *Whessowe* ; B.B. *Quesshaw* (B., C. *Wessawe*) ;
1382 Hatf. *Quesshowe.*

The second element is O.E. *hōh* (Part II), the first may
be an unrecorded Norse nickname **Hvassi,* "sharp one,"
which is perhaps found in *Hvassafell* and *Hvassahraun* in
Iceland (Kålund, i. 361, ii. 401). For the change from
a to *e,* cf. Washington *supra.*

Whickham. 1197 Pipe *Quicham* ; 1200 R.C. *id.* ;
1311 R.P.D. *Qwykham* ; 1400 D.S.T. *Qwicham,* 1507
Whycham.

---

[1] Cf. Sw. *Hvalunge,* which Hellquist takes to be a patronymic formed
from *hval* (=whale), used as a nickname.

Cf. Whittonstall *infra*. " Homestead with the quickset hedge."
Whinnetley (Haydon). 1207 Pipe *Winteleia*; 1255 Ass. *Whynneteleg, Quinteleg, Quynteley*; 1298 B.B.H. *Qwyneteley*, 1479 *Whynetle*. A difficult name. Is it possible that there was once a word *whinnet*=a clump of whin or gorse? Cf. *thick-et*.
Whirleyshaws (Guyzance). 1350 Perc. *Qwirlecharr*, 1356 *Quarlecharr*. Nothing can be made of this name. The first element may perhaps be associated with North. dial. *quarrel* (earlier *qvarel, querill*, Mod. Scots *wharrel, wharl*), " a quarry."
Whiskershiels (Elsdon). 1345 B.M. *Wyschardshell*; 1618 Redesd. *Woskershields*; 1663 Rental *Whiskersheeles*; 1672 Elsd. *Wiskersheel*. " The shiels of Wishart." *Wyschard* is N.Fr., corresponding to C.Fr. *Guiscard*.
Whitburn. *c.* 1190 Godr. *Hwiteberne*; 1292 Pat. *Wyteberme*; 1312 R.P.D. *Whitebern*; 1438 Misc. *Whittebarne*. **Whitchester** (Heddon-on-the-Wall). 1221 Pat. *Witcestre*; 1251 Ch. *Whicestre*; 1428 F.A. *Whitchestre*. **Whiteburn** (Kidland). 1233 Newm. *Whiteburne*. **Whitechapel** (Haltwhistle). 1368 Ipm. *Whitchapel*. **Whitehall** (Cramlington). *c.* 1250 T.N. *Wytelawe*; 1421 Ipm. *Whitlawe*. (Muggleswick) 1399 Acct. *Alba Aula*; 1446 D.S.T. *Whithall*. (Tribley) 1420.45 *Whithall*. **Whitehill Hall** (Chester-le-Street). 1382 Hatf. *Whytehill*. **Whitfield.** *a.* 1274 B.B.H. *Witefeld*. **Whitelees** (Nookton). *n.d.* R.P.D. *Quitteleys*. **Whiteley** (Wolsingham). 1382 Hatf. *Whitley*. **Whitemere** (Heworth). *c.* 1220 F.P.D. *Whitemere*. **Whitley** (Tynemouth). 1203 R.C. *Witelega*; 1271 Ch. (*H*)*wyteleya*. (Hexhamshire) 1349 B.B.H. *Whiteley*. **Whitlow** (Kirkhaugh). *c.* 1300 B.B.H. *Witelawe*, 1479 *Whytley*. **Whittle** (Ovingham). *c.* 1250 T.N. *Wythill*; 1428 F.A. *Whitell*. (Shilbottle) 1266 Ipm. *Vythill*; 1663 Rental *Whittle*. **Whitton** (Rothbury). 1228 Pat. *Witton*; 1275 H. 3. 2. 140 *W*(*h*)*itton*. (Grindon, Co. Durham) *c.* 1100 Allen *Wytton*.

Whitwell (nr. Durham). B.B. *Whitewell* (B., C. *Witewell*).
Whitwell Burn (Shincliffe). 1459 Acct. *Whytwellborne.*
Whitworth. *c.* 1200 B.B. *Whitworth* ; 1592 Houghton
Whitbarn.

" White barn, chester, burn, chapel, hill (= *law*), hall (2),
hill, field, ley(s) (2), mere, ley and hill (= *law*), hill (2),
farm, spring, enclosure." For Whitechapel, cf. Whit-
church, Hants. and Salop. Such were probably so called
because the outside had been whitewashed. In Whitehall,
Cramlington, *White-law* probably developed an alternative
from *White-hill*, later corrupted to *White-hall*. With reference to the application of the term " white " to ground we
may quote Hodgson's note (3. 2. 77) that " white fields " is
used in the sense of " dry open pasture ground in opposition
to woodland and black-land growing heath." Phonology,
§ 21. App. A, §§ 8, 11.

White Kirkley (Wolsingham). 1382 Hatf. *Whitekirtil-
land, Whitekirketilfeld.*

Apparently " white kirtle land," but why so called ?

Whittingham [ʍitindžəm]. *c.* 1050 H.S.C. *Hwitin-
cham* ; 1104-8 S.D. *Hwittingaham* ; 1160 Pipe *Witingeham* ;
c. 1250 T.N. *Wytingh'm* ; 1253 Ipm. *Whytincham*, 1320
Whit(t)yncham, 1327 *Whittyngeham.* **Whittington** (Corbridge). 1233 Pipe *Witynton* ; 1296 Ch. *Whytington.*

" Homestead and farm of the sons of *Hwīta* (=white)."
Cf. Whittingham, Lancs., Whicham, Cumb. (D.B. *Witing-
ham*) and Whittington, Lancs. Phonology, §§ 22, 34.

Whittonstone (Longwitton). 1292 Ass. *le Whystan.*

" Whetstone." Cf. Westernhope *supra* and Whetstone,
Leic. The modern form is corrupt.

Whittonstall (Bywell St Peter). *c.* 1150 N. vii. 178, n. 5
Quictunstal ; 1225 ib., 185 n. 3 *Cuictunstal* ; 1255 Ass.
Whittonstal ; *c.* 1250 T.N. *Quictunstal* ; 1268 Ipm. *Wyth-
tonstall*, 1270 *Whyttonstall* ; 1296 S.R. *Quikunstal, Quik-
cumstal* ; 1307 Ch. *Whittonstall.*

O.E. *cwic-tūn-steall*=farmstead (*v. steall*, Part II), with
the quickset hedge. Cf. *cwichege*, B.C.S. 207=quick-
hedge. *cwic* > M.E. *whykke* > Mod. North. Eng. *whick*.

Whitwham (Lambley). 1344 Cl. *Wytquam*; 1406 Pat. *Wytwam*; 1509-47 Dugd. vi. 306 *Whitwham*.

" White-valley." *v. hwammr*, Part II.

Wholehope alias **Holehope** (Kidland). 1233 Newm. *Holehope*; 1296 S.R. *Hollop*; 1780 Edl. *Whollop*, 1807 *Wholup*.

O.E. *hole-hop*=hollow-hope, with the same variation from *h* to *wh* as in *whole*, earlier *hool*.

Whorlton (Newburn). 1323 Pat. *Wherleton*; 1324 Cl. *Wherlton, Wherwelton*; 1724 Ponteland *Wharlton*.

Cf. Whorlton, Yorks., D.B *Wirveltun*, Kirkby's Inq. *Quereleton, Warleton, Wherleton*, Whorlton, Cumb., earlier *Wherwelton*. There is an O.E. *Hwerwyl*=Wherwell, Hants. (B.C.S. 912), a compound of O.E. *hwer*=kettle, cauldron, and *wyl*=spring. Middendorf (p. 79) takes it to mean " hot spring " and compares O.N. *Hveravellir*, the name of some hot springs in Iceland. There is an O.E. *hwyrfel*, found only in place-names, which Middendorf (p. 79) takes to be cognate with O.N. *hvirfill*, " whirlpool." An example is *wirfuldoun* (B.C.S. 867) which has become Whorwelsdown, Wilts. Neither of these words and meanings seems suited to either Whorlton, Nthb. or Yorks. In O.N. *hvirfill* is also used of the top of a hill, probably from its rounded shape, and this would suit Whorlton, Yorks., very well, for it lies on the spur of a well-rounded hill called the *Whorl*. Its aptness for the Nthb. village is not so clear, but is quite possible. The name may then be a Scand. borrowing, "farm by the rounded hill." For *whorl*, cf. *whorl* and *whirl* as dialectal forms of *whirl* (E.D.D.).

(Gainford) *c*. 1050 H.S.C. *Queornington*; 1104-8 S.D. *Cueorningtun*; 1306 R.P.D. *Querington*, 1316 *Quer(n)ington*, 1344 *Quernington*; 1360 Cl. *Quernyngton*; 1577 Barnes *Whorleton*; 1646 Map *Wharleton*.[1]

This name is very puzzling. No O.E. name *Cweorn* is known, and *cweorn* looks like the common word, " quern "= hand-mill, and *cweorning* might possibly be the *ing* or grass-

---

[1] The Rev. Professor Headlam has kindly informed me of other forms, *Quornton, Whornton*.

land where a quern is to be found, but such hybrids are doubtful. Place-names with *kvern* as the first element are fairly common in O.N. There it is used of an eddy or whirlpool, and it is possible that this sense was transferred to the English *cweorn*, and that the reference is to some eddy or pool in the Tees, on which Whorlton stands. The name would then mean " farm on the *ing* by the whirlpool." Later the first *n* was lost, and then when *cw* had become *wh* (Phonology, § 28) the name underwent complete transformation, perhaps in an attempt to distinguish it from Quarrington, which was often called Wharrington (*v. supra*). The final form may have been due to association with the dialectal *quarrel*, *wharrel*, " a quarry." There is a limestone quarry at Whorlton.

Widdrington. Type I : *c.* 1160 F.P.D. *Vuderintuna* ; 1166 R.B.E. *Wodringatone* ; 1170 Pipe *Wuderinton*, 1177 *Wudrinton* ; 1255 Ass. *Woderington* ; 1307 Ch. *id.*; 1346 F.A. *Wodryngton*, 1428 *Woddryñgton* ; 1431 D.S.T. *id.* Type II : *c.* 1180 F.P.D. *Widerintune* ; 1177 Pipe *Widerentona* ; 1295 Perc. *Widerengton* ; 1309 Ipm. *Wyderington* ; 1346 F.A. *Wedryngton* ; 1356 Perc. *Wydrington* ; 1429 Pat. *Weddryngton*, 1431 *Wederyngton* ; 1798 Corbr. *Witherington*.

" Farm of **Wuduhere* (Type I), or **Widuhere* (Type II)." Cf. *Viduarius* in Ammianus Marcellinus, as the name of a king of the Quadi, and O.N. *Viðarr* (Naumann, p. 67, and Schönfeld, p. 264). *wudu* and *wi(o)du* are variant O.E. forms of the first element of these names.

Widehope (West Auckland). 1313 R.P.D. *Wydhop*. Self-explanatory.

Wigside (Wolsingham). 1382 Hatf. *Wygesyde*. " Wicga's hill." *Wicga* and *Wigga* are fairly common in O.E.

Wilkwood (Holystone). *c.* 1230 H. 2. I. 16 n. *Wilkewde* ; 1642 Arch. 3. 4. 120 *Wilkewood*.

Cf. Wilkesley, Chesh., and Wilkesby, Lincs. All alike from O.E. *Willoc*, dimin. of *Willa.* " Little Will's wood."

Williamston (Knaresdale). 1257 Swinb. *Williameston*.

"William's farm," *William* being A.Fr.=C.Fr. *Guill-aume.*

Willimontswyke (Haltwhistle). 1279 Iter. *Wilimotes-wike*; 13th c., Swinb. *Willimoteswick*; 1542 Bord. Surv. *Willymounteswyke*; 1638 Freeh. *Willomansw'k*; 1652 Comps. *Willimoteswick*; 1663 Rental *Willimondswick.*
"Willimot's dwelling." *Wil(li)mot* is a dimin. of A.Fr. *Willeme*, as *Guillemot* (used as a pet name for the bird) is of O.Fr. *Guillaume*. Phonology, § 55.

Willington (Brancepeth). *c.* 1190 Godr. *Wyvelintun*; 1296 Halm. *Wyuelington*. **Willington Quay** (Wallsend). *c.* 1125 F.P.D. *Wiflin(c)tun*, 1203 *Wiuelington*, 1539 *Willyngtone.*
Cf. Willingham, Cambs. and Lincs. "Farm of *Wifel* (O.E.) or *Vifill* (O.N.) or his sons." Phonology, § 51.

Wilmire House (Wolviston). 1325 F.P.D. *Whyuelesmer.*
"Mere of *Wifel* (O.E.) or *Vifill* (O.N.)." App. A, § 6.

Windlestone (Auckland). 1197 Pipe *Windlesden*; 1296 Halm. *Wynelisdon*; 1304 Cl. *Wymelesdon.*
"Winel's hill." Cf. *wineles ford*, B.C.S. 769. Phonology, § 55; App. A, § 1.

Windyhaugh (Kidland). *c.* 1200 Newm. *Wyndihege.*
"Windy hay or enclosure," (M.E. *hege*=hedge), if the M.E. form is to be relied on, rather than " windy haugh " as now. App. A, § 8.

Wingate (Kelloe). *c.* 1150 Finch. *Windegat*. **Wingates** (Longhorsley). 1208 Perc. *Wyndegates.*
"Wind-gates," used of a place where the wind drives up a narrow valley or trough with special force. Cf. Wingates, Lancs., and the Winnats near Castleton, Derbys., for *Win-yats*. Phonology, § 53.

Winlaton (Ryton). *c.* 1125 F.P.D. *Winl(e)octun*; *c.* 1303 R.P.D. *Winlaweton*, 1315 *Wynlaghton*; B.B. *Wyn-laktona* (B., C. *Wynlauton*); 1316 Pat. *Wynlauton*; 1498.36 *Winlayton*; 1581 Ryton *Winlawton*, 1696 *Winlaton.*
"Farm of *Winelac*," cf. L.V.D. For the sound develop-ment cf. Laughton, Yorks., earlier *Lacton, Laghton* (Moor-man, p. 119).

Winston-on-Tees. 1091 F.P.D. *Winestona.*
"Wine's farm."

Wiserley (Wolsingham). 1382 Hatf. *Wyshill.*
Cf. Wisborough, Suss., earlier *Wiseberg* (Roberts, p. 179).
Witton, Long. 1340 Newm. *Langwotton;* 1560 N.C.W.
id. **Witton, Nether.** 1379 Ipm. *Witton by the Water.*
O.E. *wi(o)du* or *wudu-tūn*=wood farm, the wood being
doubtless the " silva de Wittun " of the foundation charter
of Newminster Abbey (Newm., p. 1). In Ipm. 1337 we have
a *Wytton Underwod.* To which place it refers is uncertain.
Phonology, § 51.
Witton Gilbert [witən džilbət]. 1275 F.P.D. *Wyttone;*
1382 Hatf. *W(h)itton;* 1479.35 *Witton gilbert;* 1636 Ryton
Witton Jelbert.
" White-farm or wood-farm," distinguished from other
W(h)ittons by the name of its one-time owner, Gilbert
de la Ley.
Witton-le-Wear. 1104-8 S.D. *Wudetun;* 1300 Pat.
Wotton in Werdale; 1313 R.P.D. *Whytton in Weredale.*
" Wood-farm by the Wear." For *le v.* Chester-le-Street.
The 1313 form shows that an *h* is not conclusive for deriva-
tion from *"white."*
Witton Rows (Witton-le-Wear). 1382 Hatf. *Wytton-*
rawe. **Witton Shiels** (Netherwitton). 1290 Ch. *Sheles.*
v. rāw, scheles, Part II.
Wolsingham (Weardale). *c.* 1150 F.P.D. *Wlsingham;*
1197 Pipe *Wulsingeham;* 1311 R.P.D. *Wolsingham;* 1336
Ipm. *Wulsingham;* 1705 Witton G. *Wisinham.*
" Homestead of (the sons of) Wulfsige." Cf. Woolsing-
ton *infra.* Reginald of Durham in the *Life of St Godric*
speaks of "Wlsingham . . . qui habitaculum Ulsi vel Lupi
habitatio seu ululatus lupi, Anglico sermone expressus,
intelligitur," an early example of inaccurate conjecture as
to the meaning of a name. Phonology, § 13.
Wolviston [wustən]. 1091 F.P.D. *Oluestona; c.* 1125
Wlueston; 1185 F.P.D. *Wulueston,* 1430 *Wolueston;* 1580
Halm. *Wolstone;* 1637 Camd. *Wuston;* 1719 Bp. M. *Woustan.*
" Wulf's farm." Phonology, §§ 39, 53.
Woodburn (Corsenside). 1265 Sc. *Wodeburn;* 1287 Ass.
Wodeburge; 1379 Cl. *Wodeburgh.*
" Stream or burh by the wood." *v.* App. A, § 10.

There is no stream of this name, but it may be an earlier name of the Lisles Burn.

Wooden (Lesbury) [u·dən]. 1237 Cl. *Wulvesdon*; 1298 Sc. *Wolvedon*; 1333 Ipm. *Wuldon*; 1663 Rental *Wooden*.

" Wolf or wolf's hill," referring either to the animal or to a man. Phonology, §§ 39, 53 ; App. A, § 1.

Woodham (Aycliffe). 1091 F.P.D. *Wodon, c.* 1150 *Wdum*; 1311 R.P.D. *Wodeham*; 1341 Cl. *Wodum*; 1539 F.P.D. *Wodhome*. **Woodhorn.** 1177 Pipe *Wudehorn*.

" Homestead and horn or corner of land by the wood, or in the latter case, with a wood on it." *Woodhorn* has by some been identified with *Wudecestre* (S.D. i. 47). App. A, § 6.

Woodhouses (W. Auckland). 1377.32 *le Wodehous*.

" House in the wood, or (less probably) of wood."

Woodifield (Bedburn). 1241-9 F.P.D. *Wdingfeud*; *n.d.* Finch. *Wudingfeld*; 1382 Hatf. *Wodingfeld*; 1446 D.S.T. *Wodefelde*.

" Field of Wuda or his sons."

Wooler. 1186 Pipe *Wullovre*; 1199 R.C. *Wllovera*, 1203 *Welloure*; 1210-2 R.B.E. *Wulovere*; 1249 Ipm. *W(i)lour*, 1250 *Wolloure*; 1255 Ass. *Wllovere, Wulloure*; *c.* 1250 T.N. *Wllovre, Willevre*; 1271 Ch. *Wolouela, Wlloure*; 1291 Ch. *Woloure*; 1296 S.R. *Wolouer*; 1311 R.P.D. *Wollouer, Wllour, Wolouere*; 1312 Ipm. *Wollouere*, 1313 *Wlhouer, Wolheuer*; 1314 Inq. a.q.d. *Wullure*; 1324 Ipm. *Wullour*; 1334 Perc. *Wolloure*; 1346 F.A. *Wellour, Wollor*; 1334 Perc. *Wolloure*; 1346 F.A. *Wellour, Wollor*; 1542 Bord. Surv. *Wouller*; 1637 Camd. *Wollovere*; 1663 Rental *Wooler*.

The second element is O.E. *ofer*, " bank or shore " (Part II). The first may be O.E. *Wulf(a)*, a personal name, or *wulf*, the animal, with very early assimilation of *lf* to *ll*. For *-er* cf. Thorner, Yorks. (Moorman, p. 188). Hence, possibly, " Wolf's bank (of the Till) or wolf-bank."

Wooley (Slaley). *c.* 1260 Perc. *Ulflawe*; 1296 S.R. *W(o)ullawe*; 1335 N. vi. 336 *Wllaw*; 1671 Arch. 2. 1. 129 *Wooley*.

O.E. *wulf-hlǣw*=wolf-hill. App. A, § 2.

Wooley Hill (nr. Billy Row).   1349-35 *Wolleys* ; 1425.45 *Wollyhall.*

"Wolf-clearing." Cf. B.C.S. 762 *to wulfa leage*, Woolley, Hunts., Woolley, Yorks.

Woolsington (Dinnington) [wisiɲtɔn].   1203 R.C. *Wulsinton* ; 1360 Ipm. *Wolsyngton* ; 1663 Rental *Wissington* ; 1798 Bothal *id.*

"Farm of Wulfsige or his sons." Phonology, § 13.

Wooperton (Eglingham) [wɔpətən], [wapətən].   1180 Pipe *Wepreden* ; 1255 Ass. *Weperdon* ; *c.* 1250 T.N. 1292 Q.W., 1331 Perc., 1346 F.A. *Weperden* ; 1346 F.A. *Weperdon*, 1428 *Weperden* ; 1498 H. 3. 2. 127 *Wyperdon* ; 1586 Raine *Weperdon* ; 1587 Bord. *Waperdon*, 1596 *Woperdon*, 1637 Camd. *Waperton;* 1663 Rental *Wopperton* ; 1671 Egling. *Woperton*, 1674 *Wopperton*, 1699 *Weeperton* ; 1746 Ingram *Wooperton*, 1811 *Wapperton.*

The first element is probably Celtic, and identical with Wepre, Flints., which Morgan (p. 160) says is from Welsh *gwybre< gwy*, water, and *bre*, hill. If so, the name is "valley of the well-watered hill." *wap-* and *wop-* are due to the infl. of initial *w*. Cf. S. Scot. *wab* and *wob* for *web* (E.D.G., p. 670). App. A, § 1.

Wreighill (Rothbury) [ri·hil].   1292 Q.W. *Werghill* ; 13th c. Newm. *Werihill, Vuerhil, Vuerchil, Vuarchil* ; 1538 Must. *Wryghyll* ; 1586 Raine *Wreghille* ; 1663 Rental *Wreghill.*

O.E. *wearg-hyll*=felon-hill. Cf. *weargedun*, B.C.S. 792, and such a compound as *wearg-rod*=cross, gallows. Possibly "gallows-hill." Phonology, § 54.

Wrekin Dike (Co. Durham).   *c.* 1135 F.P.D. *Vrakendic, c.* 1190 *Wracennhegge, c.* 1225 *Wrakendyk.*

The name of this old earth-work is certainly Celtic, but the *a* forbids our associating it with Wrekin, Salop, B.C.S. 1119 *Wreocen.*

Wydon (Haltwhistle).   1255 Ass., 1428 F.A. *Wyden.*

"Wide valley." App. A, § 1.

Wydon Eals.   *c.* 1250 H. 2. 3. 350 *le Eles. v. ele*, Part II.

Wylam-on-Tyne.   *c.* 1120 Ty. *Wylum* ; 1203 R.C. *Wilham* ; 1271 Ch. *Wylum, Wilum* ; 1326 Pat. *Wilom* ;

1380 Ipm. *Wylome*; 1428 F.A. *Wylome*; 1663 Rental *Wileham*.

" Wila's homestead."

Wynyard (Grindon). 1237 Pat. *Wyneiard*, 1238 *Wingherd*; 1311 R.P.D. *Wynhyard*, 1345 *Wyneyard*; 1421 F.P.D. *Wyneyard*.

" Wine's yard or enclosure." The climate forbids us to interpret it as O.E. *wīn-geard*=vineyard, as we can in the *Wynyards* in Ombersley, Worc. (Duignan, p. 184).

Yardhope (Holystone). 1324 Ipm. *Yerdhopp*, 1331 *Yerdhope*; 1604 Arch. 3. 4. 118 *Yardope*.

Probably the " hope " marked by a *yard* or enclosure. Cf. Earle *supra*.

Yarnspath Law (Kidland). 1233 Newm. *Hernispeth*.

" Eagle's path " (O.E. *earnes pæð*). Phonology, §§ 37, 9.

Yarridge (Hexham) [jariʃ]. 1232 Ch. *Jernerig (sic)*; 1298 B.B.H. *Yarwrigg*; 1328 *Yerurige*, 1479 *Yarowryge*; 1538 Must. *Yarath*; 1610 Speed *Yarwich*; 1663 Rental *Yarrage*.

O.E. *gearwe-hrycg*=yarrow-grass ridge. Phonology, §§ 59, 58.

Yearhaugh (Elsdon). 1312 Eccl. *Yarhalgh*; 1330 Orig. *Yarehalgh*; 1663 Rental *Yarehaugh*.

" Haugh by which there is a *yare* or fishery " (Heslop, *s.v. yare*).

Yeavering (Kirknewton) [jivrin]. *c.* 750 Bede *Ad gefrin*, *Ad gebrin*; *c.* 1000 O.E. Bede, *Aet gefrin*; *c.* 1250 T.N. *Yever*; 1296 S.R. *Yverne*; 1316 Sc. *Yeure*; 1359 Pat. *Yevere*; 1377 Ipm. *Yemrum*; 1404 Pat. *Yevern*; 1442 Ipm. *id.*; 1637 Camd. *Yeverin*, 1663 *Yeverington*; 1784 Ilderton *Evering*; 1796 ib. *Yevering*.

Clearly a Celtic name.

Yetlington (Callaly). 1186 Pipe *Yetlinton*; 1247 Ch. *Yetlington*.

Skeat derives Yattenden, Berks. (p. 29) from O.E. *Geatinga-dene*, noting the name " Godwulf *Geating* " in the W.S. genealogies. From *Geat* may have been formed a dimin. *Geatel(a)*, and Yetlington may stand for *Geatling-*

(*a*)*tun*=farm (of the sons) of *Geat(e)la*. Possibly this may be the same name as is found in Bede's *in Getlingum* (=Gilling, Yorks.). For Moorman's interesting theory with regard to this name, *v. Essays and Studies*, *u.s.*, vol. v., pp. 78 ff.

PART II

O.E. āc=oak.
Acomb, Crooked Oak, Lynesack, Pedam's Oak.

O.E. æceras (*pl.*)=pieces of tilled or arable lands, fields.
Edderacres, Farnacres, Minsteracres, Oldacres, Overacres.

M.E. bache=valley of a small stream. This probably
goes back to O.E. *bæc* often found in charters, probably
with the same meaning. Cf. Sandbach, Chesh., Debach,
Suff., Burbage, Leic. and Wilts.
Claubache (*s.n.* Cawledge).

M.E. banke<O.W.Sc. *bakki*, "ridge, eminence, hill,"
Dan. *banke*, "raised ridge of ground " (Björkman, *Scand.
Loan-Words*, p. 230). *bank* is the common North. dial. word
for the slope of a hill.
Harrowbank, Ninebanks.

O.E. berern, bern=barn.
Whitburn.

O.W.Sc. berg=hill.
Sadberge.

O.E. botl (*boðl*)=building. The first form is only found
in Nthb. and Nthts. In Scotland it appears as Buittle
(Kirkcudbr.) (Maxwell, p. 91), and *battle* in Newbattle. Elsewhere the common form is *bold* (*v.* Budle *supra*), examples
being found in Chesh., Derbys., Lancs., Lincs., Northts.,
Notts., Salop, Staffs., Warw., Worc., Yorks.[1]
The compound *boðl*+*tūn* is common in the North of

---

[1] The statements in this section, and elsewhere in Part II., are based on
a study of the documents mentioned on p. viii. of the Preface.

England, meaning "farm with a building on it," and gives later Bolton and Boulton. Nine examples have been noted in Yorks., two in Lancs. and Cumb., one in Westm., and a doubtful example in Derbys.

Budle, Harbottle, Lorbottle, Newbottle, Shilbottle, Wallbottle.

O.E. **brycg**=bridge. Usually North. Eng. *brig* has been replaced by St. Eng. *bridge.*

Corbridge, Eddys Bridge, Foulbridge House, Fowberry (Bamburgh), Piercebridge, Risebridge, Wascrow Bridge.

O.E. **burh**, dat. sg. **byrig.** The exact meaning of O.E. *burh* in its technical sense is a vexed problem, but there can be little doubt that in place-names the general idea is of some fortified place, though the fortifications may be very elementary or primitive. This element is common throughout English place-names, but is specially frequent in Herts. and Middx.

burh+tūn, meaning apparently a "fortified farm," is fairly common in English place-names as *Burton* or *Bourton.* Some forty-six have been noted, of which fifteen are found in Yorks., one in Nthb., none in East Anglia, Ess., Herts., Middx., Beds., Hunts., Surrey, Kent.

Bamburgh, Cheeseburn, Dunstanborough, Newbrough, Sockburn, Thornbrough, go back to nom. *burh.* Bradbury, Carlbury, Fowberry, Lesbury, Rothbury to dat. *byrig.*

O.E. **burna** (*m.*), **burne** (*f.*), "stream or river," originally "fountain spring." It is the regular word for a small stream throughout Nthb. and Co. Durham, and only on the modern map has it in some parts of Co. Durham been replaced by *beck.*

O.W.Sc. **bygging**=building. *bigging* is still in independent use in North. dialect for a building, especially an outhouse in contrast to the main building. It should be noted that in the case of one of the Newbiggins, *Boldon Buke* renders *bygging* by L. *villa,* and that in another, one of the MSS. has *burga* as an alternative form.

The distribution of this element is curious. Nine examples have been noted in Nthb. and Co. Durham which are not distinctively Scandinavian districts, as against eight in all the other English counties.

O.W.Sc. **býr, bœr,** Dan. *by*=farm buildings, then (in Iceland) farm, landed estate, including the farmyard and buildings, (in Norway, Sweden, and Denmark), town, village. The last is the sense that commonly prevails in England, though examples of the second are also found. It is an alternative to *burh* in a charter of King Edmund (B.C.S. 792), and *by* and *bury* are found alternatively in several place-names in D.B. and other documents. The only other suffix with which alternation has been noted is *ton*.
Aislaby, Follingsby, Killerby, Ornsby Hill, Raby, Raceby, Rumby Hill, Selaby, Ulnaby Hall.

O.E. **býre**=byre, shed, hovel.
Byers Green, Byerside, Byres, Edmundbyers.

O.E. **camb**=comb. In the sense "long narrow hill or ridge," the word is only found in place-names in Scotland and the North of England as *kame, kaim*. Skeat suggested it may also be found in a southern form in Combs, Suff.
Bingfield Comb. See also Combfield House.

M.E. **castel**=castle, fortress, stronghold.
Tedcastle, Barnard Castle.

O.E. **ceaster.** This term is generally, if not universally, applied to sites where there have been Roman encampments (L. *castra*). Normally it yields *chester*, but in parts of England where Scandinavian influence is strong, the form is more commonly *caster* (*caester*). *cester* or (in shortened form) *ster* is found in certain place-names owing to French influence.
  chester : *Aunchester, Binchester, Chester - le - Street, Chesters, Ebchester, Hetchester, Lanchester, Outchester, Rudchester, Whitchester.
  cester : Bellister, Craster, Gloster Hill.

P

O.E. **celde**=spring. Unknown except from the evidence of place-names. Cf. *to celdan*, B.C.S. 880, and *Baccanceld*= Bapchild, Kent (ib. 290). It is cognate with O.W.Sc. *kelda infra*.
Learchild.

M.E. **chace**=hunting, hunting-ground.
The earliest example of *chase*, meaning "a place for hunting," is dated 1440 in N.E.D., two hundred years later than the first example of *Chipchase*.

O.E. **clif**=a perpendicular or steep face of rock. In place-names the word is not confined to a rock overhanging the seashore, a lake or a river, but is used of a steep slope, a declivity, a sloping and cultivated escarpment.
Bewclay, Coniscliffe, Cronkley, Donkleywood, Egglescliffe, Heckley, Horncliffe, Shincliffe.

O.E. ***clōh**=ravine or valley with steep sides. This word is not found in O.E., probably because of the paucity of Anglian documents. Its M.E. equivalent *clo(u)gh* is only found in North. Eng. and Lowland Scots. In place-names it has only been noted in Derbys., Yorks., Lancs., Nthb.
Catcleugh, *Farnycleugh, Heatherley Clough, Oxcleugh, Stirkscleugh.

O.E. **cnoll**=knoll.
Butterknowle, Edge Knoll, Henknowl.

O.E. **cot**=cottage, house, dwelling. In place-names it is generally found in the plural form. It is most common in the Midlands:—Beds., Berks., Bucks., Northts., Oxon., Warw., Wilts., Glouc., Worc., and in Devon, and is very rare in East Anglia, Kent, Surrey, Sussex, and the six northern counties.
Coatham, Coatsay Moor, Coldcotes, Coldcoats, Cullercoats, Hepscott.

O.E. **croft**=enclosed field. In North Cy. dial. adjacency to a dwelling-house is usually implied.

Ancroft, Goosecroft, Greencroft, Hitchcroft, Mosscroft, Osmondcroft, Stonecroft, Woodcroft.

O.E. **cumb**=small valley, hollow. This suffix is distinctively a southern one, and is commonest in Dev., Dors., and Somerset. For its exact meaning there, *v.* N.E.D. *s.v.* It is very rare indeed in Lincs. and Yorks., only two examples have been noted. It is not found in any early forms in Lancs., Chesh., Staffs., Salop, Worc., nor in Norf., Suff., Ess., Middx., Herts., Cambs.

Escombe (?).

O.W.Sc. **dalr**=valley. O.E. *dæl*, its English cognate, is found, but the distribution of *dale* in place-names shows fairly clearly that there it is almost, if not entirely, due to Scandinavian influence. The counties in which it is commonest are Yorks., Lancs., Derbys., Norf., Lincs., and Notts. Isolated examples of its use are found in Leic., Staffs., Northts. Occasionally it is found alternating with the suffix -*dene*, but in general it is applied to the whole river-valley between its enclosing ranges of hills or high land rather than to a deep or narrow ravine.

Allendale, Glendale, Knaresdale, Redesdale, Thrundle, Tursdale.

O.E. **denu**=valley. In modern independent use as *dean* (Sc.) or *dene* (Nthb. and Durh.) it tends to be used specially of the deep, narrow, and well-wooded ravines which are so characteristic of the scenery of Nthb. and Durham and of part of the Lowlands of Scotland. This element is in fairly general use throughout England, but is commonest in Nthb. and Durham. It is very rare in Dors. and Dev., probably because of the almost universal use of *combe* in something of the same sense, and is very rare also in Warw., Worc., Staffs., and almost non-existent in Chesh., Salop, and Heref.

Acton, Ashington, Aydon, Beldon Burn, Biddlestone, Blagdon, Burdon (Bishopwearmouth), Catton, Chirdon, Dawdon, Dipton (2), Elsdon, Embleton (Sedgefield), Eppleton, Foxton (2), Gofton, Hallington, Harsondale, Haydon,

Hendon, Hinding Burn, Lysdon, Nookton, Pandon, Pigdon, Rickleton, Shawdon, Shildon (Blanchland), Togston, Tursdale, Weldon, Welton, Wydon, and several names in *-den* and *-dean*.

O.E.**dīc**=ditch, dug-out place, then bank formed by throwing the earth out of the ditch, causeway. This latter sense is denoted nowadays by *dyke* as distinct from *ditch*, but this distinction does not seem to have developed in earlier English.

Biddick, Wrekin Dike.

O.E. **dūn**=hill, down.

Berrington, Bowsden, Buston, Callerton, Cartington, Embleton, Farrington, Felkington, Glanton, Grottington, Harton, Hetton-le Hole and -le Hill, Humbleton (2), Green Leighton, Ovington, Pittington, Quarrington, Raredean, Shotton-in-Glendale, Swarden Burn, Warden, Warden Law, Windleston, Wooden, and numerous names in *-don*.

O.E. **ēa**=river.

Elvet, *Hagustaldes-ea* (*v.* Hexham).

O. North. *****ēa**=island. Bede has more than one place-name in *-eu*, and the only possible explanation of these seems to be that it is the early Nthbrian. form of W.S. *īeg*, though there are certain phonological difficulties about the form.

Heoroteu (*v.* Hartlepool).

M.E. *****ele**=a small island. This word is not found outside Nthb. place-names. It would seem to be a diminutive in *-el* of O. North. *****ēa** (*v. supra*)=an island.[1] The word *eale* is in fairly common use in Mod. Nthb. place-names. Hodgson (2. 1. 86) says, "*Eales* is the name of a hamlet on the Tyne at Knarsdale and of a portion of the haugh at Corbridge . . . Wide-*eels* and Bridge-*eels* are places on the East Allen. On North Tyne there are the

---

[1] Sedgefield believes this word to be the same as O.E. *healh* (*v. infra*), but this is always *haugh* or *hale* in Nthb. (M.L.R. vol. ix. pp. 240–1).

Eels near Wark, Bellingham *Eels*, and *Eels* in the parish of Greystead, and *Eels*-bridge on the Derwent." It is used of low grounds liable to river floods (Heslop, *s.v.*).

*Nakedale, Wydon Eals.

O.E. fald, falod=pen or enclosure for domestic animals. Stotfold, Pinfold.

O.E. feld, Mod. Eng. field. These two words differ widely in meaning. W. H. Stevenson (*Phil. Soc. Transactions*, 1895-8, p. 531) puts the case clearly when he says that O.E. *feld* was "just the opposite of our field, that is, it meant a great stretch of unenclosed land. Arthur Young uses *field land* as opposed to enclosed land." It was open land as opposed to woodland (*v. fenn, infra*). Only gradually did it come to be used of a " piece of land parted off by hedges, fences, etc." Heslop (*s.v.*) notes the relics of its older sense when he says that it means " a division of land consisting of many separate buildings, grouped together in the ancient system of cultivation for the purpose of rotation of crops. Doubtless in many of the Nthb. and Durh. names, especially those of comparatively late origin, it is used in its modern and more restricted sense.

This suffix is not found in Devonshire, and only very rarely in Somerset and Dorsetshire. It is also very rare in Lincs. There are twenty-seven names in *field* in our two counties.

M.E. flat, a derivative of the adj. *flat* (of Scand. origin), used of a piece of level ground, and common in field-names.

O.E. fenn=marshy land. In a Peterborough Charter (B.C.S. 464) land is granted to the abbey with *feld* and *wood* and *fen* thereto belonging, the three apparently covering all possible types of land.

Mason, Mousen, Pressen, and other names in *fen*.

O.E. ford=ford. This becomes in Mod. Eng. either *ford* or *forth*, and *forth* forms are common from the 14th cent. onward. Attempts have been made to explain the *forth* forms by Scand. *fjörðr*, but this does not mean a "ford," and gives English "firth" and not "forth." Further, it is much too rare a word to have had influence in producing *-forth* forms all over England. Björkman (*Scand. Loan-Words*, p. 162) agrees with Kluge-Lutz in explaining M.E. *forþ* as due to a Teut. **forþo-*, but this seems unlikely. *forth* is not found independently before the 14th cent., and a careful study of a large number of place-name forms showed one example of *-forth* from the 12th c., two from the 13th., many from the 14th, but by far the greater number from the 15th cent. onwards, pointing clearly to the *-forth* forms as a comparatively late phonological development of earlier *-ford*. (Phonology, § 30.)

Baxterwood, Flatworth, Mosswood, 31 *fords* and 5 *forths*.

O.W.Sc. **garðr** = yard, courtyard, fence. Commonly applied in North. dialect to a small piece of enclosed ground, especially near a building. It is the cognate of O.E. *geard* = yard.

Hallgarth.

O.W.Sc. **gáta** = way, road.

Cowgate, Hooker Gate.

O.E. **geard** = yard, enclosure.

Wynyard.

O.E. **geat** = "gate." This should give North. and Sc. dialectal *yet*. Modern forms in *gate* are due to the O.E. pl. form *gatu* and to the influence of the word *gate* = road (*v.* O.N. *gáta, supra*).

Eastgate, Leadgate, Portgate, Stotgate, Westgate, Wingate(s).

M.E. **grene** = "a grassy spot," and later "a piece of public or common grassy land in or near a town or village" (N.E.D.).

Dewsgreen, Town Green.

O.E. **hām** = 'a village or town, a collection of dwellings, a vill with its cottages," "a dwelling-place, house, abode"

(N.E.D.). It is probable that the former is the meaning in the names in *ingham*, which all seem to be of early origin. The latter is the sense in the names of more recent formation.

Alnham, Bellingham, Bolam, Billingham, Carham, Chillingham, Cleatlam (?), Deanham, Downham, Ealingham, Edlingham, Eglingham, Ellingham, Eltringham, Fenham (?), Fleetham, Greatham, Harnham, Headlam, Hexham, Higham Dykes, Ingram, Leam (2), Lyham, Bp. Middleham, Neasham, Newham (2), Ovingham, Norham, Polam, Seaham, Stamfordham, Streatlam, Ulgham, Ulwham, Warnton, Whickham, Wolsingham, Wylam.

It should be noted that, except in the somewhat doubtful cases of Ulgham and Wylam, *ham* is only compounded with a patronymic, a river-name, or some name descriptive of the position or soil of the homestead. It is never compounded with a personal name pure and simple.

The suffix *ham* is found throughout England. It is most frequent in Norf., Suff., and Cambs., in Surr. and Suss., and then in Essex and Middlesex, and Kent. It is rarest in Glouc., Worc., Warw., Salop, Staffs., Derbys., and Leic.

It is difficult to be sure whether M.E. forms go back to O.E. *hām* or to O.E. *hamm* in either of the senses given below, but as the first sense is very rare in place-names, and the second is unknown in the dialect of Norf., Suff., Cambs., or Essex, where *ham* names are most numerous, it is probable that any calculations of the relative frequency of the suffix *hām* are not seriously affected by the existence of the suffix *hamm*. Its presence in M.E. can, moreover, often be detected by a -*hom(me)* or -*hamme* form.

O.E. **hamm, homm,** is found in place-names in two senses, the etymology differing accordingly. They are (1) a plot of pasture-ground, meadow-land ; (2) a piece of ground shaped like the human " ham," i.e. the hollow or bend at the back of the knee. This suffix is only found once in Nthb. and Durh., and as sense (1) is unknown in N. England, it must be interpreted under (2).

Farnham.

O.E. hēafod=head, then, highest point of a field, a stream or a hill. Cf. Wyld, pp. 344-5.
Consett, Gateshead, Greenhead, Hartside, Wearhead.

O.E. healh. The nom. of this word has given Nthb. *haugh* [ha·f], used of a " piece of flat alluvial land by the side of a river, forming part of the floor of the river-valley " (N.E.D.). Such land is common within the bend of a river. The oblique case forms *h(e)ale* have given M.E. *hale*, " corner, nook, secret place." Dialectal *hale* is used in the same sense as "haugh," and also of a triangular corner of land, a bank or strip of grass separating lands in an open field " (Lincs.). It is very common as a suffix in place-names, sometimes doubtless with the meaning of *haugh*, but at others the sense is less certain. For a full discussion *v.* Wyld, pp. 340-1.

Beadnell, Bothal, Cornhill, Dinsdale, Etal, Featherstone, Finchale, Henshaw, Hepple, Houghall, Howtel, Oxenhall, Ramshaw Hall, Ricknall, Seghill, Snabdaugh, Stokoe (?), Tughall, and 17 *haughs*.

This suffix is almost unknown in the three western counties—Dev., Dors., and Som. It is remarkably common in the group Chesh., Staffs., Salop, Derbys., and also in Norf. and Suff. The nom. form is rare except in Nthb.

O.W.Sc. heimr. The Scandinavian equivalent of O.E. *hām* (*v. supra*).
Skirningham.

O.E. helm=helmet, then crown, top or summit of anything. For its use of a hill cf. *helm-cloud*, i.e. a cloud which forms over a mountain top.
Bensham, Helme, Helme Park.

O.E. hlāw, hlǣw = rounded hill, barrow, tumulus. *law* is still in independent use in North England and Scotland. In the Midl. it generally appears as *low*. *law* is very common in Nthb. and Durham, *low* is a characteristic Derbyshire suffix, probably because of its well-rounded hills, and is fairly common in Heref., Staffs., and Salop. It is unknown in Dors., Dev., Som., Hants., I. of Wt. and Surr., and only occurs once in Sussex. No example has been noted in Herts., Middx., Kent., Norf., Northts., Rutl. Barley Hill, Brenkley, Crawley, Dewley, Fairnley, Harlow Hill, Hartley, Hauxley, Hunterley Hill, Kearsley, Kellah, Kelloe, Kirkley, Moorsley, Sheddon's Hill, Shellbraes, Slingley, Softley (Auckland), Stanley, Stickley, Thornley (Kelloe), Throckley, Wheatley Hill, Whitehall, Whitlow, Wooley (Slaley), and twenty-nine names in *law*.

O.E. **hlynn**="torrent," reinforced by Gaelic *linne*, of similar meaning (N.E.D.), and used in M.E. both of a cascade and of a pool.

Kipperlynn, Lowlynn, Lyne.

O.E. **hōh, hō**=a projecting ridge of land, a promontory. It is probably identical with *hōh*, "a heel," hence "point of land formed like a heel and projecting into more level ground."

The nom. *hōh* has given Sc. and North. dial. *heugh*, *heuch* [hjuf], "precipitous or hanging descent, a craggy or rugged steep" (N.E.D.). In Sacriston Heugh (*v. supra*), it is rendered by Lat. *clivus*. The form *hō* has given Hoo, as in Hooe, Kent., Luton Hoo, Beds. The oblique case form *hōge*, M.E. *howe*, has given later English *how* and *hoe*, as in Morthoe, Dev. In those counties where Scand. influence is prevalent, it is very difficult to distinguish this suffix from O.N. *haug*, M.E. *howe*=burial mound.

The distribution of this suffix is curious. It is remarkably common in Beds. and Northts., and fairly so in Ess., Herts., Suff., and Bucks. Elsewhere it is common only in Nthb. and Durham.

Nom. sg. : Ten names in *heugh* and Cambo (?).

Dat. sg. : Belsay, Cornsay, Duddo, Ingoe, Kyo, Rivergreen, Sandoe, Shaftoe, Wharnley, Whessoe, and five names in *ho(e)*.

O.W.Sc. **holmr**=islet, meadow on the shore, but used also in Norwegian place-names of a "grass plot in a field,"

and in other senses.   It is used in England, chiefly in
Scotland and North England, and there has the sense, "a
piece of flat, low-lying ground by a river or stream, sub-
merged or surrounded in time of flood " (N.E.D.).
Broomyholme, Salt Holme.

O.E. hop=" a small enclosed valley, especially a smaller
opening branching out from the main dale, the upland part
of a mountain valley, a blind valley " (N.E.D.).   It is distinctively Scottish and North-East English in its distribution, being common only in Nthb. and Co. Durham.   It is
comparatively rare in Yorks., and is only common elsewhere
in Derbys., Heref. and Salop.

It is possible that in Ryhope, and in one or two other of
these names, we have to do with O.E. *hop* (possibly a distinct
word), meaning " a piece of enclosed land " (cf. *hope* N.E.D.).

Blenkinsopp, Cassop, Hoppen, Philip, Temple Heap,
Pontop, Snope, and forty-four names in *hope*.

O.E. horn=horn, then, a projecting piece of land, a
promontory.
Woodhorn.

O.E. hrycg = back of a man or animal, long and
narrow stretch of elevated ground.   The proper North.
Eng. form is *rigg*.   (Phonology, § 27.)

This suffix is specially common in Nthb., and practically
unknown in East Anglia.
Aldin Grange, eleven names in *ridge*, and two in *rigg*.

O.E. hūs=house.
Eleven names in -*house*, Woodhouses.
The dat. pl. (*æt þǣm*) *niwan hūsum*=(at the) new houses,
gives, in later English, *Newsham*.   One example is found
in Nthb., one in Co. Durh., four in Yorks., two in Lincs.
Variant forms are Newsam, Newsholme (3), Newsome (3) in
Yorks.   It does not seem to be found outside these counties.

Woodhouse(s) is only found in Yorks. and the North
Midlands.

O.W.Sc. **hvammr.** Used in Norway, according to Rygh (*Indledning*, p. 57), of "a short valley or depression surrounded by high ground, but in such a way that there is an opening on one of the sides." Cf. *Hvítar hvammar*, in Iceland (Kálund I., 190). From this comes the dial. *wham*, used in Scotl., Nthb., Cumb., and Yorks. (E.D.D.) to denote (1) a marshy hollow, (2) a hollow in a hill or mountain. Cf. Goodall, p. 297.

There is a word *hwomm* (*hwamm*) used in the O.E. Vespasian Psalter to gloss Lat. *angulus*, which is evidently the English cognate, but associations of meaning are a good deal stronger in the case of the Norse word.

Wham, Whitwham. Cf. also Whamlands.

O.E. **hyll**=hill. This element is found in place-names throughout England. In Ess. and Herts. alone is the suffix very rare : only one example has been noted in each.

Beal, Bearl, Burnigill, Clennell, Cockle Park, Earle, Fairley, Ogle, Redmarshall, Ryal, Ryle, Whittle (2), and nine names in *hill.*

O.E. **hyrst**=copse, wood.

Hesleyhurst, Hirst, Keyhirst, Longhirst, Moralhirst.

O.W.Sc. **kiarr**=copsewood, brushwood. Norw. *kjerr*= swamp, marsh. Mod. dial. *car*=pool, hollow place, low-lying ground.

Byker, Walker.

O.W.Sc. **krókr**=crook, bend, M.E. *cr(o)uke* (North Cy.), used of a small piece of ground of a crooked shape, an odd corner or nook. It is very common in field-names.

Coppy Crook, Crawcrook, Crook, Crookham, Crookhouse, Crooks, Darncrook.

O.E. **land.** In place-names the suffix *land* seems in O.E. to be used definitely of the soil or ground with some qualifying first element describing its tenure (e.g. *bócland*), its cultivation or lack of it (e.g. *eyrðlond, wudulond*), its crops (e.g. *lĩnland*). It is never used with a personal name as the first element.

Blanchland, Copeland, and Coupland (cf. *ceaplond*, B.C.S. 1020), Mayland, Newland (4), Sunderland, and Swarland, Tone, belong to one or other of these types. Buteland, Dotland, Hoppyland, Kidland, Warland are all probably of comparatively late origin, and in them *land* may be used in one or other of its later developments (*v.* E.D.D. and N.E.D.), e.g. it may be used of one of the strips into which a corn-field or pasture-field has been ploughed.

O.E. lēah, m. and f., dat. lēage=a tract of open ground, either meadow, pasture, or arable land, used primarily of land from which forest has been cleared away. *ley(e)* forms are clearly from the dat. Cf. Wyld, p. 368.

Baydale, Brotherlee, Callaly, Cleatlam, Dally Castle, Fallowlees, Fawnlees, Garretlee, Glantlees, Greenlee, Hanging Leaves, Hawkuplee, Karswelleas, Kyloe, Leas and Lee Hall, Lees, Longlee Moor, Morralee, Pokerly, Raylees, Ridlees, St John Lee, Stella, Stobbilee, Tinely, Whitelees, Whitelee, and one hundred-and-six names in *ley*.

M.E. leche, lache=a small stream. Mod. Nthb. *letch* is used of "a long narrow swamp in which water moves slowly among rushes and grass." Further, *v.* Wyld, p. 365.
Cawledge Park. *v.* also Cong Burn.

O.E. luh is used in the Lindisf. Gospels as a gloss of L. *fretum* and *stagnum*, and in M.E.=lake or pool.
*Lowes.

O.E. (ge)mǣre= boundary, landmark, may possibly be found in one or two of the Nthb. names in M.E. *mere*, as in Greymare Hill, but it is more probable that, as a rule, we have to do with the next word.

O.E. mere=mere, standing water, lake, also used in dialect of low, marshy ground.
Boulmer, Thackmires, Whitemere, Wilmire. See also Black Lough and Greenlee Lough.

O.E. **mersc**=marsh. This forms the first element in the numerous Marstons throughout England. Cowpen Marsh, Homers Lane, Owmers.

O.E. **mōr**=waste or marshy land, but used also of "moors" in the modern sense.

O.E. **mūþ, mūþa**=mouth, estuary. Jesmond, Blyth, and nine names in *mouth*.

O.W.Sc. **mýrr**=mire, swamp.

O.E. **ofer**=shore, margin, bank. Wooler.

O.E. **pæð**. Except for three examples in Oxf. and Warw. this suffix has not been noted outside Nthb. and Co. Durham. Here it is fairly common, and its presence is explained by the use of Nthb. *pæð*, North. M.E. and Mod. Nthb. and Scots *peth*, of a hollow or deep cutting in a road, and also of a steep road or path. N.E.D. quotes from the Old Northumbrian version of the Gospels, where *vallis* in "Every valley shall be filled" is glossed *pæð vel dene*. Soppit, five names in *peth*, and four in *path*.

O.E. **pōl**=pool, deep or still place in a river.

O.E. **rǣw, rāw**=row. From the second of these forms comes Sc. and North. *rawe* and *rowe*, used of a number of houses standing in a line. Six names in *row* and two in *raw*.

Late O.E. **sǣte** < O.W.Sc. **sæti**=seat. For this element *v*. Wyld, p. 380. It is not clear whether this suffix in place-names means "seat," referring to the shape of the ground, or to the fact that a settler was "seated" or stationed there. Allerside, Bebside (?), Causey Hall, Corsenside, Earlside, Gibside, Simonside (2), Tarsett.

O.E. **sceaga**=thicket, small wood, copse or grove.

The only county in which this suffix is found with any frequency is Lancs. Elsewhere it only occurs sporadically. No examples have been noted in the group Derbys., Lincs., Notts., Northts., Beds., Cambs., Hunts., Suff., Ess., Middx., in Kent or Suss., Dors. or Som. Six are found in our counties.

O.E. **scēat**=corner, quarter, region, lit. that which *shoots* forth.

Bebside (?), Eshott.

M.E. **schele,** the English cognate of O.W.Sc. *skáli*, Cumb. and Westm. *scale*. It is used first of a temporary building, such as a shepherd's summer hut, then of a small house, cottage, hovel. See a full discussion in Lindkvist, pp. 189-90. Its use is confined to Nthb. and Durham, and S. Scotland. The derivative *shieling* is used occasionally. Cf. Shelley *supra.*

Agarshill, Axwell, Eshells, fifteen names in *shiel(s)*, and nine in *shield*.

O.E. **scīr**=province, district. For its use in these counties *v.* Introd. § 1.

M.E. **side** (O.E. *sīde*)=slope of a hill or bank, especially one extending for a considerable distance (N.E.D.). This suffix is much more common in Nthb. and Co. Durham than anywhere else. Three examples have been noted in Yorkshire. It is found also in Lancs., but no old forms have been noted. There is no evidence for the use of O.E. *sīde* in this sense, but that may only be due to the scanty records of O. Northumbrian.

Gallow Hill, and twenty-one names in *side*.

O.E. **slæd**=slope, hollow. Mod. Eng. *slade* has the various meanings " valley, dell, dingle, an open space between banks and woods, a forest glade, a strip of greensward or boggy land."

Weedslade.

O.E. stān=stone, rock.

Dunstan (2), Falstone, *Forston, Fourstones, Greystones, Holstone Ho, Holystone, Iveston, *Rodestane, Settling-stones, Spindleston, Whittonstone.

O.E. steall=standing place for horses or cattle, cattle-shed, cow-house.

Hamsteel, Tunstall (2), Whittonstall.

O.E. stede=place, position, site. This suffix is found in most English counties, though no examples have been noted in Dev., Som., Derbys., Chesh., Heref., Oxon., Warw. It is rare throughout the west of England. The counties in which it is commonest are Berks., Ess., Herts., Suff., Kent, Surr., Suss., and I. of Wt., Essex and Herts standing well above the rest. Its presence in Cumb. and Westm. is doubtful. Here it is hard to distinguish it from the O.N. cognate staðr, and so also in Yorks. there are several doubtful cases. It is not found in Co. Durham, and there is no evidence that those names in Nthb. which are formed with this suffix are at all early in origin.

stede is often found in combination with O.E. *hām*. Such compounds, giving later *hampstead*, are specially common in Herts. and Berks. They are unknown in the North.

Newstead (2), Barneystead.

O.W.Sc. steinn = stone, rock.
Middlestone.

O.E. stigu=sty, pen for cattle.
Housty, Houxty.

O.E. stōw=place, not apparently with any particular significance.

This element is comparatively rare in place-names, and only occasionally is it found in independent use, as in Stow-on-the-Wold. It is not found in Nthb., Cumb., Westm.,

Lancs., Chesh., Staffs, Warw., Worc., nor in Dors., Hants.,
I. of Wt., Surr., Suss.
Westoe.

M.E. strother. This element is in common use in
Northern England and Scotland. W. H. Stevenson (*Phil.
Soc. Trans.*, 1845-8, p. 531) quotes from the Scotch Rolls
Series, No. 1650, "*una marresia* (marsh) *vulgariter nuncupata
a strudire*," which fixes its meaning as a marsh or swamp.
He takes it to be a derivation of O.E. *strōd*, with the same
meaning, which is very common in the forms Strode, Strood
and Stroud in England, and notes that it is the name given
by Chaucer to the village " far in the North " of his Reeve's
tale.
Broadstruthers Burn, *Coldstrother, Haughstrother,
Strother (2).

O.W.Sc. toft (*topt*)=homestead, site of a house and its
outbuildings.
Bruntoft, Toft House.

O.E. tūn=an enclosed place or piece of ground, enclosed
land surrounding or belonging to a single dwelling, a manor,
an estate, the enclosed land of a village community, " a
small group or cluster of dwellings or buildings." It is in
one or other of these senses that O.E. *tūn* has given rise to
the numerous Mod. Eng. place-names in -*ton*. In none of
them is there that contrast with the smaller and more
dependent village that is implied in its modern derivative
" *town*." It is the commonest of all English place-name
suffixes, Essex and Herts. being the counties in which it
is distinctly least common, Shropshire and Herefordshire
those in which it is commonest. Some 4500 examples in
all have been noted.
Coldtown, Molesdon (?), Nelson Newtown (2), Walltown.
and 213 names in -*ton*.

O.E. twisla=a point or part at which anything divides
into branches, a fork.
Haltwhistle, Twizel (3).

O.E. þorp, O.W.Sc. þorp = hamlet, village or small town. In O.E. glossaries it is found as an alternative to *tūn* and is used to gloss L. *fundus* and *villa*.

This suffix is clearly found in many places of purely English origin, often disguised in such forms as *throp, drop*. That Scandinavian influence has been at work in increasing its frequency is clear from the fact that Rutl., Leic., Lincs., Notts., Northts., Norf., and Yorks, are the counties in which it is most frequent. It is unknown in Cambs., Middx., Kent, Suss., I. of Wt., Dev., Heref., Salop, Nthb., and almost so in Dors., Som., Staffs., Worc., Surr., so that it would seem not to have been used in the West of England, in the West Midlands, or in S.E. England, and was probably not in use in Northumbria.

Fulthorpe, Thorpe (3).

O.E. þyrne = thorn-bush.
Caistron.

O.E. **wæsce** is found in the compound *sceap-wæsce*= sheep-washing place. *wæter-gewæsc* is used of alluvium, land formed by the washing up of earth.
Allerwash, Sheepwash.

O.E. **weg**=road.
Garmondsway.

O.E. **weorþ, worþ, wyrþ** = an enclosed homestead, a habitation with surrounding land. The Lat. equivalents are *villa, villula, viculus*. Cf. Bosworth-Toller, *s.v. worþ*.

Chesterwood, Clarewood, Ewart, Ewart's Hill, Lils-wood, Pauperhaugh, Pegswood, Shoreswood, Staward, and twenty-two names in *worth*.

O.E. **wīc**=(1) dwelling-place, abode, residence, and is used as the equivalent of L. *villa* and *mansio*, (2) a collection of small houses, a village, is equated with *lytel port* and used as a gloss of L. *castellum, vicus*.

This suffix, in the form *wick* or, less commonly, *wich*, is

Q

in wide use in place-names, and the word is often found by itself in place-names such as *Wick* and *Wyke*, and possibly *Week*. The suffix *wick* is specially common in Cambs., Herts. and Beds., *wich* in Worc. and Staffs. Two compounds of *wick* are very common, viz. *Berwick* and *Hardwick*. The former is found as *Berwick, Barwick*, or *Borwick* in Ess., Norf., Kent, Suss., Oxon., Wilts., Som., Salop, Lancs., Yorks., Nthb. The latter is found in the Midlands—Beds., Bucks., Hunts., Leic., Lincs., Cambs., Northts., Derbys., Notts., Warw., Oxon., Glouc., Worc., also in Norf. and Yorks. and Co. Durham.

Anick, Carrick, Gatherick, Rosebrough, Willimontswyke, and thirty-two names in *wick*.

O.E. **wielle**=well, spring, stream, fountain.
Twenty-four examples of names in -*well*(s).

O.E. **wisce**=a piece of meadow. *v.* W. H. Stevenson in *Phil. Soc. Trans.*, 1895-8, p. 542. It is found several times in O.E. charters, and is frequent in Sussex as *wish*. Cf. L.Ger. *wische*, meadow < *wīska*, related to O.H.G. *wisa*, meadow.
Sledwish.

O.E. **wudu**=wood, forest.
Twenty-five names in *wood*.

PART III

O. E. Names.

Abba	Abshiels.
Aca	Acton (?).
Æccel	Acklington.
Ælf	Elsdon.
Ælfgār	Agarshill (?).
Ælfhere	Allerdean, Elrington (?).
Ælfsige	Elswick.
Ælfwine	Elstob (?).
Ælla, Ella	Eldon, Elford (?), Ellingham, Ellington, Elton (?), Elwick (?).
*Æppel	Eppleton.
Æsc	Ashington.
*Æscel	Eslington.
*Æspheard	Esperley, Esper Shiels.
Æðelred	Edderacres, Etherley (?).
Æðelwine	Anick.
Alubeorht or Aloburh (f.), (L.V.D.)	Abberwick.

Bab(b)a	Bavington.
Bacca	Backworth.
Bacga	Bagraw.
*Baða	Baydale.
Bǣda or Bēada	Bedburn, Biddick (?).

[1] All names are masc. unless otherwise stated. An asterisk denotes a hypothetical restoration of a lost name. Names marked L.V.D. and Nthb. are not found elsewhere than in L.V.D. and distinctively Northumbrian documents.

Bǣre	Berrington.
*Bēaghere	Barmoor, Byermoor.
Bebba, Bæbba (f.)	Bamburgh, Bebside (?).
Becca	Beckley.
*Bēdel	Bedlington.
Bedwine (L.V.D.)	Beadnell.
Bell(a), Beola	Beldon, Belford, Bellingham, Bellister, Belsay, Benton (?).
Be(o)nna	Benwell (?).
Beonnic (?)	Bensham (?).
Beorhtwine	Birchope.
*Beorma	Barmpton.
Beorn	Barmston.
*Berela	Birling.
Bibba	Bebside (?).
Bicca	Beechburn.
*Bidel, *Bydel	Biddlestone (?).
Bilheard or *Bilhere	Bildershaw.
Billa	Billingham, Billy Row, Bilton. Bingfield (?).
Billing	Billy Mill.
*Bodel	Bollihope.
Boll(a)	Bowsden, Boldon (?).
Bōta, Bōte (f.)	Bothal, Buteland.
Brannoc	Branxton.
*Bregn, Bregwine	Brainshaugh.
Brocc	Broxfield.
Brūn	Burnigill.
Brynca (L.V.D.)	Brenkley (?), Brinkburn (?).
Bucc(a)	Buckton.
Butel	Buston, Butsfield.
Bynna, Byni	Binchester (?).
Bytel	Biddlestone (?).
Cæcca	Catch Burn.
Casa	Cassop.
Catta	Catton (?).
Caua (L.V.D.)	Caw Burn, Cawledge Park.
Ceatta	Chatton (?).

Ceappa (Ceabba)	Choppington, Chopwell.
Cedd	Sedgefield.
Cēna	Keenlyside.
*Ceofel	Chillingham.
Cēolferþ	Chollerton.
Ceorra	Charlaw, Chirdon, Chirton.
Cetta	Chatton (?).
Cietel	Chattlehope (?).
Cifa, Ceofa	Cheveley, Chevington.
Cilla	Chibburn, Chilton (?).
Cippa	Chipchase.
Cissa	Sessinghope (?).
Cniht	Knitsley (?).
Cnyt(t)el	Hazeldean (?).
Cocc(a)	Coastley, Coxhoe, Cock-field, -law, -le (?).
Coenwald	Kenner's Dene.
Collan	Coanwood.
Coppa	Copley (?).
Corn (?)	Cornsay (?).
Cotten	Cottonshope.
Cotta	Cottingwood.
Crāwa, Crawe (f.)	Craster (?), Crawcrook (?), Crawley (?).
Crin	Kearsley.
Crossan (Celtic)	Corsenside (?).
Cuneca	Chester-le-Street, Consett.
Cunda	Coundon (?).
Cybbel	Kibblesworth.
Cȳda (L.V.D.)	Kidland.
Cylla	Killingworth.
*Cymel	Kimblesworth.
Cynemǣr	Kimmerston.
Dealla	Dally (?).
*Dīca	Dissington, Ditchburn (?)
Docc	Doxford.
Dodda, Dudda	Doddington.
*Ducc	Dukesfield.

Duda	Duddo, Duddoe.
Dunn(a)	Dinnington, Dunsheugh, Duns Moor.
Ĕadgȳð (f.)	Eddys Bridge (?).
Ĕadhere	Etherley (?).
*Eadmann	Edmondsley (?).
Ĕadmund	Edmundbyers, Edmondsley (?).
Ĕadred	Adderstone, Etherley (?).
Ĕadwine	Edge Knoll.
Ĕadwulf	Edlingham.
Ealda	Aldin Grange, Aldeworth (?).
Ealdgār	Agarshill (?).
Ealdgȳth	Old Shield (?).
Ealdwine	Old Shiel (?).
*Eard(a)	Ardley, Earsdon.
Eata, Eota	Etal.
Ebbi (L.V.D.)	Ebchester.
Ecga	Edgewell (?).
Ecgwulf	Eglingham.
Edda, Æddi	Eddys Bridge (?), Escombe (?).
Ehha	Eighton.
*Elm (?)	Embleton (?).
*Emel	Emblehope (?), Embleton (?).
*Eofel	Ealingham.
Esi	Easington.
*Fær	Farrington.
Filica, *Feoleca	Felkington.
Fisc	Fishburn (?).
*Fisel (?)	Fiselby.
*Framel	Framlington.
Gārmund	Garmondsway.
*Geatela	Yetlington.
*Geoc	Yokesley (v. Nubbock).
Gildwine	Gilden Burn.
Golda	Golden Pot.

*Gor	Gorfen (?).
*Grot(t)a	Grottington.
Hadd	Hadston.
Hæðe (L.V.D.)	Headworth.
*Hæðhere	Hetherington, Heatherley Clough, Heatherslaw, Hetherslaw, Black Hedley.
*Heardgār	Haggerston.
Heafoc	Hauxley, Hawkwell (?).
Hēah(a)	Heighington.
*Heddel	Headlam.
Hegær (L.V.D.)	Harelaw-in-Glendale (?).
*Heort(a)	Harton.
Herefrið or *Herefær	Harraton.
*Herela	Kirkharle.
Hēring (Hæring)	Herrington.
Hicca	Hitchcroft.
Hidd(i) (L.V.D.)	Hedgeley.
Hildeburh (f.)	Filbert Haugh (?).
Hlothere	Lutterington (?).
Hlyda	Lysdon (?).
*Hol(l)	Howsdon Burn.
Hodd, Hudd	Hudspeth.
Hræfn	Ramshope, Ravensworth, Ravensfield.
Hrisa	Risebridge.
Hrōca	Rookhope (?).
*Hrōð(a)	Rothley.
Hūn	Humshaugh, Hunwick.
Hūnstān	Hunstanworth.
Hwala	Whalton (?).
Hwīta	Whittingham, Whittington.
Īda (Nthb.)	Edington.
Ifa	Ivesley (?), Iveston (?).
Illa	Elilaw, Ellishaw.
Ini	Isehaugh.
Ing(a)	Ingoe.

Ladda	Ladley.
*Lǣce	Lesbury.
Lēodhere	Lutterington (?).
Lēofhere	Learchild, Learmouth (?), Lorbottle.
Lida	Lysdon (?).
Lihtwine	Greenleighton (?).
Lilla	Lilburn, Lilswood.
Līn	Lynesack.
*Līnel	Linnolds.
Lippa	Lipwood (?).
Lud(d)a	Ludworth.
*Mægen	Mainsforth.
*Mǣr(a)	Merrington.
Mǣðhelm (L.V.D.)	Medomsley (?).
Mǣðhere	Matfen (?).
*Merc	March Burn, Marchingley.
*Mearð	Mason.
*Miloc (?)	Melkington.
Moll (L.V.D.), Nthb.	Molesdon.
Mōr	Moorsley.
Mucel	Muggleswick.
Mūl	Mousen.
*Netel	Nettlesworth.
Ocga	Ogle (?).
Ofa	Ovingham, Ovington, Owton.
Ōsa	Ewesley, Usworth, Usway Ford.
Ōsburh (f.)	Rosebrough.
Ōsmund	Osmond Croft.
*Pælloc	Paston.
Paga (Nthb.)	Pawlaw Pike.
Pecg	Pegswood.
Penda	v. Old and New Moor.
Pitta, Pytta	Pittington.
*Plaga (?)	Plawsworth (?).
Pruda	Prudhoe.

*Ræda	Redworth (?).
Ricola	Rickleton.
Ricwine	Ricknall (?).
Ridda	Ritton.
*Rim	Rimside (?).
*Rippel	Riplington.
Rugga	Rugley.
Sceaft(a)	Shaftoe (?).
*Sceldwine	Sheddons Law (?).
*Scīrmund	*Shirmonden.
Scorra	Shoresworth.
Scot	Shoreston, Shotton (?).
*Scufel	Shilburnhaugh, Shilmore.
Scurfa	Sheraton.
Scylf(a)	Shildon, Shilvington.
Scyne	Shincliffe.
*Scyttel	Shitlington, Chesters, Shittle-heugh (?), Shittlehope (?).
Sigemund	Simonburn, Simonside (2).
Sigewine	Sewing Shiels.
Sigga, *Siga	Seghill.
Sledda	Sledwick.
Snoter	Snotterton.
*Snytre (?)	Snitter (?).
*Socca (?)	Sockburn (?).
*Styfel	Stillington
Sunna	Sunnyside.
Syla	Sills.
Tada	Tedcastle (?).
Tīr	Tarsett.
Titel, Tytel	Titlington.
Tocg(a)	Togston.
Tolla	Tone (?), Tow Law (?).
Tot(t)a	Todburn (?).
*Trema	Trimdon.
*Tucga	Tughall.
Tudda	Tudhoe.

Tyrhtel	Tritlington.
*Đocer (?)	Thockrington.
Đrocc	Throckley.
Đūrwine	Thrundle (?), Thrunton.
*Ulca, Ulga (?)	Ulgham, Ulwham.
*Ur(a) (?)	Urpeth.
Wacer	Wackerfield.
*Wǣra	Warland, Warton.
Wala	Waldridge (?).
Wassa	Washington.
Wealh	Wallington.
*Wealt(a)	Welton.
Wearda	Wardley.
We(o)rc or Werce (f.)	Warkworth.
Wicg(a)	Wigside.
Wif(a)	Westoe.
Wifel	Willington (2), Wilmire.
*Wiggel	v. Greenlee Lough.
*Wila	Wylam.
Willoc	Wilkwood.
Wine	Winston, Wynyard.
Winel	Windleston
Winelac	Winlaton.
Wuda	Woodifield.
*Wuduhere, *Widuhere	Widdrington.
Wulf	Wolviston, Wooden, Wooler.
Wulfsige	Wolsingham, Woolsington.

M.E., including names of French and Low German origin.

Aleyn	Allensford, Allenshiel.
Androwe	Andrews House.
Anton (?)	Anton Field (?).
Bernard	Barnard Castle, Barneystead.
Blenkin	Blenkinsopp.

Chater	Chatterley (?).
*Cramel	Cramlington.
Crane	Crane Row.
Crawe (= Crow)	Crowsfield.
Davie	Davyshiel.
Emmot	Emmethaugh (?).
Erle (= Earle)	Earlshouse (?), Earlside (?).
Falder	Fortherley.
Feldyng (= Fielding)	Fielden Gate.
Flemyng	Flemingfield.
Gerard	Garretlee, Garretshiels.
Gibbe	Gibside (?).
*Goffe	Gofton.
*Greneson	Girsonsfield.
*Hamstre (?)	Hamsterley.
Harper	Harperley.
*Hebbe	Hepscott
Heppo (?)	Hepden Burn (?), Hetton (?).
*Hesse, *Hasse	Haswell.
Heued (= Head)	Headshope.
Hopper	Hopperclose.
Ivo	Ivesley (?), Iveston (?).
*Kepe	Keepwick.
Ket	Ketton.
*Kever	Keverstone.
*Lame	Lamesley.
Morel	Morleston.
Nele (= Nigel)	Nelson.

P(i)ers	Piercebridge.
*Poid	Podge Hole.
Pollard	Pollard's Lands.
*Prende	Prendwick.
Roger	Rogerley.
*Schakel	v. Snook Bank.
*Schimpel	Shilbottle.
Scot	Scots House.
*Scrimer	Scremerston.
Slater	Slatyford.
Smale	Smales (?).
*Spening	Spennymoor.
Spayn	Spain's Field.
Stokker	Stockerley.
*Thrastere, *Thristere	Thirston.
William	Williamston.
Willimot	Willimontswyke.
Wischard	Whiskershiels.

Scandinavian.

Ásgeirr	Angerton (?), Ingram (?).
Ásketill, Ásketin	Aislaby (?), Eshells.
Áslákr	Aislaby.
Bleikr, Bleiki	Blakeston.
Boltr	Bolts Law.
Brandr	Brancepeth.
Bróðir	Brotherlee, Brotherwick.
O.Sw. **Dote**, O.Dan. **Dota**	Dotland.
Egill	Eggleston.
Eymundr	Amerston Hall.

Fleinn	Plainfield (?).
Freyviðr [1]	Farrow Shiels.
*Fulliði	Follingsby.
Gamall	Gamelspath (?).
Garpr [1]	Carp Shield (?).
Gagni or Gegnir [1]	Gainford (?).
Gellir [1]	Gellesfield Hole.
*Glante, *Glente	Glantlees, Glanton.
Gunnvarðr(m) or	Gunnerton.
Gunnvǫr [1]	
Háki	Hackford (?).
Hallþórr	Eltringham.
Hestr	Hisehope Burn (?).
Heðinn	Henshaw.
Hild (f.)	Ilderton.
Holti [1]	Howtel (?), Houtley (?).
Horni	Horncliffe (?).
Hreinn [1]	Rainton (?).
*Hreiðr	Raceby.
Hrómundr	Rumby.
*Hvassi	Whessoe.
Hvelpr [1]	Kirkwhelpington.
Ingjaldr	Ingleton.
Karlr	Carlbury.
Ketilvarðr	Killerby.
Kiartan [1] (O.Ir.)	Cartington.
Kjarni	Kearsley (?).
*Kjǫt	Ketton (?).
Klakkr	Claxton.
Knǫrr [1]	Knaresdale.
Krókr	Croxdale.
Lumi [1]	Lumley.

[1] Scandinavian names not hitherto recorded as found in England.

Náttfari	Nafferton.
Ormr	Ornsby Hill.
Pampi	Pandon.
Rauði	Rothbury, Rudchester (?).
Rögnvaldr	Rennington, Rainton (?).
*Sæliði	Selaby.
Silki [1]	Silksworth.
*Skraffinnr	Scrainwood.
Skúli	School Aycliffe.
Slöngr	Slingley Hall.
Snörtr [1]	Snitter (?).
Steinn	Stannington, Stainton, Stoney Burn.
*Stykki	Stickley (?).
*Styrkolr	Stirkscleugh.
Sveinn	Swainston.
Titlingr [1]	Titlington (?).
Toli	Tone (?), Tow Law (?).
Trani	Tranwell.
Þor, Þur	Throston.
Þorsteinn	Thrislington.
Þorviðr [1]	Trewitley.
Úlfr	Ouston, Ushaw.
Úlfheðinn [1]	Ulnaby.
Úlfketill	Ouston (2), Ulgham (?).
*Útfari	Offerton.
Vagn	Wainhope (?).
Vakr [1]	Wackerfield (?).
Vífill	Willington (?), Wilmire (?).

[1] Scandinavian names not hitherto recorded as found in England.

PHONOLOGY

N.B.—The treatment of phonology attempted in these paragraphs does not attempt to be exhaustive, but an attempt is made to present as fully as possible all points of definite dialectal interest, and certain general points not hitherto observed.

Vowels.

a (æ).

§ 1. O.E. **a (æ)** in a closed syllable, apart from the influence of a neighbouring consonant > Nthb. and Durbh. [a] or [e], e.g. *a* in *glass, e* in *ash, path.*
Names in *Ael-* show forms in *e* only in *Ellingham, Ellington, Elsdon, Elstob, Elwick* (Nthb.). In *Elswick, Elvet, Elwick* (Co. Durh.) we have early forms in *A-, Ae-, Ai-,* but later always *El-*. In *Allerdean,* forms in *All-* have prevailed under the influence of M.E. *aller,* Nthb. [alr]=alder.
æsc gives *e-* forms alone in Esh. In Ashington St. Eng. *Ash-* has prevailed. *Washington* shows early and late forms in *wesh-* in agreement with dial. [weʃ] for *wash.*
æspe (espe) gives *esp,* except in quite late forms of Esper Shields, where the St. Eng. *asp* makes itself felt. *Espley* shows fluctuations in early forms, but settles down to *Esp-*.

§ 2. O.E. **hæs(e)1** > Nthb. and Durh. [hezl] for *hazel.*
This has been replaced by the St. Eng. form in Hazeldean, Hazelrigg. Similarly *Aeccel* > *Ackle-* and *Eckle-* in Acklington. Cf. North. *kekkyll* and *shekyll* for *cackle* and *shackle.*
A similar *e* (< M.E. *ā*) is heard in Nthb. [pepə] for *paper.*
Cf. Pauperhaugh.

§ 3. O.E. **a** before **ld** remains in North. M.E. and becomes Nthb. and Durh. [ad], [a·d] though forms like [kauld], [kould], [ko·d] are also heard. *Cold, Old* are universal on the modern map.

§ 4. O.E. a before mb > *ā* in North. M.E. *came* > Nthb. [kiəm], [kjem]. Forms in *comb* (e.g. Combfield) due to St. Eng.

§ 5. O.E. **an** is in M.E. often represented by *aun* under the influence of French words with nasalised *a* before *n* (cf. Blanchland). How far this is purely scribal it is difficult to say, but the persistence of *au, aw* in the forms of *Brancepeth* suggests that in some names at least a definite sound-change took place. Ultimately the *a(u)n* of Eng. and Fr. words alike > Nthb. [a]. Similarly *am* > *aum* in Bamburgh.

§ 6. O.E. **ang**, as in *lang*, remains in North. M.E. and Mod. Nthb. and Durh. On the map St. Eng. *long* appears as a rule.

e

§ 7. O.E. **e** is represented in M.E. forms very frequently by *i*, fluctuation between *e* and *i* forms being so common in some names that in words of doubtful etymology it is difficult to be certain what is the original form. Modern forms show *e* in some names, *i* in others.

Morsbach (§ 109) notes such spellings with *i*, some sporadic, some regular. They are found before (*a*) *Dentals* ; *d*, Biddick, Ridlees, Whittonstone; *n*, Grindon, Hinding Burn, *Rinnington* for Rennington; *s*, Hisehope; *l*, Chillingham, Shilford; (*b*) *Palatals*. Bitchfield, Fitches. Note [jivrin] for Yeavering.

Slingley and Sting Head show the same sound-development as *England*. (Cf. Jespersen, 3.113).

Bebside and Trimdon are very uncertain.

§ 8. O.E. **er** > M.E. **ar** > [a·] before a following consonant as in English generally, e.g. Barford, Darncrook, Hardwick, so also in *Farn-*. Cf. Nthb. [fa·n]=fern, where St. Eng. has a spelling pronunciation. In *Derwent* a spelling pronunciation now prevails.

In Nthb. *er* > [ar] before a vowel also. Cf. Nthb. and Durh. [vari] for *very*, e.g. Barrington.

ir > *ar* > [aˑ] in Tarsett, but is unchanged in Chirdon and Chirton.

In [jɔˑzn] for Earsdon we have the characteristic S. Nthb. [ɔˑ] heard in [tʃɔˑtʃ], [tɔˑn] for *church, turn*.

§ 9. Initial **e** > Nthb. [je] in Earsdon, Yarnspath Law, *Yerlesset* for Earlside, *Yelderton* for Ilderton, cf. E.D.G., § 248.

i

§ 10. O.E. **i** has often become [*e*] in Mod. Nthb. (E.D.G., § 68), and there is evidence that this lowered variety of *i* was common also in M.E. In place-names *e* and *i* prevail in about equal proportions in Mod. Nthb. and Durh.

Morsbach (§ 114), in dealing with the sporadic appearance of *e* for *i* in M.E., notes its special frequency before certain sounds. Using his grouping, we have examples before (*a*) nasals, e.g. Benwell, Brenkley, *Brenk-* forms for Brinkburn, *Fenkle* for Finchale, Hendon, *Kem-* forms for Kimblesworth; (*b*) labials, e.g. Bebside, Hepburn, Cheviot; (*c*) *l*. *Bell-* forms for Billingham, Felton, *Mel-* forms in Milbourne, Relley.

Edington, Hedgeley, Seghill, Spredden, Westoe, and certain forms of Fiselby, Pittington, and Riddlehamhope fall outside his grouping.

Heddon is due to the neighbouring Heddon-on-the-Wall.

Detchant shows Nthb. [detʃ] for *ditch*.

Medlem for (Bp. Middleham) seems to be quite modern.

o

§ 11. O.E. **o** > Nthb. [u] in Budle, Hulam, Hulne, and early forms of *strother* and Osmond Croft.

§ 12. O.E. **ord** > Nthb. [uəd], [uərd], [urd]. Cf. Ord and names in *-ford* for spellings indicative of this. *Dultries*= Dortrees may show the same change, but the history of *door* (O.E. *duru* and *dor*) is obscure.

R

u

§ 13. O.E. u > [i] in Crimden, Dinnington, Shilburn-haugh, Shilmore, Shittlehope, Shittleheugh, *Wissington* for Woolsington, Witton, and similarly M.E. *ŭ* < O.E. *ū* (§ 21) becomes *i* in Dinley, Philip. Cf. E.D.G., § 100, and [ʃil] Ayr., [ʃiul] Durh., for *shovel.*

ā

§ 14. O.E. ā > North. M.E. *ā* > Nthb. and Durh. [iə]. In a good many cases, at least on the map, the modern form shows St. Eng. *o.* Contrast *Acomb* and *Lynesack* with *Oakwood* and *Pedams Oak,* and note *Crowsfield.*

§ 15. O.E. ār > North. M.E. *ār* > Nthb. and Durh. [eə(r)], [eˑr], e.g. Harsondale.

§ 16. O.E. āw > North. M.E. *āw* > Nthb. and Durh. [a] and [aˑ], though [ou] is also heard, and *o* written under the influence of St. English. Cf. Bagraw and Cranerow.
M.E. *āw* < O.E. *āg* develops similarly, ẹ.g. Fawdon.

§ 17. Initial ā > Nthb. and Durh. [ja] and [je] as in the local pronunciation of Alne, Alnham, Aycliffe and Acomb.

ō

§ 18. O.E. ō > North. M.E. [yˑ] > Nthb. [in], [iə] or, under the influence of St. Eng., [uˑ]. See forms under Broom, Bewclay, Buteland, Ewesley, Pooltree, Rookhope.

ū

§ 19. O.E. ū has been shortened in *Lucker.* Cf. E.D.G., § 172 for examples of this change in monosyllables.

§ 20. O.Fr. eau has a two-fold development :—
(1) > *eu* > [juˑ]. Cf. *Bewmys* under Beamish, *Beure-pair* under Bear Park.
(2) > *eu* > [iˑ]. Cf. Beamish and *Beatreby* for Butterby

and the local pron. of Bear Park, *v.* Wyld, *Short English Grammar*, § 172. The modern forms are often influenced in form or pronunciation by Mod. Fr. *beau.*

§ 21. Shortening of long vowel before consonant group.

M.E. *ā*: e.g. Acton, Brafferton, Stamfordham, *Ackewode* under Oakwood.

M.E. *ē*: e.g. Bedburn, *Chesborne* under Cheesburn, Meldon, Pespool.

M.E. *ī*: e.g. Ditchburn, Pigdon, Swinburn.

M.E. *ō*: e.g. Bothal, Gosforth, Rothley.

M.E. *ū*: e.g. Fulford, Hunwick, Prudhoe, *Utchester.*

In some names, e.g. Cheesburn, the influence of the independent word, e.g. *cheese,* has ultimately prevailed and led to the use of a form with long vowel.

§ 22. Shortening of the first long vowel in words of three syllables. Cf. Wyld, *u.s.,* § 176.

M.E. *ā*: e.g. Hallington, Stannington.

M.E. *ē*: e.g. Berrington, Kenners Dene, Rennington.

M.E. *ī*: e.g. Dissington, Edington, Linacres.

M.E. *ō*: e.g. Bockenfield, Morralee, Ovingham.

M.E. *ū*: e.g. Burnigill.

§ 23. Shortening in an unstressed syllable, e.g. M.E. *ā* in Lynesack.

Consonants

b

§ 24. Intervocalic **b** > **v** in Bavington and *Averwick* (*s.n.* Abberwick). Cf. Pavenham, Beds., D.B. *Pabeneham* and the forms of Baverstock, Wilts. (Ekblom, p. 20). In all cases the development seems quite modern.

c

§ 25. O.E. **c** (palatal) > Nthb. and Durh. *ch.,* e.g. Chatton. The only cases of *k* are Birkenside, Lightbirks, Picktree, Finchale. *birk* is Nthb. for *birch.* For *pick*=pitch, *v.*

Heslop, p. 533. *Kirk* is found as the first element in a small group of names. *kirk* itself is never now used independently in Nthb.

§ 26. M.E. **ch** > Nthb. (ʃ) in Chatton and Chillingham. Heslop (p. 684) doubts if this was ever more than a rustic joke based upon some family peculiarity, the shibboleth being, " The children of Chillingham gied to the children o' Chatton a chain to sit on," or some other such sentence. Cf. Heslop, pp. 85, 147.

§ 27. O.E. **cg** in *brycg* and *hrycg* gives place-name forms with and without palatalisation in M.E., and there was probably the same uncertainty in M.E. that there is in the present-day dialect. On the map *rigg* survives in Hazlerigg alone.

§ 28. O.E. **cw** > M.E. *qu(h)* > Nthb. [hw], [ʍ] in Whaggs, Wharmley, Whickham, Whittonstall, Whorlton. See forms under Quarrington, also cf. E.D.G., § 241, Nthb. [hwik] and Durh. [wik] for *quick*, and Quernmore, Lancs. [waˑrmər], 1575 *Wharnemores* (Wyld, p. 213).

d

§ 29. Intervocalic **d** followed by *r* in the next syll. > [ð] regularly, though in some words, under the influence of words with ð—*r* which tend to become *d*—*r* (§ 41), the change has not persisted (E.D.G., § 297). Cf. Etherley, Gatherick, Fortherley, [eðəsən] for Adderstone, and the history of Widdrington.

§ 30. Final **rd** tends to become *rth* in the suffix *-ford*. This change is not found in any words in independent use in Mod. Eng. or its dialect, the word *ford* itself always having *d*, though a form with *th* is in rare and comparatively late use in M.E. The change may in part be due to lack of stress in the second element, in part to confusion with words in *rd* from *rth* (§ 43), e.g. *afford* for M.E. *afforthe*.

§ 31. ds > [dz] > [dž] in Edge Knoll, Hedgeley, Podge Hole, Sedgefield, and forms under Leadgate, *v.* M.L.R vol. xiv., p. 342. For further examples cf. § 40.

§ 32. M.E. *schele* often becomes Nthb. *shield.* *d* is occasionally developed in dialect after *l*, but no example has been noted in Nthb. or Durh. Confusion with the common word *shield* may have been at work.

g

§ 33. O.E. **g** (palatal) in *geat* > *y* in M.E. *yate*, in Portgate, Stotgate, Eastgate, Westgate. Later this was replaced by St. Eng. *-gate.* Cf. the similar fluctuation between [giət], [giat], [geət], [gjet] on the one hand, and [jat], [jet], [je·t], [jit] on the other, in Nthb. and Durh. dial. forms of *gate.*

§ 34. A distinctive feature of Northumbrian is the pronunciation, with one exception (viz. Chillingham), of *-ingham* as [indžəm]. This development of O.E. *ingaham* (Introd., p. xxiv.) is found also in South Scotland, where it is often spelled *ingehame,* but is unknown in Co. Durham. Traces of it are found sporadically elsewhere, as in Bengeo, Herts., Cowlinge, Suff. Such names must go back to O.E. patronymics with *ja-* stems in place of the more usual *a-* stems. Nom. pl. **ingjoz,* gen. pl. **ingja,* would naturally yield M.E. *inge* [indžə]. M.E. spellings in *-incham, -ingjam* show attempts to represent the pronunciation, the latter being a cross between a historic and a phonetic spelling. It is difficult to see why Chillingham does not show this change, except on some principle of distant dissimilation.[1]

h

§ 35. h has been lost initially in Aydon, Eltringham, Ewarts Hill, Ilderton, Ousterley. The loss is the more noteworthy in that initial *h* is never dropped in Nthb. dialect.

[1] Since this paragraph was written my attention has been called to an article by Zachrisson in Herrig's *Archiv.,* 1915, 348 ff., which deals very fully with the history of these names.

§ 36. h has been lost medially through lack of stress in numerous names in *-hill, -hale, -head, -hope, -ho(e), -ham*, e.g. Whittle, Bothal, Hartside, Pontop, Ingoe, Leam, Snope. See also forms under *Coxhoe*.

§ 37. h has been added initially in certain M.E. spellings of names, e.g. Adderstone, Earsdon, Edington, Elsdon. The *h* is probably due to N.Fr. scribes, and in no case has it affected the pronunciation.

§ 38. h has been added medially before a suffix beginning with a vowel in certain M.E. spellings of Crooked Oak, Pedam's Oak, Wallsend. In the last it is curiously persistent. Cf. B.C.S. 458 *heuedakerhende*, R.P.D. *Wodeshende*, and the personal name *Townshend*=Town's end.

1

§ 39. O.E. l is lost before following *b, d, k, m, p, s, t*, but modifies the preceding vowel.

a > [ɔ·], [a·], cf. spellings *s.n.* Abberwick, Wallbottle, Daldon, Auckland, Doepath, Paston, Falstone, Causey, Dalton (2), Walltown.

o > [au] or [ou]. E.D.G. (§ 86) gives the former for N. Nthb., and the latter for S. Nthb. and Co. Durh. Cf. Bowsden [bauzən] in Nthb. and Boldon [boudən] in Co. Durham. Other names are Bowmont, Homer's Lane, Owmers, Howtel. See also forms under *Colepike Hall*.

ulb > *owb* in Fowberry; *ulm* > [u·m] in Bulmer [bu·mə]. Cf. Hulme [hju·m] in Chesh. and Lincs; *uln* > *oun* (? = [u·n]) in *Ounbie* (*s.n.* Ulnaby); *uls*>*ous* in Ouston.

Irregulars are the developments, *alg, ald* > *ag, ad* in Agarshill and Aydon Shiel, *alb* > *av* or *abb* in Abberwick.

t

§ 40. ts > [tʃ] in Pytchley, Northts., earlier *Pihtesle*. So in Knitsley, Nthb., we get *Knitchley* for earlier *cnihteslea*. The modern form has reverted to the original *ts*. In *Rateswood* (*s.n.* Wretchwood) we have the reverse process, *ts* being written for *ch*. Cf. M.L.R. vol. xiv., p. 342, and Wyld, *History of Colloquial English*, p. 292.

þ, ð

§ 41. Intervocalic ð followed by r in the next syllable > d in the dialects of Nthb. and Durham (E.D.G., § 314). Cf. Edderacres, and forms s.n. Brotherwick, Etherley, Heatherslaw, Heatherwick, Netherton.

§ 42. Continuant ð > stop d before l, m, s, w in Baydale, Bedlington, Budle, Hedley, Medomsley, Weedslade, Headworth, [sudik] for Southwick. Cf. Nthb. and Durh. [faðəm] for fathom (E.D.G., § 315), and M.E. wurdli for wurðli (Horn, § 200).

§ 43. Medial rth > rd in Mason. Cf. Nthb. fardin for farthing (E.D.G., § 315). Final rth > rd > rt in Ewart, Staward, and Heward (s.n. Heworth). Cf. Nthb. erd for earth, and M.E. stalwart, Jeddart for stalworth, Jedworth.

§ 44. th between vowels is lost in Bolton, Henshaw. Cf. Jespersen, 2.612.

M.E. v

§ 45. v is lost medially between vowels before following l in Dilston, Ealingham, Shilburn, Shilmore. Cf. Nthb. and Durh. [di·l] for devil, [ʃul] and [ʃiul] for shovel. Similarly before r in Clarewood, Learchild, Learmouth. In both cases the change may, in part at least, have been due to assimilation of v to following l or r, after syncopation of the unstressed vowel.

§ 46. v is lost before following b in Coe Burn, d in Cleadon, l in Slaley (cf. Coaley, Glouc., earlier Coveley), n in Scrainwood (cf. Denshire, Daintry for Devonshire, Daventry, Jespersen 2.532). The change is doubtless largely due to assimilation.

§ 47. v has become vocalic u, and then gives rise to a diphthong in Owton, earlier Oveton. Cf. Owsden, Cambs. earlier Ovesden, Uvesden.

w

§ 48. **w** is lost initially in (*W*)*ooden*. It is kept in Nthb. *wood*, but lost in (*w*)*ool* and (*w*)*ound*. The loss is due to absorption by the following *u*. *v*. Wyld, *u.s.* p. 296.

§ 49. **w** is lost medially at the beginning of an unstressed syllable. This is specially common in the case of the suffixes -*wick*, -*well*, and O.E. names in -*wine*. Cf. Allen, Anick, [kɔləl] for Colwell, and forms under Edge Knoll.

Assimilation.

§ 50. **Progressive.** *Unvoicing:* *d* > *t* after *f, k, s, t*, e.g. Gofton, Nookton, Shoreston, Catton; [tj] > [tç] Portgate. *Miscellaneous:* *ld, ln, lv* > *ll*, e.g. *Illerton* (*s.n.* Ilderton), Kilham, Allerdean; *rð, rv* > *rr*, e.g. Norham, Harraton; *sf* > *ss*, e.g. Pressen; *tl* > *tt* in *Butteston* (*s.n.* Buston).

§ 51. **Regressive.** *Unvoicing:* *b* > *p* before *s, t*, e.g. Hepscott, *Lampton* (*s.n.* Lambton); *d* > *t* before *f, s, t*, e.g. Mitford, *Weetsleatt* (*s.n.* Weedslade), Whitton; *g* > *k* before *s*, e.g. Houxty. *Voicing:* *k* > *g* before *b, d*, e.g. *Aslagby* (*s.n.* Aislaby), Blagdon; *t* > *d* before *m*, e.g. Pedam's Oak. *Miscellaneous:* *df, dl, dr* > *ff, ll, rr*, e.g. Brafferton, Bollihope, Relley; *hb* > *bb* in Hebburn; *kb* > *bb* in Aislaby; *kp* > *pp* in Soppit; *lb, ls* > *bb, ss* in Chibburn and *Wissington, s.n.* Woolsington[1]; *m* > *n* or [ŋ] before *d, k, s, t*, e.g. Brandon, Cronkley, Ornsby, Branton; *n* > *m* before *b, f, m*, e.g. Amble, Stamford, Kimmerston; [ŋ] > *n* before *ch, sh, t*, e.g. Lanchester, *Lanshaes* (*s.n.* Langshaws), Lanton; *pt* > *tt* in Hetton; *th* is assimilated to *t, d, l*, e.g. Hetton, Heddon, and *Helley* (*s.n.* Hedley); *vl* > *ll*, e.g. Chillingham.

§ 52. **Dissimilation.** *ns* > *ms* in Barmston, *Edomsley* (*s.n.* Edmondsley); *mf* > *nf* in Tanfield; *nl* > *ml* in Wharmley; *nm* > *lm* in *Kilmerston* (*s.n.* Kimmerston). For the last cf. Zachrisson, pp. 132-3.

---

[1] Or possibly under § 39.

§ 53. Simplification of Consonant Groups.

Loss of medial consonant, e.g. *l(d)f* in Shilford, *l(v)d* in Shildon, *dr(d)s* in Adderstone.

Loss of initial consonant, e.g. *(b)nb* in Bamburgh, *(l)df* in Shadfen, *(v)st* in *Iseton, Isley (s.n.* Iveston, Ivesley). Loss of final consonant, e.g. *ls(t)* in Nelson.

§ 54. Metathesis

is most common in the case of the consonant *r*, e.g. Bruntoft, Burnigill, Sturton, but we have also *cs > sc* in Coastley; *dns > nsd* in Dinsdale; *lk > kl* in Aislaby; *lm > ml* in Embleton; *nl > ln* in [bi·dlən] for Beadnell, *Keednall (s.n.* Kidland); *rdg > grd* in Haggerston.

§ 55. Epenthesis.

The following cases are illustrated :—
ml > mbl, mpl, e.g. Embleton, Humbledon, and *Framp(ling)ton (s.n.* Framlington); *lf, lr, ls > ldf, ldr, lds,* e.g. Holdforth, Bildershaw, Shields (cf. E.D.G., § 298); *nf, nl > ndf, ndl* in *Taundfeld (s.n.* Tanfield), Spindlestone; *ms, mt > mps, mpt,* e.g. *Rampshaw (s.n.* Ramshaw), *Brampton (s.n.* Branton); *nl > ntl* in Parmentley. *v.* Wyld, *u.s.* p. 309.

An intrusive *n* is found in an unstressed syllable in Auckland, Edmondhills, Lemmington, Willimontswyke, and some forms *s.n.* Emmethaugh. The *n* is the same as that found in *messenger,* and other words discussed by Jespersen (2.429).

§ 56. Loss of Final Consonant.

d in *Akell (s.n.* Akeld), Hazon, *Kidlin (s.n.* Kidland), Warren Burn; cf. *Englan(d)* and *erran(d)* in Nthb. and Durh. (E.D.G., p. 235), and the common loss of *d* after *l* in Scotland (ib. § 307).

f in Bewclay, [jakli] for Aycliffe, *Cunsly (s.n.* Coniscliffe), and other names. Cf. *bailie* and *hussy* for *bailiff* and *housewife* (Jesperson, 2.534).

n after *l* in [jel] for Alne, Ayle, [hul] for Hulne, Till. Cf. [kil] for *kiln* in Scots., Nthb., Durh. (E.D.G., § 271).

t after *n* in Allen, *Alwen (s.n.* Alwent), *Darwen (s.n.* Derwent) and [detʃən] for Detchant. This is specially common in Scots. (E.D.G., § 295). *v.* Wyld, *u.s.* pp. 303-4.

§ 57. **Unvoicing of final d.** This is found in Bowmont, Consett, Detchant, Garretlee and Garret Shiel, and cf. *Barnettsteed, Newton Hanset* (*s.n.* Barneystead, Newton Hansard), E.D.G., § 303. *v.* Wyld, *u.s.* p. 313.

§ 58. Unvoicing of final [dž] to [tʃ] or [ʃ]. See forms under Melkridge, Todridge, Waldridge, Yarridge. These developments are found sporadically in N. Eng. and Scots. dialect, *v.* E.D.G., § 366.

Miscellaneous.

§ 59. **Loss of an unstressed syllable.** This is common, and accounts e.g. for such forms as *Cramelton* (Cramlington), *Chetlup* and *Chestrop* (Chettlehope and Chesterhope), *Cunsley* (Coniscliffe), *Farnton* (Farrington), *Darnton* (Darlington), [dintən] for Dinnington, [emləp] for Embleton, Hexham from *Hextildham*, [framptən] for Framlington.

§ 60. **-end > -and** probably under the influence of the North. Pres. Part. in *-ande*. Cf. forms under Detchant and Hazon.

§ 61. Initial **c** and **g** often interchange. How far the change is an orthographic blunder, due to Anglo-Norman scribes, or represents a real phonetic change, it is difficult to say. *c > g* occasionally in Catch Burn, Cresswell, Carp Shiel, Kirkley, Knaresdale, *g > c* in some forms of Gamelspath. See further Zachrisson (pp. 137-8), and Wyld (p. 34), where we have several examples from Lancashire of initial *g > c*.

APPENDIX A

§ 1. O.E. **dene, dūn,** and **tūn.**

dene > ton. This change is specially common after an unvoiced consonant, when, by assimilation, *den > ten,* and is in the unstressed syllable easily confused with *ton.*

Acton, Ashington, Biddleston, Catton, Dipton (2), Embleton, Eppleton, Foxton (2), Gofton, Hallington, Nookton, Togston, Welton, Wooperton.

dene > don. In unstressed syllables these suffixes are identical in pronunciation, map-makers are more familiar with the latter, and confusion is the more easy in that wherever there is a " dene " there is probably a " down."

Aydon (2), Beldon Burn, Blagdon, Burdon, Chirdon, Dawdon, Elsdon, Hendon, Lysdon, Pandon, Shawdon, Shildon, Weldon, Wydon, and *v.* forms under Biddleston, Marden, Roseden.

don > dene is less common, as might be expected from the relative frequency of these words in place-nomenclature generally.

Bowsden, Swarden Burn, Warden, Wooden, Raredean, and *v.* forms under Meldon, Pittington, Pelton, Hetton, Windleston.

don > ton usually by assimilation of *d* to preceding unvoiced consonant, and under the influence of the more common suffix.

Berrington, Buston, Callerton, Cartington, Embleton, Felkington, Glanton, Harton, Hetton (2), Humbleton, Pittington, Windleston, Quarrington, Langton, *Rareton (s.n.* Raredean).

ton > don in isolated forms under Lemmington, Riplington.

In some cases variation existed from the earliest times.

Thus, between *ton* and *don* in Farrington, Leighton, Molesdon, Shoreston, and between *den, don,* and *ton* in Horden, Rare Dean, Melkington.

§ 2. **ley** (O.E. *lēah, lēage*) and **law** (O.E. *hlǽw, hlāw*). Nthb. *law* has often given place, at least on the map, to the more common suffix *ley*, though traces of the earlier form remain in local pronunciations such as [krala] for Crawley.

Brenkley, Dewley, Hartley, Fairnley, Hauxley, Stickley, Thornley, Throckley, Woolley, Wheatley. See also *s.n.* Cocklaw, Highlaws. In Barley Hill and Hunterley Hill, pleonastic *hill* has been added after the change of suffix.

§ 3. **worth** (O.E. *weorþ*) and **wood** (O.E. *wudu*). **worth** > **wood** in Chesterwood, Clarewood, Lilswood, Pegswood. See also *s.n.* Hunstanworth, Shoresworth. This is an example of the type dealt with under § 8. There was probably an intermediate stage, in which *worth* > *word* (Phonology, § 43). The reverse change, as might be expected, is very rare, and **wood** > **worth** only in certain forms under Broadwood, Stobswood.

§ 4. **ford** (O.E. *ford*) > **worth** (O.E. *weorþ*) in Flatworth. See also *s.n.* Doxford, and cf. Longworth, Heref., with *-ford* as late as 1781, Duxford, Cambs., with *-worth* as late as 1662. Confusion has arisen through alternative forms in *forth* and *ford* on the one hand (Phonology, § 30), and *word* and *worth* on the other (ib. § 43).

§ 5. **ford** > **wood** in Mosswood, Baxterwood, probably with an intermediate *worth* (*v.* §§ 4, 3), cf. *Gosworth, s.n.* Gosforth.

§ 6. Certain changes are due to the phonetic identity of some unstressed suffixes or to their close similarity. This is already illustrated under §§ 1-5. Further examples are **field** > **fold**, see *s.n.* Stotfield ; **hale** (O.E. *hēale*) > **hill** in Cornhill, Seghill ; and **hill** > **hale**, see *s.n.* Redmarshall ; **hale** > **hall** in Bothal, Houghall, Oxenhall, Ricknall, Tughall ; **hill** > **hall**, see *s.n.* Whitehall (?) ; **haugh** > **(h)oe** in Stokoe ;

heugh > **haugh**, see *s.n.* Redheugh, Shittleheugh; **mere** > **mire** in Thackmire, Wilmire.

In Eastgate, Hooker Gate, Leadgate, Portgate, Stotgate, Westgate there has been confusion between *gate* (Dial. *yet*), an opening, and *gate*, a road.

Pl. in -es > (h)ouse. Crookhouse, Harbourhouse.

Dat. pl. in -um > (h)am. Carham, Bolam, Crookham, Downham, Fenham, Kilham, Newsham, Roddam, Woodham. Some of these may be examples of original unstressed -(*h*)*am* written as -*um*.

§ 7. Many changes of suffix are due to a misdivision of the word, consequent as a rule upon some phonological development. Thus *Agar-shele* > Agars-hill, *Ak-sheles* > *Ak-sels* > Ax-well, *Alding-ridge* > *Aldern-edge* > Aldin Grange, *Bade-ley* > Bay-dale, *Belles-ho* > *Bel-shaugh* (*s.n.* Belsay), *Be-repar* > Bear-park, *Biddles-ton* > Biddle-stone, *Burning-*(*h*)*ill* > Burni-gill, *Col-pighill* > Colepike-hall, *Cons-*(*h*)*ed* > *Con-side* (*s.n.* Consett), *Didens-hale* > Dins-dale, *Fore-stone* > Fors-ton (?), *Gates-head* > *Gate-side* (*s.n.* Gateshead), *Harts-head* > Hart-side, *Hens-halgh* > Hen-shaw, *Henne-dene* > Hind-ing, *Ive-stan* > Ives-ton, *Kevers-ton* > Kever-stone, *Kyl-*(*h*)*oe* > *Kil-ley* (*s.n.* Kyloe), *Rams-hale* > Ram-shaw, *Scremer-stone* > Scremers-ton, *Stan*(*f*)*ord-*(*h*)*am* > *Staner-den* (*s.n.* Stamfordham), *Stel-ley* > *Stel-*(*h*)*oe* (*s.n.* Stella), *Warend-ham* > *Warn-don* (*s.n.* Warenton). Names in *clif* which lose final *f* (Phonology, § 56) develop a suffix *ley*, e.g. *Croun-clef* > Cronk-ley.

§ 8. There is a tendency to replace a rare suffix by a more familiar one closely resembling it in sound (§ 6).

barn (O.E. *bern*) > **burn** in Whitburn; **berg** (O.E. *beorg*) > **bury** in Sadberge; **helm** > **ham** in Bensham; **hay** (O.E. *hege*) > **haugh** in Windyhaugh (?); **set(e)** > **side** in Allerside, Bebside, Corsenside, Gibside, Holmside, Simonside (2); **shet(e)** (O.E. *scēat*) > **sete** > **side** in Bebside (?); **tail** > **dale** in Croxdale; **wish** > **wick** in Sledwick.

§ 9. An unstressed suffix, whose meaning has become disguised, is given a fresh form giving a more definite meaning, e.g. *Akell > Akehill* (*s.n.* Akeld), *Cornell >* Corn-hill.

§ 10. **burgh** (O.E. *burh*) and **burn** seem sufficiently distinct both in meaning and form, but they often interchange in Nthb. and Co Durham. Elsewhere the interchange has only been noted in Yorkshire.

burgh > burn in Cheesburn, Sockburn. See also *s.n.* Bamburgh, Thornbrough.

burn > burgh. See *s.n.* Hartburn, Brinkburn, Woodburn.

In Newburn it is difficult to say whether *burn* or *burgh* is the older.

§ 11. Occasionally one suffix is replaced by another of similar meaning. *Harson-den* and *Trelles-den >* Harsondale and Turs-dale; *Ward-law > War-don* (*s.n.* Wardon Law), **law > hill** (see *s.n.* Whitehall), **law > braes** in Shell-braes, **law > side** in *Heforside* (*s.n.* Hefferlaw), **side > hill** in Gallow-hill, **mere > lough** in Black Lough, **wick > worth** (see *s.n.* Muggleswick), **worth > town** (see *s.n.* Pegswood), New-**burg**(a) is once given for New-**biggin**.

§ 12. Miscellaneous changes are: *Binwall >* Benwell; *Conside >* Consett; *Dudden >* Duddoe; *Foul-brigg >* Fowberry; *Gates-(h)ende* (*s.n.* Gateshead); *Har-low* for *Har-law* (*s.n.* Harlow), with S. and Mid. Eng. *low* for North *law*, and, similarly, *Kellaw > Kel-low > Kell-ow > Kell-(h)oe*; *Hengandleys >* Hanging Leaves.

ADDENDA

P. 59, *s.n.* **Dalton Piercy.** The place was also known as *Dalton in Herternesse* (Cl. 1316). *Herternesse* must be O.W.Sc. *hjartarnes* = hart's headland, found in *Hjartenes* (N.G. xii. 398), Swedish *Hiortanæs* (No. B., 1917, p. 180), Icelandic *Hjarðarnes* (Jónsson, p. 492). *Herternesse* would seem to be the earlier and fuller name of *Hart*. If this is so the explanation of that name given on p. 103 must be abandoned, and we must believe it to be a shortened form of a name given to the place by some Viking settler.

P. 97, *s.n.* **Gubeon.** Mr C. B. Lewis, of St Andrews, calls my attention to the fact that *gudgeon* is from O.Fr. *goujon*, from Lat. *gobionem*. The use of the form *Gudgeon* for *Gubeon* must be explained as due to the suggestion of some 17th century antiquary who was aware of their ultimate identity.